Lecture Notes in Physics

For information about Vols. 1–151, please contact your bookseller or Springer-Verlag.

Vol. 152: Physics of Narrow Gap Semiconductors. Proceedings, 1981. Edited by E. Gornik, H. Heinrich and L. Palmetshofer. XIII, 485 pages. 1982.

Vol. 153: Mathematical Problems in Theoretical Physics. Proceedings, 1981. Edited by R. Schrader, R. Seiler, and D.A. Uhlenbrock. XII, 429 pages. 1982.

Vol. 154: Macroscopic Properties of Disordered Media. Proceedings, 1981. Edited by R. Burridge, S. Childress, and G. Papanicolaou. VII, 307 pages. 1982.

Vol. 155: Quantum Optics. Proceedings, 1981. Edited by C.A. Engelbrecht. VIII, 329 pages. 1982.

Vol. 156: Resonances in Heavy Ion Reactions. Proceedings, 1981. Edited by K.A. Eberhard. XII, 448 pages. 1982.

Vol. 157: P. Niyogi, Integral Equation Method in Transonic Flow. XI, 189 pages. 1982.

Vol. 158: Dynamics of Nuclear Fission and Related Collective Phenomena. Proceedings, 1981. Edited by P. David, T. Mayer-Kuckuk, and A. van der Woude. X, 462 pages. 1982.

Vol. 159: E. Seiler, Gauge Theories as a Problem of Constructive Quantum Field Theory and Statistical Mechanics. V, 192 pages. 1982.

Vol. 160: Unified Theories of Elementary Particles. Critical Assessment and Prospects. Proceedings, 1981. Edited by P. Breitenlohner and H.P. Dürr. VI, 217 pages. 1982.

Vol. 161: Interacting Bosons in Nuclei. Proceedings, 1981. Edited by J.S. Dehesa, J.M.G. Gomez, and J. Ros. V, 209 pages. 1982.

Vol. 162: Relativistic Action at a Distance: Classical and Quantum Aspects. Proceedings, 1981. Edited by J. Llosa. X, 263 pages. 1982.

Vol. 163: J.S. Darrozes, C. Francois, Mécanique des Fluides Incompressibles. XIX, 459 pages. 1982.

Vol. 164: Stability of Thermodynamic Systems. Proceedings, 1981. Edited by J. Casas-Vázquez and G. Lebon. VII, 321 pages. 1982.

Vol. 165: N. Mukunda, H. van Dam, L.C. Biedenharn, Relativistic Models of Extended Hadrons Obeying a Mass-Spin Trajectory Constraint. Edited by A. Böhm and J.D. Dollard. VI, 163 pages. 1982.

Vol. 166: Computer Simulation of Solids. Edited by C.R.A. Catlow and W.C. Mackrodt. XII, 320 pages. 1982.

Vol. 167: G. Fieck, Symmetry of Polycentric Systems. VI, 137 pages. 1982.

Vol. 168: Heavy-Ion Collisions. Proceedings, 1982. Edited by G. Madurga and M. Lozano. VI, 429 pages. 1982.

Vol. 169: K. Sundermeyer, Constrained Dynamics. IV, 318 pages. 1982.

Vol. 170: Eighth International Conference on Numerical Methods in Fluid Dynamics. Proceedings, 1982. Edited by E. Krause. X, 569 pages. 1982.

Vol. 171: Time-Dependent Hartree-Fock and Beyond. Proceedings, 1982. Edited by K. Goeke and P.-G. Reinhard. VIII, 426 pages. 1982.

Vol. 172: Ionic Liquids, Molten Salts and Polyelectrolytes. Proceedings, 1982. Edited by K.-H. Bennemann, F. Brouers, and D. Quitmann. VII, 253 pages. 1982.

Vol. 173: Stochastic Processes in Quantum Theory and Statistical Physics. Proceedings, 1981. Edited by S. Albeverio, Ph. Combe, and M. Sirugue-Collin. VIII, 337 pages. 1982.

Vol. 174: A. Kadić, D.G.B. Edelen, A Gauge Theory of Dislocations and Disclinations. VII, 290 pages. 1983.

Vol. 175: Defect Complexes in Semiconductor Structures. Proceedings, 1982. Edited by J. Giber, F. Beleznay, J.C. Szép, and J. László. VI, 308 pages. 1983.

Vol. 176: Gauge Theory and Gravitation. Proceedings, 1982. Edited by K. Kikkawa, N. Nakanishi, and H. Nariai. X, 316 pages. 1983.

Vol. 177: Application of High Magnetic Fields in Semiconductor Physics. Proceedings, 1982. Edited by G. Landwehr. XII, 552 pages. 1983.

Vol. 178: Detectors in Heavy-Ion Reactions. Proceedings, 1982. Edited by W. von Oertzen. VIII, 258 pages. 1983.

Vol. 179: Dynamical Systems and Chaos. Proceedings, 1982. Edited by L. Garrido. XIV, 298 pages. 1983.

Vol. 180: Group Theoretical Methods in Physics. Proceedings, 1982. Edited by M. Serdaroğlu and E. İnönü. XI, 569 pages. 1983.

Vol. 181: Gauge Theories of the Eighties. Proceedings, 1982. Edited by R. Raitio and J. Lindfors. V, 644 pages. 1983.

Vol. 182: Laser Physics. Proceedings, 1983. Edited by J.D. Harvey and D.F. Walls. V, 263 pages. 1983.

Vol. 183: J.D. Gunton, M. Droz, Introduction to the Theory of Metastable and Unstable States. VI, 140 pages. 1983.

Vol. 184: Stochastic Processes – Formalism and Applications. Proceedings, 1982. Edited by G.S. Agarwal and S. Dattagupta. VI, 324 pages. 1983.

Vol. 185: H.N. Shirer, R. Wells, Mathematical Structure of the Singularities at the Transitions between Steady States in Hydrodynamic Systems. XI, 276 pages. 1983.

Vol. 186: Critical Phenomena. Proceedings, 1982. Edited by F.J.W. Hahne. VII, 353 pages. 1983.

Vol. 187: Density Functional Theory. Edited by J. Keller and J.L. Gázquez. V, 301 pages. 1983.

Vol. 188: A.P. Balachandran, G. Marmo, B.-S. Skagerstam, A. Stern, Gauge Symmetries and Fibre Bundles. IV, 140 pages. 1983.

Vol. 189: Nonlinear Phenomena. Proceedings, 1982. Edited by K.B. Wolf. XII, 453 pages. 1983.

Vol. 190: K. Kraus, States, Effects, and Operations. Edited by A. Böhm, J.W. Dollard and W.H. Wootters. IX, 151 pages. 1983.

Vol. 191: Photon Photon Collisions. Proceedings, 1983. Edited by Ch. Berger. V, 417 pages. 1983.

Vol. 192: Heidelberg Colloquium on Spin Glasses. Proceedings, 1983. Edited by J.L. van Hemmen and I. Morgenstern. VII, 356 pages. 1983.

Vol. 193: Cool Stars, Stellar Systems, and the Sun. Proceedings, 1983. Edited by S.L. Balliunas and L. Hartmann. VII, 364 pages. 1984.

Vol. 194: P. Pascual, R. Tarrach, QCD: Renormalization for the Practitioner. V, 277 pages. 1984.

Lecture Notes in Physics

Edited by H. Araki, Kyoto, J. Ehlers, München, K. Hepp, Zürich
R. Kippenhahn, München, H. A. Weidenmüller, Heidelberg
and J. Zittartz, Köln
Managing Editor: W. Beiglböck, Heidelberg
Fondazione C. I. M. E., Firenze
Adviser: Roberto Conti

228

Thermodynamics and Constitutive Equations

Lectures Given at the 2nd 1982 Session of the
Centro Internationale Matematico Estivo (C. I. M. E.)
Held at Noto, Italy, June 23 – July 2, 1982

Edited by G. Grioli

Springer-Verlag
Berlin Heidelberg GmbH

Editor

Giuseppe Griolo
Dipartimento di Matematica, Università degli Studi di Padova
Via Belzoni 7, 35100 Padova, Italy

ISBN 978-3-540-15228-6 ISBN 978-3-540-39394-8 (eBook)
DOI 10.1007/978-3-540-39394-8

Originally published by Springer-Verlag Berlin Heidelberg New York Tokyo in 1985

2153/3140-543210

PREFACE

This volume collects the series of lectures given by Professors B.D. Coleman, C.M. Dafermos, and I. Müller and four seminars given by Professors A. Bressan, D. Graffi, A. Morro, and T. Ruggeri.

The lectures and seminars analyze problems and modern aspects of the interaction between thermodynamics and the mechanics of continua with finite deformations.

In the book are treated on the one hand the general framework and the constitutive relations, on the other hand analytical problems and specific important theories, for instance mixture theory, electromagnetism, and the theory of wave propagation.

Giuseppe Grioli

TABLE OF CONTENTS

On Thermodynamics and Constitutive Relations

Bernard D. Coleman

Carnegie-Mellon University
Pittsburgh, Pennsylvania

This course of eight lectures was based on the two articles printed here.

The first, entitled Actions for General Thermodynamical Systems is a preliminary version of an appendix which B. D. Coleman and D. R. Owen have been preparing for the 1984 reprinting of C. Truesdell's book Rational Thermodynamics. The reader of this article should bear in mind that when the general thermodynamical theories treated here are applied to materials with fading memory, such as viscoelastic substances or electromagnetic media that can show absorption and dispersion, a state is identified with an appropriate history.

The second article, entitled Thermodynamics and the Constitutive Relations for Second Sound in Crystals is an English version (here printed for the first time) of a paper prepared in Italian by B. D. Coleman, M. Fabrizio, and D. R. Owen. That paper on heat transfer in dielectric crystals exhibiting second sound illustrates the usefulness of Coleman and Owen's interpretation of the second law of thermodynamics as an assertion that a particular action has the Clausius property.

It is hoped that the two articles together provide an up-to-date view of research on two problems: (i) that of finding sufficiently general forms of the second law of thermodynamics to cover the systems of interest in modern continuum physics and (ii) that of finding the implications of the second law for the constitutive relations of non-classical materials.

Actions for General Thermodynamical Systems

Bernard D. Coleman and David R. Owen

Department of Mathematics
Carnegie-Mellon University
Pittsburgh, Pennsylvania

The principles of thermodynamics have found application in many branches of science. These principles have been employed to understand the efficiency of heat engines, the electromotive force of galvanic cells and thermal junctions, the dependence of chemical equilibrium on temperature and pressure, the properties of phase transitions, and the thermomechanics of continuous bodies.

Although thermodynamics is the science of heat and temperature, its principles are often usefully applied to experiments in which heat is not flowing (e.g., those involving poor thermal conductors or insulated reaction chambers) or others in which temperature is not changing (because, say, the object under study is a good thermal conductor in contact with an isothermal environment). One recognizes a thermodynamical argument by its reference to consequences of either the first or the second Law. Every student of physics or chemistry has been taught that the first law is an assertion about the balance of heat and work and that the second law is an assertion about the rate of increase of entropy that, in some sense, is equivalent to a denial of the existence of certain perpetual motion machines, or to a denial of the existence of cycles in which heat is absorbed at some temperatures without emission at others, or to an assumption about the sign of the sum over a cycle of the ratio of the heat absorbed to the temperature at which it is absorbed.

We here describe some recent work toward a precise formulation of the second law as a general principle whose implications can be derived with rigor. We do not believe that the results of this work can be dismissed as "mere axiomatics". The development in the 1960's of the thermodynamics of materials with memory raised questions whose resolution required a careful examination of the mathematical foundations of thermodynamics. In its original presentation, the theory of the thermodynamics of materials with fading memory rested on the Clausius-Duhem inequality, i.e., on the assumption that for each substance there is a function of state, called the entropy, whose difference at two states dominates the ratio of the heat absorbed to the absolute temperature along each process taking one state to the other. The question was raised: does each substance have such a function of state with the properties of regularity needed to derive the now known consequences of the Clausius-Duhem inequality? Of course, the question is meaningful only if one has a statement of the second law that does not presuppose the presence of entropy as a function of state. A statement of this type can be obtained by making mathematical the ideas behind the familiar assertion that _the sum along a cycle of the ratio of the heat gained to the absolute temperature at which it is gained cannot be positive_. However, to be useful for materials with gradually fading memory and for other substances with few nontrivial cycles, the statement must be formulated in such a way that it has meaning for "approximate cycles". We have obtained such a formulation of the second law and have used it to study various questions, including the existence, uniqueness, and regularity of entropy functions.

In this recent work [1974,1] [1975,1], a careful distinction is made between the general structure of thermodynamical systems and the equations defining special classes of systems. The concept of a _system_ employed is one in which a system is a pair (Σ,Π) of sets with the following mathematical structure: Σ is a topological space whose elements are the _states_; Π is the set of _processes_; associated with each process is a continuous function ρ_P, mapping a non-empty open subset $D(P)$ of Σ onto a subset $R(P)$ of Σ; ρ_P is called the _transformation induced by_ P and its value at a state σ in $D(P)$ is denoted by $\rho_P\sigma$; to each pair (P'',P') of processes for which R(P') intersects D(P") there is assigned a process P"P' called the process resulting from the successive application of (first) P' and (then) P". It is assumed that (I) for each σ in Σ, the set of states accessible from σ, i.e., the set of states of the form $\rho_P\sigma$ with P in Π, is dense in Σ, and (II) if P"P' is the result of the successive application of P' and P", then the transformation $\rho_{P"P'}$ induced by P"P' is the composition of $\rho_{P'}$, and $\rho_{P"}$, i.e., is the function defined on $D(P"P') = \rho^{-1}_{P'}(D(P"))$ by the equation $\rho_{P"P'}\sigma = \rho_{P"}\rho_{P'}\sigma$.

The mathematical concept that renders precise and general the idea of a "sum along a process" is that of an action. An action is a function that assigns a number $\underline{a}(P,\sigma)$ to each pair (P,σ) with P in Π and σ in D(P); $\underline{a}(P,\sigma)$ is called the supply of $\underline{a}$ on going from σ to $\rho_P\sigma$ via the process P. Two properties are required of an action: (i) additivity in the sense that if P is the result of the successive application of P' and P", then for each σ in D(P) the supply of $\underline{a}$ obtained by going from σ to $\rho_P\sigma$ via P is the sum of the supplies of $\underline{a}$ obtained by going from σ to $\rho_{P'}\sigma$ via P' and from $\rho_{P'}\sigma$ to $\rho_P\sigma = \rho_{P"}\rho_{P'}\sigma$ via P", i.e.,

$$\underline{a}(P,\sigma) = \underline{a}(P',\sigma) + \underline{a}(P",\rho_{P'}\sigma), \tag{1}$$

and (ii) continuity in the sense that for each P in Π, the function $\underline{a}(P,\cdot)$ is continuous on D(P).

A process P and a state σ° are said to form a cycle (P,σ°) if σ° is in D(P) and $\rho_P\sigma^\circ = \sigma^\circ$. One may consider taking the second law to be the assertion that an appropriate action $\underline{s}$ (which, of course, must be specified) is not positive when its argument is a cycle, i.e., is such that

$$\rho_P\sigma^\circ = \sigma^\circ \text{ implies } \underline{s}(P,\sigma^\circ) \le 0. \tag{2}$$

For materials with gradually fading memory, the class of cycles (P,σ°) is too small for (2) to have the full implications expected of the second law. To obtain an extension of (2) to "approximate cycles", we have employed the following concept: We say that $\underline{s}$ has the Clausius property at a state σ° if, for each

$\varepsilon > 0$, $\sigma°$ has a neighborhood $\mathcal{O}_\varepsilon(\sigma°)$ for which

$$\rho_P\sigma° \in \mathcal{O}_\varepsilon(\sigma°) \quad \text{implies} \quad s(P,\sigma°) < \varepsilon. \tag{3}$$

It is clear that if (3) holds and $\rho_P\sigma° = \sigma°$, we have $s(P,\sigma°) < \varepsilon$ for every $\varepsilon > 0$, and hence (2) holds; i.e., if $\underline{s}$ has the Clausius property at $\sigma°$ then $\underline{s}(P,\sigma°)$ is not positive when $(P,\sigma°)$ is a cycle.

If an action has the Clausius property at a state $\sigma°$, then it has the property at each state in a set $\Sigma°$ that is dense in Σ and contains all states σ that are accessible from $\sigma°$; this set $\Sigma°$ may be defined as follows: For each state $\sigma°$ in Σ let $\mathcal{S}(\sigma°,\sigma)$ be the collection of all the open subsets $\mathcal{O}$ of Σ that contain σ and are such that the sets $\underline{s}\{\sigma° \to \mathcal{O}\}$, defined by

$$s\{\sigma° \to \mathcal{O}\} = \{s(P,\sigma°) \mid P \in \Pi, \; \rho_P\sigma° \in \mathcal{O}\} \tag{4}$$

are individually bounded above, i.e., have

$$\sup \underline{s}\{\sigma° \to \mathcal{O}\} < \infty; \tag{5}$$

$\Sigma°$ is the set of states for which

$$m(\sigma°,\sigma) := \inf_{\mathcal{O} \in \mathcal{S}(\sigma°,\sigma)} \sup \underline{s}\{\sigma° \to \mathcal{O}\} \tag{6}$$

is finite, i.e.,

$$\Sigma° = \{\sigma \mid \sigma \in \Sigma, \; m(\sigma°, \sigma) > -\infty\}. \tag{7}$$

In [1974,1] we interpreted the second law as the statement that <u>a particular action $\underline{s}$ has the Clausius property at least at one state</u>. To prove that such a statement implies the existence of an entropy function that enters a relation with the form of the Clausius-Duhem inequality, we there introduced the concept of an upper potential.

A real valued function S on a dense subset $\underline{D}(S)$ of Σ is called an <u>upper potential</u> for an action $\underline{s}$ if for each pair of states σ_1, σ_2 in $\underline{D}(S)$ and each $\varepsilon > 0$ there is a neighborhood $\mathcal{O}_\varepsilon(\sigma_1,\sigma_2)$ of σ_2 such that whenever $\rho_P\sigma_1$ is in $\mathcal{O}_\varepsilon(\sigma_1,\sigma_2)$

there holds

$$S(\sigma_2) - S(\sigma_1) > \underline{s}(P,\sigma_1) - \varepsilon. \tag{8}$$

In the special case in which σ_2 is accessible from σ_1, i.e., in which σ_2 has the form $\sigma_2 = \rho_P\sigma_1$, this relation holds for all $\varepsilon > 0$ and hence implies that the supply of $\underline{s}$ on going from σ_1 to σ_2 is dominated by the difference $S(\sigma_2) - S(\sigma_1)$:

$$S(\rho_P\sigma_1) - S(\sigma_1) \geq \underline{s}(P,\sigma_1). \tag{9}$$

It is easily seen that an action that has an upper potential has the Clausius property at each state in the domain of the upper potential. Although far less trivial to show, the converse is also true: the assumption that there are states at which s has the Clausius property implies that s has an upper potential, in fact one that is upper semicontinuous. If we identify $\underline{s}(P,\sigma)$ with the sum of the heat added divided by the temperature at which it is added as the system is taken from the state σ to the state $\rho_P\sigma$ by the process P, then in (9) the upper potential S is playing the role played by entropy in the Clausius-Duhem inequality. Thus, the existence of an upper potential for $\underline{s}$ is tantamount to the existence of entropy as a function of state.

Our construction of an entropy function employs the observation that if $\underline{s}$ has the Clausius property at $\sigma°$ and if $S°$ is defined on $\Sigma°$ by

$$S°(\sigma) = m(\sigma°,\sigma), \tag{10}$$

then not only is the domain $\Sigma°$ of $S°$ dense in Σ, but $S°$ is an upper potential for s and is upper semicontinuous on $\Sigma°$.

Under the assumptions stated up to this point, we can say no more about the regularity of upper potentials for $\underline{s}$ than that there is one that is semicontinuous. This should not be surprising, for we have so far assumed very little about systems and actions. At this level of generality, the collection Σ of states has a topology but not the vector-space or manifold structure required to make meaningful the concept of a differentiable function on Σ. When more is assumed about the system (Σ,Π) and the action $\underline{s}$, one expects to be able to prove more about entropy as a function of state.

We have not yet assumed any special properties for the state σ° with which we start when we construct Σ° as shown in equations (4) - (7). We would like to take this state as a "standard state" and be able to normalize entropy functions S on Σ° so that

$$S(\sigma^\circ) = 0. \tag{11}$$

Although the hypotheses we made so far imply that Σ° is dense in Σ, they are not strong enough to imply that σ° is in Σ°. It does suffice, however, to assume that this selected state σ° is equilibrated with respect to $\underline{s}$ in the sense that there is at least one process P° in Π for which

$$\rho_{P^\circ}\sigma^\circ = \sigma^\circ \quad \text{and} \quad \underline{s}(P^\circ,\sigma^\circ) = 0. \tag{12}$$

In fact, we have the following theorem: *Suppose that σ° is equilibrated with respect to $\underline{s}$ and is a state at which $\underline{s}$ has the Clausius property. Then (1) σ° is in Σ°; (2)*

$$m(\sigma^\circ,\sigma^\circ) = 0, \tag{13}$$

and hence the upper potential S° defined in (10) vanishes at σ°, i.e.,

$$S^\circ(\sigma^\circ) = 0; \tag{14}$$

moreover (3) S° is the smallest entropy function that is normalized in this way: if S is an upper potential for s that is defined on Σ° and obeys (11), then for each state σ in Σ°,

$$S^\circ(\sigma) \leq S(\sigma). \tag{15}$$

If, in addition, $m(\sigma,\sigma^\circ)$ [defined by interchanging the roles of σ and σ° in the relations (4) - (6)] is finite for each σ in Σ°, then the function $S_\circ$ defined on Σ° by

$$S_\circ(\sigma) = -m(\sigma,\sigma^\circ) \tag{16}$$

also is an upper potential for $\underline{s}$ and not only obeys the normalization (11) but is the largest entropy function that does; i.e., each upper potential for $\underline{s}$ that is defined on $\Sigma°$ and obeys (11) has the bounds:

$$S°(\sigma) \leq S(\sigma) \leq S_°(\sigma). \tag{17}$$

The set of entropy functions on $\Sigma°$ normalized according to (11) is convex: if S_1 and S_2 are two such entropy functions, then so also is each function of the form $\alpha S_1 + (1-\alpha)S_2$, $0 < \alpha < 1$. We have just observed that if $\underline{s}$ has the Clausius property at $\sigma°$, if $\sigma°$ is equilibrated with respect to $\underline{s}$ at $\sigma°$, and if $m(\sigma,\sigma°)$ is finite whenever $m(\sigma°,\sigma)$ is, then $S_°$ is the maximal and $S°$ is the minimal element of this convex set of normalized entropy functions. Clearly, then, this set reduces to a singleton if and only if $S° = S_°$. That is, under these hypotheses about the reference state $\sigma°$, in order that there be only one entropy function S on $\Sigma°$ obeying (11) it is necessary and sufficient that

$$m(\sigma°,\sigma) + m(\sigma,\sigma°) = 0. \tag{18}$$

This condition, although met by elastic materials and viscous materials, is not met in general. Among the exceptions are certain elastic-plastic materials, materials with fading memory, and certain materials with internal state variables. We have seen that if one takes the second law to be the assertion that an action $\underline{s}$ has the Clausius property, one can deduce the existence of entropy as a function of state and obtain information about the regularity and uniqueness of entropy. We have left open the questions: (i) Which of the many actions one can formulate for a system should be assumed to have the Clausius property? (ii) Can information about the form of $\underline{s}$ be deduced from a statement of the second law that makes precise an assertion to the effect that there can be no cycles in which heat is only absorbed? (iii) In what sense is "absolute temperature" a distinguished measure of hotness?

Recent papers of Serrin have shed light on these questions. Basic to his theory [1979,2,3] is the concept of the hotness manifold H, introduced by Mach

[1896,1] and assumed by Serrin to be a continuous, oriented, one-dimensional manifold whose points L are called levels of hotness, or for short, hotnesses. It is assumed that the orientation of H induces a total strict order "$\prec$" on hotnesses, with "$L_1 \prec L_2$" read "L_1 is a lower level of hotness than L_2", or "L_1 is below L_2", or "L_2 is above L_1". Serrin's theory [1979, 2, 3] does not rest on a concept of "state", but does refer to objects that we may here identify with cycles, i.e., with pairs (P,σ) in $\Pi \times \Sigma$ with σ in $D(P)$ and $\rho_P \sigma = \sigma$. Let us define a classical thermodynamical system to be a set $\mathbb{P}_c$ of cycles and a real-valued function Q on $\mathbb{P}_c \times H$ called the accumulation function; the value $Q(\underline{P},L)$ of Q at a point $(\underline{P},L)$ in $\mathbb{P}_c \times H$ is called the net heat absorbed by the system at levels of hotness at or below L in the cycle $\underline{P}$. It is assumed that $Q(\underline{P},L)$ varies only over a bounded interval in H in the sense that for each cycle $\underline{P}$ there are levels of hotness $L^\ell = L^\ell(\underline{P})$, $L^u = L^u(\underline{P})$, with $L^\ell \prec L^u$, such that

$$\left.\begin{aligned} Q(\underline{P},L) &= 0 \text{ for } L \prec L^\ell, \\ Q(\underline{P},L),L &= Q(\underline{P},L^u) \text{ for } L^u \preceq L. \end{aligned}\right\} \tag{19}$$

For each $\underline{P}$, the function $Q(\underline{P},\cdot)$ generates a finitely additive set function q_P for H whose value $q_{\underline{P}}(I) = Q(\underline{P},L_2) - Q(\underline{P},L_1)$ on the set $I = \{L \mid L_1 \prec L \preceq L_2\}$ is the net heat absorbed by the system at levels of hotness in I, (i.e., above L_1 and at or below L_2); (19) implies that this set function has compact support. The number $Q(\underline{P},L^u)$ is called the overall net gain of heat (by the system) in the cycle $\underline{P}$.

Serrin develops a language for discussing the effect of the operation of two or more systems or the repeated operation of a single system: If $\underline{S}' = (\mathbb{P}_c', Q')$ and $\underline{S}'' = (\mathbb{P}_c'', Q'')$ are two thermodynamical systems (which may or may not be the same), their union $\underline{S}' + \underline{S}''$ is the thermodynamical system $\underline{S} = (\mathbb{P}_c, Q)$ with

$$\mathbb{P}_c = \mathbb{P}_c' \times \mathbb{P}_c'' \tag{20}$$

and

$$Q((\underline{P}',\underline{P}''),L) = Q'(\underline{P}',L) + Q''(\underline{P}'',L) \tag{21}$$

for each pair (P',P'') in $\mathbb{P}_c$ and each L in H.

Serrin assumes that a collection $\underline{U}$ of classical thermodynamical systems, closed under the union operation, has been given. His statement of the second law is: <u>If P is a cycle of a system</u> ($\mathbb{P}_C$,Q) <u>in U with</u> $Q(\underline{P},L) \geqslant 0$ <u>for every</u> L, <u>then</u> $Q(\underline{P},L) = 0$ <u>for every</u> L. In other words, in a cycle for which the net heat absorbed at or below each hotness level is not negative, the accumulation function is identically zero and, in particular, the overall met gain of heat in that cycle is zero.

To show that his statement of the second law permits the construction for each system of a function $\underline{s}$ obeying (2), Serrin assumes that the collection $\underline{U}$ of classical thermodynamical systems contains at least one special system that is the mathematical embodiment of an elastic or viscous substance. His proofs take their simplest form if the distinguished systems are <u>ideal gases</u>; these are systems for which each cycle $\underline{P}$ can be represented as a closed (oriented) curve γ_C in the first quadrant of a coordinate plane with one coordinate, V, interpreted as the volume of the gas, and the other, θ, a coordinate indicating the level of hotness L in the gas. The number θ is related to L by a strictly increasing, positive-valued, continuous function ϕ on the manifold H. (Use of the coordinate system ϕ on H corresponds to measurement of hotness with an "ideal gas thermometer".) For each process $\underline{P}$ of an ideal gas and each level $\bar{L}$ of hotness, $Q(P,\bar{L})$ equals the integral of a differential form, $c(V,\theta)d\theta + p(V,\theta)dV$, over the portion $\gamma_C(\bar{L})$ of γ_C on which $L \leqslant \bar{L}$, i.e., on which the coordinate $\theta = \phi(L)$ is equal to or less than $\bar{\theta} = \phi(\bar{L})$; p is the pressure in the gas and is given by the formula,

$$p(V,\theta) = \frac{r\theta}{V}, \tag{22}$$

with r a constant; c is the heat capacity of the gas and is given by a function of θ alone:

$$c(V,\theta) = \tilde{c}(\theta). \tag{23}$$

(The relations (22) and (23) distinguish an ideal gas from other homogeneous fluid bodies.) Thus, for an ideal gas,

$$Q(\underline{P},\overline{L}) = \int_{\gamma_{c(\overline{L})}} \left(\tilde{c}(\theta)d\theta + \frac{r\theta}{V}\,dV\right), \tag{24}$$

i.e.,
$$Q(\underline{P},\phi^{-1}(\overline{\theta})) = \int_{\gamma_c(\phi^{-1}(\overline{\theta}))} \left(\tilde{c}(\theta)d\theta + \frac{r\theta}{V}\,dV\right). \tag{25}$$

Serrin shows that his form of the second law implies that the functions ϕ corresponding to two distinct ideal gases must be proportional, and hence that ideal gases determine, to within a constant factor, a distinguished coordinate system on H. In [1979,2] it is shown that the presence in $\underline{U}$ of elastic, or even viscous substances far more general than ideal gases determines the same coordinatization of H. This coordinatization, which is unique if, as in practice, one preassigns the value of the difference in the coordinates of the hotness levels of two phase transitions at a standard pressure, such as the freezing and boiling points of water at one atmosphere, is called the <u>absolute temperature scale</u>.

Serrin's principal result is that his statement of the second law is equivalent to asserting that <u>for every cycle P of each system in U</u>

$$\int_0^\infty \frac{Q(\underline{P},\phi^{-1}(\theta))}{\theta^2}\,d\theta \leq 0. \tag{26}$$

The importance of this relation, called the <u>accumulation inequality</u>, lies in its generality: it refers only to the absolute temperature scale ϕ and the "accumulation" $Q(\underline{P},\cdot)$ of the (countably additive) heat measure q_p on the hotness manifold H ; it is independent of the "space-time structure" or the concepts of "body" and "force" used in the specific physical theory to which the thermodynamical concepts of heat and hotness may be applied. When the function $\hat{Q}_p$, defined by $\hat{Q}_{\underline{P}}(\theta) = Q(\underline{P}, \phi^{-1}(\theta))$, is of bounded variation, for each small $\delta>0$, an integration by parts yields, in view of (19),

$$\int_0^\infty \frac{Q(\underline{P},\phi^{-1}(\theta))}{\theta^2}\,d\theta = \int_{\phi(L^\ell)-\delta}^{\phi(L^u)} \frac{\hat{Q}_{\underline{P}}(\theta)}{\theta^2}\,d\theta + \int_{\phi(L^u)}^{\infty} \frac{\hat{Q}_{\underline{P}}(\phi(L^u))}{\theta^2}\,d\theta$$

$$= \int_{\phi(L^\ell)-\delta}^{\phi(L^u)} \frac{d\hat{Q}_{\underline{P}}(\theta)}{\theta} - \frac{\hat{Q}_{\underline{P}}(\theta)}{\theta}\Bigg|_{\phi(L^\ell)-\delta}^{\phi(L^u)} + \frac{\hat{Q}_{\underline{P}}(\phi(L^u))}{\phi(L^u)}$$

$$= \int_{\phi(L^\ell)}^{\phi(L^u)} \frac{d\hat{Q}_{\underline{P}}(\theta)}{\theta} = \int_0^\infty \frac{d\hat{Q}_{\underline{P}}(\theta)}{\theta}\,. \tag{27}$$

Thus the accumulation inequality does gives a mathematical form to the assertion that "the sum along a cycle of the ratio of heat gained to the absolute temperature at which it is gained cannot be positive".

The integral in the accumulation inequality plays the role of the action s in (2). It is clear that Serrin's form of the second law and his derivation of the accumulation inequality go a long way toward the resolution of problems (i), (ii), and (iii). In our recently published research with Serrin [1981,1], we have extended Serrin's form of the second law and the accumulation inequality so that they are meaningful for "approximate cycles". In this research, by combining definitions and methods of the papers [1974,1][1975,1], and [1979,2,3], we answer the questions (i)-(iii) in a way that supplies not only identification of the action s in (2) but also a derivation of the implication (3). We consider thermodynamical systems (Σ,Π,Q) that are systems (Σ,Π) in the sense explained above and possess an accumulation function Q that assigns a number $Q(P,\sigma,L)$ to each triple (P,σ,L) with P in Π, σ in D(P), and L in the hotness manifold H; $Q(P,\sigma,L)$ is called the net heat absorbed by the system at levels of hotness at or below L in the process P starting at the state σ. Let $\mathcal{P}$ be the set of pairs (P,σ) with P in Π and σ in D(P). In addition to a mild regularity condition for Q, it is assumed that, for each (P,σ) in P [even if (P,σ) is not a cycle, i.e., does not have $\rho_P\sigma = \sigma$] there are hotness levels $L^\ell = L^\ell(P,\sigma)$ and $L^u = L^u(P,\sigma)$, with $L^\ell \prec L^u$, such that, in analogy with (19),

$$\left.\begin{array}{l} Q(P,\sigma,L) = 0 \text{ for } L \prec L^{\ell}, \\ \\ Q(P,\sigma,L) = Q(P,\sigma,L^{u}) \text{ for } L^{u} \prec L. \end{array}\right\} \quad (28)$$

We say that a pair (P,σ) in $\mathbb{P}$ is <u>absorptive</u> if $Q(P,\sigma,L) \geq 0$ for all L in H; i.e., if the heat absorbed by the system at or below each level of hotness is not negative.

The <u>union</u> $\underline{S}' + \underline{S}''$ of two thermodynamical systems $\underline{S}' = (\Sigma',\Pi',Q')$ and $\underline{S}'' = (\Sigma'',\Pi'',Q'')$ is taken to be the system $\underline{S} = (\Sigma,\Pi,Q)$ with $\Sigma = \Sigma' \times \Sigma''$, $\Pi = \Pi' \times \Pi''$, $\mathbb{P} = \mathbb{P}' \times \mathbb{P}''$, and with

$$\rho_{(P',P'')}(\sigma',\sigma'') = (\rho_{P'}\sigma', \rho_{P''}\sigma'') \quad (29)$$

and

$$Q((P',P''),(\sigma',\sigma''),L) = Q'(P',\sigma',L) + Q''(P'',\sigma'',L), \quad (30)$$

for each $((P',P''),(\sigma',\sigma''))$ in $\mathbb{P} = \mathbb{P}' \times \mathbb{P}''$ and each L in H. As in [1979,2,3], it is assumed that a collection $\underline{U}$ of thermodynamical systems, closed under the union operation, is given and that $\underline{U}$ contains at least one special system that corresponds to an elastic or viscous substance. Again, the discussion of the second law takes its simplest form if the distinguished systems are ideal gases. Each state σ of an ideal gas is represented as a point (V,θ) in the first quadrant of a coordinate plane; each process P_t of the gas is a piecewise continuous function on an interval $[0,t)$ with values $(\dot{V}(\tau),\dot{\theta}(\tau))$ that, for each τ in $[0,t)$, can be interpreted as the rates of change of V and θ at time τ; a pair (P_t,σ°), with

$\sigma^\circ = (V^\circ, \theta^\circ)$, is in $\mathcal{P}$ if, for each s in $[0,t]$, $(V(s), \theta(s))$, with

$$\left.\begin{aligned} V(s) &= V^\circ + \int_0^s \dot{V}(\tau)d\tau, \\ \theta(s) &= \theta^\circ + \int_0^s \dot{\theta}(\tau)d\tau, \end{aligned}\right\} \tag{31}$$

is in Σ, i.e., has $V(s) > 0$ and $\theta(s) > 0$; in such a case,

$$\rho_{P_t}\sigma^\circ = \big(V(t),\ \theta(t)\big). \tag{32}$$

It is again part of the definition of an ideal gas that θ is given by a coordinate system ϕ on H, and no more is assumed about ϕ than that it is a strictly increasing, positive-valued, continuous function. For an ideal gas the function, Q has the form

$$Q(P,\sigma^\circ,\mathsf{L}) = \int_{M(P_t,\sigma,\mathsf{L})} \left(\tilde{c}\big(\theta(s)\big)\dot{\theta}(s) + \frac{r\theta(s)}{V(s)}\, \dot{V}(s)\right) ds, \tag{33}$$

where r is a number, $\tilde{c}$ is a function characteristic of the gas, and

$$M(P_t,\sigma,\mathsf{L}) = \left\{ s \mid 0 \leqslant s < t,\ \theta(s) \lessgtr \phi(\mathsf{L}) \right\}. \tag{34}$$

When the curve with the parameterization (31) on $[0,t]$ is a closed curve, and hence $V(t) = V^\circ$, $\theta(t) = \theta^\circ$, (32) yields $\rho_{P_t}\sigma^\circ = \sigma^\circ$, and the pair (P_t,σ°) is a cycle; in such a case , if we write $\underline{P}$ for (P_t, σ°), the equations (33) and (24) become the same. The equation (33), which can be written in the line-integral notation used in equations (24) and (25), is an extension of these equations from pairs (P_t, σ°) that are cycles to pairs (P_t,σ°) with σ° in $\underline{D}(P_t)$, i.e., from $\mathcal{P}_c$ to $\mathcal{P}$.

We take the second law to be the following statement that pertains to each system $\underline{S} = (\Sigma,\Pi,Q)$ in $\underline{U}$: <u>For each level L of hotness and each $\varepsilon > 0$, each state σ has a neighborhood $\mathcal{O}_\varepsilon(\sigma,\mathsf{L})$ in Σ for which</u>

$$\rho_P\sigma \in \mathcal{O}_\varepsilon(\sigma,\mathsf{L}),\quad (P,\sigma) \text{ absorptive},\ L^u(P,\sigma) \prec \mathsf{L} \ \text{ implies } \ 0 < Q(P,\sigma,L^u(P,\sigma)) < \varepsilon. \tag{35}$$

In terms more suggestive but less precise: the overall net gain of heat is small in an approximate cycle that is absorptive and operates at or below a fixed level Γ of hotness.

In (35), the relation $\rho_P\sigma \in \mathcal{O}_\varepsilon(\sigma,\Gamma)$ indicates an "approximate cycle" and the relation $L^u = L^u(P,\sigma) < \Gamma$ is the assertion that a pair (P,σ) "operates at or below the fixed level Γ of hotness". The relation $0 < Q(P,\sigma,L^u)$ is true for any absorptive pair (P,σ). The relation $Q(P,\sigma,L^u) < \varepsilon$, however, is the assertion that the "overall net gain of heat is small" and is the important conclusion of the implication (35).

It is a consequence of this law that the hotness manifold again has a distinguished coordinate system ϕ that is unique to within a constant multiple, and this coordinate system is that employed in the formula (33) for the accumulation function of an ideal gas. The principal results obtained in [1981,1] are of the following type: <u>The second law is equivalent to the assertion that for every</u> Γ <u>in</u> H <u>and every thermodynamical system</u> (Σ,Π,Q) <u>in</u> $\underline{U}$, <u>each state</u> $\sigma°$ <u>has, for each</u> $\varepsilon > 0$, <u>a neighborhood</u> $\mathcal{O}_\varepsilon(\sigma°,\Gamma)$ <u>in</u> Σ <u>for which</u>

$$\int_0^\infty \frac{Q(P,\sigma°,\phi^{-1}(\theta))}{\theta^2}\, d\theta < \varepsilon \tag{36}$$

<u>whenever</u> $(P,\sigma°)$ <u>is in</u> $\mathbb{P}$, $L^u(P,\sigma) < \Gamma$, <u>and</u>

$$\rho_P\sigma \in \mathcal{O}_\varepsilon(\sigma,\Gamma). \tag{37}$$

In other words: the second law holds if and only if each system in $\underline{U}$ is such that its accumulation integral is approximately negative on approximate cycles. In particular, the second law implies that when

$$\underline{s}(P,\sigma°) = \int_0^\infty \frac{Q(P,\sigma°,\phi^{-1}(\theta))}{\theta^2}\, d\theta \tag{38}$$

and $(P,\sigma°)$ operates at or below Γ, the implication (3), with $\mathcal{O}_\varepsilon(\sigma°) = \mathcal{O}_\varepsilon(\sigma°,\overline{L})$, holds for each system in $\underline{U}$, each state $\sigma°$, and each $\varepsilon > 0$.

In this essay we have discussed the problem of characterizing the action $\underline{s}$ in a general, essentially context-free, manner in which only the concepts of heat and hotness need be mentioned . Of course, in the thermomechanics of continuous media, formulae for $\underline{s}$ have long been known. Of primary interest to researchers in that field will be the ideas and theorems presented in the first part of our discussion, namely in the paragraphs containing the relations (1) - (18). The concepts set forth there give us an approach to thermodynamics in which the existence and regularity of entropy (and of free energy) as a function of state is to be deduced rather than assumed. In our first paper employing this new approach [1974,1], we examined the problem of finding the restrictions that the second law places on the constitutive equations of elastic and viscous materials, materials with internal state variables, and materials with fading memory, and we found that the assumption that $\underline{s}$ has the Clausius property yields restrictions on the response functions (or functionals) that give such experimentally observable quantities as stress, heat flux, and internal energy (or temperature) agreeing perfectly with restrictions obtained in the treatments that start with a differentiable entropy, or free energy, function and employ the Clausius-Duhem inequality [1963,1,2],[1964,1-3],[1967,1,2]. For each of these materials more was known about $\underline{s}(P,\sigma)$ than its continuity in σ or its general representation as an accumulation integral, and consequently more was proven about entropy and free energy than semi-continuity. In each case it was shown that, <u>starting with an appropriate expression for s in terms of the response functions or functionals for stress, heat flux, and internal energy, and assuming that s has the Clausius property, one can construct an entropy function (or functional) with the properties of differentiability needed to derive the principal results of the earlier studies.</u> The failure of the entropy and free energy of a material with fading memory to be unique does not invalidate the earlier studies based on the Clausius-Duhem inequality. The implications of the earlier work for the response functionals for stress, heat flux, and internal energy, required only the existence of an appropriately smooth free-energy functional, not its uniqueness.

It has been found that for some materials one can separate the problem of finding the class of entropy functions from that of deriving the thermodynamical restrictions on response functions relating experimentally accessible quantities. For example, the local thermomechanics of a unidimensional elastic-plastic material (with its elastic behavior linear and its plastic behavior perfect) is described by giving the elastic modulus μ and the yield strain α as functions of the temperature θ, and the heat capacity κ, the latent elastic heat Λ_e and the latent plastic heat Λ_p as functions of the elastic strain λ_e, the plastic strain λ_p, and the temperature:

$$\mu = \mu(\theta), \quad \alpha = \alpha(\theta),$$

$$\kappa = \kappa(\lambda_e,\lambda_p,\theta), \qquad \Lambda_e = \Lambda_e(\lambda_e,\lambda_p,\theta), \qquad \Lambda_p = \Lambda_p(\lambda_e,\lambda_p,\theta).$$

Without mention of entropy or free energy, one can derive relations among the functions μ, σ, κ, Λ_e, Λ_p that are necessary and sufficient for compliance with the laws of thermodynamics (see [1976,2],[1979,1]). One may separately find conditions on these functions sufficient for the entropy function, normalized as in equation (11), to be unique and, in cases where entropy and free energy are not unique, the class of such functions can be described with precision [1975,1],[1979,1].

We are grateful to James Serrin for the opportunity to work with him on the theory of the accumulation inequality.

The preparation of this essay was supported in part by the U.S. National Science Foundation and the Italian National Council for Research.

References

1896 [1] Mach, E.: Die Prinzipien der Wärmelehre, Historisch-kritisch entwickelt. Leipzig, Barth.

1963 [1] Coleman, B.D., & V.J. Mizel: Thermodynamics and departures from Fourier's law of heat conductions. Arch. Rational Mech. Anal. 13, 245-261.

[2] Coleman, B.D., & W. Noll: The thermodynamics of elastic materials with heat conduction and viscosity. Arch. Rational Mech. Anal. 13, 167-178.

1964 [1] Coleman, B.D.: Thermodynamics of materials with memory, Arch. Rational Mech. Anal. 17, 1-46.

[2] Coleman, B.D.: Thermodynamics, strain impulses, and viscoelasticity. Arch. Rational Mech. Anal. 17, 230-254.

[3] Coleman, B.D., & V.J. Mizel: Existence of caloric equations of states in thermodynamics. J. Chem. Phys. 40, 1116-1125.

1967 [1] Coleman, B.D., & M.E. Gurtin: Equipresence and constitutive equations for rigid heat conductors. Z.A.M.P. 18, 199-208.

[2] Coleman, B.D., & M.E. Gurtin: Thermodynamics with internal state variable. J. Chem. Phys. 47, 597-613.

1972 [1] Boyling, J.B.: An axiomatic approach to classical thermodynamics. Proc. R. Soc. London A 329, 35-70.

[2] Day, W.A.: The Thermodynamics of Simple Materials with Fading Memory. Springer Tracts in Natural Philosophy, Vol. 22. Berlin, etc.: Springer.

[3] Noll, W.: A new mathematical theory of simple materials. Arch. Rational Mech. Anal. 48, 1-50.

1974 [1] Coleman, B.D., & D.R. Owen: A mathematical foundation for thermodynamics. Arch. Rational Mech. Anal. 54, 1-104.

1975 [1] Coleman, B.D., & D.R. Owen: On thermodynamics and elastic-plastic materials. Arch. Rational Mech. Anal. 59, 25-51.

1976 [1] Coleman, B.D., & D.R. Owen: On thermodynamics and intrinsically equilibrated materials. Annali Mat. Pura applicata (IV) 108, 189-199.

[2] Coleman, B.D., & D.R. Owen: Thermodynamics of elastic-plastic materials. Rend. Accad. Naz. Lincei (VIII) Classe di Scienze, fis. mat. nat. 61, 77-81.

1977 [1] Truesdell, C., & S. Bharatha: The Concepts and Logic of Classical Thermodynamics as a Theory of Heat Engines, Rigorously Developed upon the Foundation Laid by S. Carnot and R. Reech, New York, etc.: Springer.

1978 [1] Silhavy, M: On the Clausius Inequality. Abstracts, Czechoslovak Academy of Sciences and Skoda, National Corporation, Plzen, p. 68.

1979 [1] Coleman, B.D., & D.R. Owen: On the thermodynamics of elastic-plastic materials with temperature-dependent moduli and yield stresses. Arch. Ratonal Mech. Anal. **70**, 339-354.

[2] Serrin, J.: Lectures on Thermodynamics, University of Naples.

[3] Serrin, J.: Conceptual analysis of the classical second laws of thermodynamics. Arch. Rational Mech. Anal. **70**, 355-371.

[4] Truesdell, C.: Absolute temperatures as a consequence of Carnot's General Axiom. Arch. History Exact Sci. **20**, 357-380.

1980 [1] Silhavy, M.: On measures, convex cones, and foundations of thermodynamcis, I. Systems with vector-valued actions; II. Thermodynamic Systems. Czech. J. Phys. B **30**, 841-861, 961-991.

[2] Truesdell, C.: The Tragicomical History of Thermodynamics, 1822-1854, New York, etc.: Springer.

1981 [1] Coleman, B.D., D.R. Owen, & J. Serrin: The second law of thermodynamics for systems with approximate cycles, Arch. Rational Mech. Anal. **77**, 103-142.

1982 [1] Feinberg, M., & R. Lavine: Thermodynamics based on the Hahn-Banach Theorem: the Clausius inequality, Arch. Rational Mech. Anal., **82**, 203-293. (Pagination added in proof; paper appeared in 1983).

[2] Owen, D.R.: The second law of thermodynamics for semi-systems with few approximate cycles, Arch. Rational Mech. Anal. **80**, 39-55.

[3] Šilhavý, M: On the Clausius Inequality, Arch. Rational Mech. Anal. **81**, 221-243.

Thermodynamics and the Constitutive Relations for Second Sound in Crystals+

Bernard D. Coleman*, Mauro Fabrizio**, and David R. Owen*

This is an English language version of the paper: Il secondo suono nei cristalli: Termodinamica ed equazioni costitutive, by Bernard D. Coleman, Mauro Fabrizio, and David R. Owen [Rend. Sem. Mat. Univ. Padova, 68, 206-227 (1982)], dedicated to Giuseppe Grioli on his 70th Birthday.

+ This research was supported by the U.S. National Science Foundation and the Consiglio Nazionale delle Ricerche.

* Address: Department of Mathematics, Carnegie-Mellon University, Pittsburgh, Pennsylvania.

** Address: Istituto Matematico, Università degli Studi, Ferrara, Italy

Summary. A derivation is given of implications of the second law of thermodynamics for the constitutive equations of materials for which the heat flux vector $\underset{\sim}{q}$ and the temperature θ obey the relation

$$\underset{\sim}{T}(\theta)\dot{\underset{\sim}{q}} + \underset{\sim}{q} = -\underset{\sim}{K}(\theta) \operatorname{grad} \theta , \qquad (+)$$

with $\underset{\sim}{T}(\theta)$ and $\underset{\sim}{K}(\theta)$ non-singular second-order tensors that, as functions of θ, depend on the material under consideration. The relation (+), which is a natural generalization to anisotropic media of the relation of Cattaneo, has been used by Pao and Banerjee to describe second sound in dielectric crystals. It is here shown that when (+) holds the specific internal energy e depends not only on θ but also on $\underset{\sim}{q}$; that is:

$$e = e_0(\theta) + \underset{\sim}{q}\cdot\underset{\sim}{A}(\theta)\underset{\sim}{q} ,$$

where e_0 is the classical or "equilibrium" internal energy, and $\underset{\sim}{A}$ is determined by $\underset{\sim}{K}$ and $\underset{\sim}{T}$:

$$\underset{\sim}{A}(\theta) = -\frac{\theta^2}{2}\frac{d}{d\theta}\left(\frac{\underset{\sim}{Z}(\theta)}{\theta^2}\right) , \qquad \underset{\sim}{Z}(\theta) = \underset{\sim}{K}(\theta)^{-1}\underset{\sim}{T}(\theta) .$$

It is also shown that the second law implies that $\underset{\sim}{Z}(\theta)$ is a symmetric tensor and that $\underset{\sim}{K}(\theta)$ is positive definite. It is observed that if $\underset{\sim}{Z}(\theta)$ and $\underset{\sim}{Z}(\theta)^{-1}\underset{\sim}{A}(\theta)$ are positive definite and $\partial e/\partial\theta$ is positive, a temperature-rate wave, i.e., a singular surface across which there is a jump in $\dot{\theta}$, will travel faster if it propagates opposite to, rather than parallel to the heat flux.

Introduction

In the classical theory of heat conduction, it is assumed that the heat flux vector $\underset{\sim}{q}$ and the spatial gradient $\underset{\sim}{g}$ of the temperature θ, i.e.,

$$\underset{\sim}{g} = \operatorname{grad}_{\underset{\sim}{x}} \theta(\underset{\sim}{x},t) \ , \tag{1}$$

are related by the constitutive equation,

$$\underset{\sim}{q} = -\underset{\sim}{K}(\theta)\underset{\sim}{g} \ . \tag{2}$$

Thermodynamical arguments (vid., e.g., [1]) imply that the temperature-dependent second-order tensor $\underset{\sim}{K}(\theta)$, called the thermal conductivity, is positive semi-definite, and since $\underset{\sim}{K}(\theta)$ is, in practice, an invertible tensor, it is positive definite.

In a now frequently cited paper published in 1948, Cattaneo [2] used a rough model yielding results having some properties in common with a result of Maxwell in the kinetic theory of gases* to suggest that the relation (2) should be replaced by one which, for the isotropic materials (namely gases) considered by Cattaneo, has the form

$$\tau\dot{\underset{\sim}{q}} + \underset{\sim}{q} = -\kappa\underset{\sim}{g} \tag{3}$$

(with τ and κ positive functions of θ). Cattaneo pointed out that his constitutive relation yields field equations for θ and $\underset{\sim}{q}$ that are free from the "paradox of instantaneous propagation of thermal distrubances" known to be associated with the relation (2). In 1963 Chester [4] observed that current theories of the physics of heat conduction in pure dielectric crystals at low temperatures suggest that there can be a range of temperatures in which a relation of the form (3) holds approximately with the order of magnitude of κ/τ equal to $\frac{1}{3}cV^2$, with c the heat capacity at constant volume and V an

---

* For a modern presentation and extension of Maxwell's result see Chapters XIII and XVII of the treatise of Truesdell and Muncaster [3].

average value of the phonon velocities, which depend, in general, on frequency and polarization as well as direction.* Pao and Banerjee [17] (see also, Banerjee and Pao [18]) have remarked that for anisotropic media the natural generalization of the relation (3) is

$$\underset{\sim}{T}(\theta)\dot{\underset{\sim}{q}} + \underset{\sim}{q} = -\underset{\sim}{K}(\theta)\underset{\sim}{g} , \tag{4}$$

with $\underset{\sim}{K}(\theta)$ as in (2) and with $\underset{\sim}{T}(\theta)$, like $\underset{\sim}{K}(\theta)$, a temperature-dependent, positive definite, second-order tensor.** When $\dot{\underset{\sim}{q}} = \underset{\sim}{0}$, equation (4) reduces to equation (2), and hence, in the theory of (4), one may call $\underset{\sim}{K}(\theta)$ the <u>steady state thermal conductivity tensor</u>. We call $\underset{\sim}{T}(\theta)$ the <u>tensor of relaxation times</u>.

In an article [19] for the Archive for Rational Mechanics and Analysis, we have derived the restrictions that the second law of thermodynamics places on the constitutive relations of a class of materials that includes those considered by Cattaneo and Pao and Banerjee. We there show that the relation (4) with the tensors $\underset{\sim}{T}(\theta)$ and $\underset{\sim}{K}(\theta)$ non-singular is compatible with thermodynamics only if $\underset{\sim}{K}(\theta)$ is positive definite, the tensor

---

* Many authors have proposed modifications and extensions of the relation (3). Of particular relevance is the derivation, for dielectric crystals, given by Guyer and Krumhansl [5]. Also of interest are the earlier articles on second sound by Ward and Wilks [6], Dingle [7], Sussman and Thellung [8], Griffin [9], Prohofsky and Krumhansl [10], and Guyer and Krumhansl [11], and the more recent articles of Enz [12], Kwok [13], and Hardy [14]. The problem of formulating constitutive relations that yield a finite velocity for the propagation of thermal disturbances has been discussed from another point of view by Gurtin and Pipkin [15] and Morro [16].

** It is clear that the relations (3) and (4) are not invariant under time-dependent changes of frame, but this lack of invariance is not important for the problems we treat. A modification of (4) that <u>is</u> invariant under all changes of frame is $\underset{\sim}{T}(\theta)(\dot{\underset{\sim}{q}} - \underset{\sim}{W}\underset{\sim}{q}) + \underset{\sim}{q} = -\underset{\sim}{K}(\theta)\underset{\sim}{g}$ with $\underset{\sim}{W}$ either the velocity or the vorticity tensor, i.e., the skew part of the velocity gradient. (As our discussion is confined to rigid bodies, in any motion of the materials we consider the velocity gradient is skew.) Pao and Banerjee [17][18] considered non-rigid bodies and discussed the relations between thermal and acoustical waves in the framework of the linear theory of infinitesimal elastic deformations.

$$\underset{\sim}{Z}(\theta) = \underset{\sim}{K}(\theta)^{-1}\underset{\sim}{T}(\theta) \tag{5}$$

is symmetric, i.e.,

$$\underset{\sim}{Z}(\theta)^T = \underset{\sim}{Z}(\theta) , \tag{6}$$

<u>and</u> the specific internal energy e (per unit volume), the specific entropy η, and the specific Helmholtz free energy $\psi = e - \theta\eta$ are <u>not</u> given by functions of θ alone, but are instead given by functions $\tilde{e}$, $\tilde{\eta}$, and $\tilde{\psi}$ of the form*

$$e = \tilde{e}(\theta,\underset{\sim}{g}) = e_0(\theta) + \frac{1}{\theta}\,\underset{\sim}{g}\cdot\underset{\sim}{Z}(\theta)\underset{\sim}{g} - \frac{1}{2}\,\underset{\sim}{g}\cdot\frac{d}{d\theta}\,\underset{\sim}{Z}(\theta)\underset{\sim}{g} , \tag{7a}$$

$$\eta = \tilde{\eta}(\theta,\underset{\sim}{g}) = \eta_0(\theta) + \frac{1}{2\theta^2}\,\underset{\sim}{g}\cdot\underset{\sim}{Z}(\theta)\underset{\sim}{g} - \frac{1}{2\theta}\,\underset{\sim}{g}\cdot\frac{d}{d\theta}\,\underset{\sim}{Z}(\theta)\underset{\sim}{g} , \tag{7b}$$

$$\psi = \tilde{\psi}(\theta,\underset{\sim}{g}) = \psi_0(\theta) + \frac{1}{2\theta}\,\underset{\sim}{g}\cdot Z(\theta)\underset{\sim}{g} ; \tag{7c}$$

these functions obey the relations

$$\partial_\theta\tilde{\psi} = -\tilde{\eta} \quad \text{and} \quad \theta\partial_\theta\tilde{\eta} = \partial_\theta\tilde{e} , \tag{8}^{**}$$

which imply the familiar formulae,

$$d\psi_0/d\theta = -\eta_0 \quad \text{and} \quad d\eta_0/d\theta = c_0/\theta , \tag{9}$$

in which c_0 is the "equilibrium heat capacity", i.e.,

---

* (Footnote added to the English version, May, 1983): We have recently seen a paper by P.J. Chen and M.E. Gurtin [On second sound in materials with memory, Zeits. Angew. Math. Phys., 21, 232-241 (1970)] in which they extend part of the theory of Gurtin and Pipkin [15] to deformable media and discuss specializations of that theory to unidimensional cases. It is clear from the example treated in their §6 that, despite differences in language, methods, and initial assumptions, in the special case in which $\underset{\sim}{T}$ and $\underset{\sim}{K}$ are independent of θ and the heat flow is unidimensional, Chen and Gurtin's theory intersects ours and agrees with it in the conclusion that an equation equivalent to our equation (7c) can hold. Chen and Gurtin do not observe that equation (7c) is <u>implied</u> by equation (4), i.e., that (7c) is the <u>only</u> free energy function compatible with (4). Of course, the papers of Gurtin and Pipkin [15] and Chen and Gurtin predate not only our own work on second sound but that of Pao and Banerjee [17][18] on second sound in deformable media.

$$c_0(\theta) = de_0(\theta)/d\theta \; . \tag{10}$$

In the absence of both deformation and a supply of heat by radiation, the law of balance of energy takes the form

$$\dot{e} + \operatorname{div} \underset{\sim}{q} = 0 \; . \tag{11}$$

If we let

$$\underset{\sim}{A}(\theta) = \frac{1}{\theta}\underset{\sim}{Z}(\theta) - \frac{1}{2}\frac{d}{d\theta}\underset{\sim}{Z}(\theta) = -\frac{\theta^2}{2}\frac{d}{d\theta}\left(\frac{\underset{\sim}{Z}(\theta)}{\theta^2}\right) , \tag{12}$$

then equation (7a) becomes

$$\tilde{e}(\theta,\underset{\sim}{q}) = e_0(\theta) + \underset{\sim}{q}\cdot\underset{\sim}{A}(\theta)\underset{\sim}{q} \; , \tag{13}$$

and, clearly, $\dot{e}$ in (11) is <u>not</u> given by the classical formula, $\dot{e} = c_0(\theta)\dot{\theta}$ (which was, of course, employed in references [2], [4], [5], [14], [17], and [18]), but is instead given by the expression

$$\dot{e} = [c_0(\theta) + \underset{\sim}{q}\cdot\underset{\sim}{B}(\theta)\underset{\sim}{q}]\dot{\theta} + 2\underset{\sim}{q}\cdot\underset{\sim}{A}(\theta)\dot{\underset{\sim}{q}} \tag{14}$$

with

$$\underset{\sim}{B}(\theta) = \frac{d}{d\theta}\underset{\sim}{A}(\theta) \; . \tag{15}$$

Thus, the evolution of the heat flux and temperature fields is governed by a pair of partial differential equations,

$$\left.\begin{aligned} \underset{\sim}{T}(\theta)\dot{\underset{\sim}{q}} + \underset{\sim}{q} + \underset{\sim}{K}(\theta)\operatorname{grad}\theta &= \underset{\sim}{0} \; , \\ \operatorname{div}\underset{\sim}{q} + c_0(\theta)\dot{\theta} + \underset{\sim}{q}\cdot B(\theta)\,\underset{\sim}{q}\,\dot{\theta} + 2\underset{\sim}{q}\cdot\underset{\sim}{A}(\theta)\dot{\underset{\sim}{q}} &= 0 \; , \end{aligned}\right\} \tag{16}$$

for which the tensorial coefficients $\underset{\sim}{A}(\theta)$ and $\underset{\sim}{B}(\theta)$ in the second equation are determined by the temperature-dependence of the coefficients $\underset{\sim}{T}(\theta)$ and $\underset{\sim}{K}(\theta)$ in the first. As the relations (12) and (15) do not, in general, yield $\underset{\sim}{A} = \underset{\sim}{B} = \underset{\sim}{0}$, the second equation in (16) is non-linear in $\underset{\sim}{q}$.

Below we give a derivation of the relations (6)-(8) that has certain advantages over that which we gave in reference [19]. Both derivations are set in the framework of Coleman and Owen's general theory of thermodynamical systems [20]

[21]. Here we take a state to be a pair $(\theta,\underset{\sim}{q})$, rather than $(e,\underset{\sim}{q})$, * and we avoid the assumption that the function $\theta \mapsto \underset{\sim}{e}(\theta,\underset{\sim}{q})$ is invertible at fixed $\underset{\sim}{q}$. In future papers Coleman and Owen will discuss circumstances under which such invertibility does not hold, and, for fixed non-zero values of $\underset{\sim}{q}$, $\partial_\theta \tilde{e}(\theta,\underset{\sim}{q})$ changes sign from positive to negative as θ decreases toward 0 .**

---

* In [19] we treat at length a general class of materials for which states are pairs $(e,\underset{\sim}{\alpha})$, or $(\theta,\underset{\sim}{\alpha})$, whose second members, $\underset{\sim}{\alpha}$, become $\underset{\sim}{q}$ only in certain special cases such as that in which (4) holds.

** (Footnote added to the English version, May, 1983): The first of these papers has been printed: B.D. Coleman and D.R. Owen, On the nonequilibrium behavior of solids that transport heat by second sound, Comp. Math. with Appls. 9, 529-546 (1983). The abstract reads as follows:

> In earlier work Coleman, Fabrizio, and Owen observed that if a constitutive equation for the heat flux $\underset{\sim}{q}$ employed in theories of second sound in dielectric solids is assumed to hold for large values of $\underset{\sim}{q}$, then thermodynamical principles require that the equation of state for the internal energy contain a term that is quadratic in $\underset{\sim}{q}$. Here the resulting field equations for unidimensional heat flow are studied. It is shown that they are compatible with the occurrence of oscillatory traveling waves containing discontinuities in temperature and heat flux. The class of waves constructed contains limiting cases in which the heat flux and temperature exhibit small amplitude, high frequency, fluctuations about their mean. Such a traveling wave would appear to an experimenter as a strong and nearly constant flow of heat in a nearly uniform temperature field.

Derivation of Thermodynamical Relations

We are concerned with materials for which the state at a point (or material element) can be described by giving the local temperature θ and the local heat flux $\underset{\sim}{q}$. The collection Σ of all states of an element is here a set of the form $\underline{I} \times \underline{A}$ with $\underline{I}$ an open interval of positive numbers and $\underline{A}$ an open connected subset of $V^{(3)}$ containing the zero vector $\underset{\sim}{0}$. ($V^{(3)}$ is the three-dimensional Euclidean vector space in which $\underset{\sim}{q}$ lies.) We write either σ or $(\theta,\underset{\sim}{q})$ for the members of Σ, and, when we discuss such concepts from analysis as derivatives and line integrals, we regard Σ as a subset of the vector space $R \oplus V^{(3)}$ with the inner product

$$\langle(\alpha,\underset{\sim}{a}),(\beta,\underset{\sim}{b})\rangle = \alpha\beta + \underset{\sim}{a}\cdot\underset{\sim}{b} \tag{17}$$

("$\cdot$" is the inner product on $V^{(3)}$).

With each material there are associated three continuously differentiable functions:

$$\left.\begin{aligned} \underset{\sim}{T}: &\quad \underline{I} \to \mathrm{Lin}^{\neq}(V^{(3)}), \\ \underset{\sim}{K}: &\quad \underline{I} \to \mathrm{Lin}^{\neq}(V^{(3)}), \\ \tilde{e}: &\quad \Sigma \to R, \end{aligned}\right\} \tag{18}$$

with $\mathrm{Lin}^{\neq}(V^{(3)})$ the set of all invertible linear transformations of $V^{(3)}$ into $V^{(3)}$. Each process of an element of the material is a piecewise continuous (p-C^0) function P_t mapping an interval $[0,t)$, with $t > 0$, into $R \oplus V^{(3)}$. The values of P_t are pairs $(\zeta(\xi),\underset{\sim}{g}(\xi))$ in which $\zeta(\xi)$ is the time-derivative of θ and $\underset{\sim}{g}(\xi)$ is the spatial gradient of θ. The set Π of all processes of a given element is defined as follows: For each p-C^0 function $P_t = (\zeta,g): [\underset{\sim}{0},t) \to R \oplus V^{(3)}$, let $D(P_t)$ be the set of states $\sigma_0 = (\theta_0,\underset{\sim}{q}_0)$ for which the equations

$$\left.\begin{aligned} \dot{\theta} &= \zeta, \\ \dot{\underset{\sim}{q}} &= -\underset{\sim}{T}(\theta)^{-1}\underset{\sim}{q} - \underset{\sim}{T}(\theta)^{-1}\underset{\sim}{K}(\theta)\underset{\sim}{g}, \end{aligned}\right\} \tag{19}$$

with the initial conditions

$$\theta(0) = \theta_0 \,, \qquad \underset{\sim}{g}(0) = \underset{\sim}{g}_0 \,, \tag{20}$$

have a solution $\xi \mapsto (\theta(\xi), \underset{\sim}{g}(\xi))$ whose values lie in Σ for all ξ in $[0,t]$; if $D(P_t)$ is not empty, then P_t is a process, i.e.,

$$\Pi = \{\, P_t\colon [0,t) \to R \oplus V^{(3)},\ \text{p-}C^0 \mid D(P_t) \neq \phi \,\} \,. \tag{21}$$

For each pair (P_t, σ_0) with P_t in Π and $\sigma_0 = (\theta_0, \underset{\sim}{g}_0)$ in $D(P_t)$, the solution $\xi \mapsto (\theta(\xi), g(\xi))$ of (19) obeying (20) is called the parameterized trajectory in Σ corresponding to (P_t, σ_0). The final point $(\theta(t), \underset{\sim}{g}(t))$ of this trajectory is interpreted as the "state $\sigma_t = (\theta_t, \underset{\sim}{g}_t)$ at the instant of completion of the process P_t". The dependence of σ_t upon the "initial state" σ_0 is indicated by writing

$$\sigma_t = \rho_{P_t} \sigma_0 \,, \tag{22}$$

and the operator ρ_{P_t} so defined is called the state-transformation function induced by the process P_t. Familiar theorems in the theory of differential equations tell us that, for each P_t in Π, the domain $D(P_t)$ of ρ_{P_t} is an open subset of Σ, and ρ_{P_t} is not only single-valued but also continuous on $D(P_t)$.

A pair (P_t, σ_0) for which σ_0 is in $D(P_t)$ and

$$\rho_{P_t} \sigma_0 = \sigma_0 \tag{23}$$

is called cyclic.

If P_t is in Π, ν is in $(0,t)$, and P_ν is the restriction of P_t to $[0,\nu)$, then P_ν is also in Π, and $D(P_\nu)$ contains $D(P_t)$ as a subset. Moreover, if P_{t_1} and P_{t_2} are in Π, and if the range $R(P_{t_1})$ of $\rho_{P_{t_1}}$ intersects the domain $D(P_{t_2})$ of $\rho_{P_{t_2}}$, then the function $P_{t_1+t_2}$, defined on $[0, t_1 + t_2)$ by the formula

$$P_{t_1+t_2}(\xi) = \begin{cases} P_{t_1}(\xi), & \xi \in [0,t_1) \,, \\ P_{t_2}(\xi - t_1) \,, & \xi \in [t_1, t_1+t_2) \,, \end{cases} \tag{24}$$

is in Π, has $D(P_{t_1+t_2}) = \rho_{P_{t_1}}^{-1}(D(P_{t_2}) \cap R(P_{t_1}))$, and $\rho_{P_{t_1+t_2}}\sigma = \rho_{P_{t_2}}\rho_{P_{t_1}}\sigma$ for each σ in $D(P_{t_1+t_2})$; $P_{t_1+t_2}$ is called <u>the process resulting from the successive application of</u> (first) P_{t_1} <u>and</u> (then) P_{t_2}.

Let c be an oriented $p-C^1$ (piecewise continuously differentiable) curve that lies in Σ, and let the function $\xi \mapsto (\theta(\xi),\underset{\sim}{q}(\xi))$, from $[0,t]$ into Σ, be a $p-C^1$ parameterization of c. Clearly, the function $P_t = (\zeta,g)$, defined on $[0,t)$ by the equations

$$\left.\begin{aligned} \zeta(\xi) &= \dot{\theta}(\xi) , \\ \underset{\sim}{g}(\xi) &= -\underset{\sim}{K}(\theta(\xi))^{-1}\underset{\sim}{T}(\theta(\xi))\dot{\underset{\sim}{q}}(\xi) - \underset{\sim}{K}(\theta(\xi))^{-1}\underset{\sim}{q}(\xi) , \end{aligned}\right\} \tag{25}$$

is $p-C^0$, and if this function is substituted into equation (19), the solution of (19) with initial value σ_0 will be precisely the function $\xi \mapsto (\theta(\xi),q(\xi))$. It follows that P_t is in Π, $\xi \mapsto (\theta(\xi),q(\xi))$ is the parameterized trajectory in Σ corresponding to the pair (P_t,σ_0), and $\rho_{P_t}\sigma_0$ equals $(\theta(t),q(t))$, the final point of c. In summary, we here assert, as in [19],

<u>Remark</u> 1. If c is an oriented $p-C^1$ curve lying in Σ, then each $p-C^1$ parameterization of c is the parameterized trajectory corresponding to a unique pair (P_t,σ_0) with P_t in Π and σ_0 in $D(P_t)$; σ_0 is the initial point of c; $\rho_{P_t}\sigma_0$ is the final point of c. If c is a closed curve, then $\rho_{P_t}\sigma_0 = \sigma_0$, and the pair (P_t,σ_0) is cyclic.

Because each pair of points in Σ can be joined by a $p-C^1$ curve, Remark 1 implies the validity of the following assertion:

<u>Remark</u> 2. For each pair (σ',σ'') of states in Σ, there is a process P_t in Π with σ' in $D(P_t)$ and $\sigma'' = \rho_{P_t}\sigma'$; in other words, the set $\{\sigma \in \Sigma \mid \sigma = \rho_{P_t}\sigma_0\}$ of states "accessible" from any given state σ_0 is equal to all of Σ.

It follows from these observations that the material elements under consideration are <u>systems</u> in the sense in which the term is used in the general theory of references [20] and [21].

We turn now to the function $\tilde{e}$. The value $e = \tilde{e}(\theta,q)$ of $\tilde{e}$ is the specific internal energy. The rate of change of e along the parameterized trajectory $\xi \mapsto (\theta(\xi),q(\xi))$ corresponding to a pair (P_t,σ) is, for each ξ in $[0,t]$,

$$\dot{e}(\xi) = \partial_\theta \tilde{e}(\theta(\xi),\underset{\sim}{q}(\xi))\dot{\theta}(\xi) + \partial_{\underset{\sim}{q}} \tilde{e}(\theta(\xi),\underset{\sim}{q}(\xi))\cdot\dot{\underset{\sim}{q}}(\xi) \ , \tag{26}$$

and, according to the first law of thermodynamics,

$$\dot{e}(\xi) = h(\xi) \ , \tag{27}$$

where h is the rate at which heat is absorbed at the material element. This rate is determined by the heat flux $\underset{\sim}{q}$ and the rate r of supply of heat by radiation from sources external to the element:

$$h = r - \operatorname{div} \underset{\sim}{q} \ . \tag{28}$$

Our assumption that e is given by a function $\tilde{e}$ of state is clearly compatible with the first law of thermodynamics. It has been customary in this subject to assume that the function $\tilde{e}$ reduces to a function of θ alone, i.e., to assume that e is independent of $\underset{\sim}{q}$, but, as we shall show below, such an assumption is not compatible with thermodynamics. Using the method of Coleman and Owen [20],* we shall show that the second law implies that $\tilde{e}$ must have the form shown in equation (7a), with $\underset{\sim}{Z}$ as in equation (5).

Along the parameterized trajectory $\xi \mapsto (\theta(\xi),\underset{\sim}{q}(\xi))$ corresponding to a pair (P_t,σ), one may calculate the integral

$$s(P_t,\sigma) = \int_0^t \left(\frac{r}{\theta} - \operatorname{div} \frac{\underset{\sim}{q}}{\theta} \right) d\xi \ ; \tag{29}$$

---

* The principal advantage of this method over that proposed by Coleman and Noll [1] and Coleman and Mizel [22] [23], and employed in the investigations of Gurtin and Pipkin [15], Coleman and Gurtin [24], and Morro [16], is that it does not require a priori assumptions about the regularity or even existence of entropy (or free energy) as a function of state. The method we use, i.e., that of reference [20], also avoids a situation encountered by Morro [16], who observed that his starting assumptions are not in accord with equipresence.

for, by (28) and the relation

$$\operatorname{div} \frac{q}{\theta} = \frac{1}{\theta} \operatorname{div} q - \frac{1}{\theta^2} q \cdot g \,, \tag{30}$$

there holds

$$s(P_t,\sigma) = \int_0^t \left(\frac{h(\xi)}{\theta(\xi)} + \frac{q(\xi)\cdot g(\xi)}{\theta(\xi)^2} \right) d\xi \,, \tag{31}$$

and hence (26) and (27) yield

$$s(P_t,\sigma) = \int_0^t \left(\frac{\partial_\theta \tilde{e}(\theta(\xi),q(\xi))}{\theta(\xi)} \dot{\theta}(\xi) + \frac{\partial_q e(\theta(\xi),q(\xi))}{\theta(\xi)} \cdot \dot{q}(\xi) + \frac{q(\xi)\cdot g(\xi)}{\theta(\xi)^2} \right) d\xi \,. \tag{32}$$

We here take this last expression as the definition of s; it makes obvious the fact that s is well defined for <u>every</u> pair (P_t,σ) with P_t in Π and σ in $D(P_t)$. It is not difficult to show that, for each process P_t, the function $\sigma \mapsto s(P_t,\sigma)$ is continuous on $D(P_t)$. Moreover, if $P_{t_1+t_2}$ is the result of the successive application of P_{t_1} and P_{t_2}, then for each σ in $D(P_{t_1+t_2})$,

$$s(P_{t_1+t_2}, \sigma) = s(P_{t_1}, \sigma) + s(P_{t_2}, \rho_{P_{t_1}} \sigma). \tag{33}$$

Hence, s is an <u>action</u> for the system (Σ,Π) in the sense of Definition 2.2 of reference [20]. In accord with Definition 3.1 of that paper, we say that s has the <u>Clausius property</u> at a state σ_0 if, for each $\varepsilon > 0$, σ_0 has an open neighborhood, $\mathcal{O}_\varepsilon(\sigma_0)$, for which

$$P_t \in \Pi, \quad \sigma_0 \in D(P_t), \quad \rho_{P_t}\sigma_0 \in \mathcal{O}_\varepsilon(\sigma_0) \text{ implies } s(P_t,\sigma_0) < \varepsilon \,; \tag{34}$$

i.e., $s(P_t,\sigma_0)$ is approximately negative whenever the trajectory in Σ determined by the pair (P_t,σ_0) is approximately closed.

As in [19], [20], [21], we here take the second law of thermodynamics to be the following assertion:

Second Law. <u>The action s has the Clausius property at least at one state in Σ.</u>

It follows from Remark 2 above and Remark 3.1 of reference [20] that <u>the second law here implies that</u> s <u>has the Clausius property at every state in</u> Σ.

If a pair (P_t,σ) is cyclic, i.e., if $\rho_{P_t}\sigma = \sigma$, then $\rho_{P_t}\sigma$ is in every neighborhood of σ, and, if s has the Clausius property at σ, $s(P_t,\sigma)$ is less than every $\varepsilon > 0$, which means that $s(P_t,\sigma) \leq 0$. Hence the second law has the following implication: <u>If the pair</u> (P_t,σ) <u>is cyclic</u>, $s(P_t,\sigma)$ <u>is not positive</u>; i.e., <u>for each</u> σ <u>in</u> Σ,

$$P_t \in \Pi\,, \quad \sigma \in D(P_t)\,, \quad \rho_{P_t}\sigma = \sigma \quad \text{implies} \quad s(P_t,\sigma) \leq 0\,. \tag{35}$$

Let c be an oriented p-C^1 curve lying in Σ. Each p-C^1 parameterization of c is, by Remark 1, the parameterized trajectory $\xi \mapsto (\theta(\xi),\underset{\sim}{g}(\xi))$ corresponding to a pair (P_t,σ) with σ in $D(P_t)$, and for this pair the equations (32) and (19) yield the formula,

$$s(P_t,\sigma) = \int_0^t \Big(\frac{\partial_\theta \tilde{e}(\theta,\underset{\sim}{q})}{\theta}\,\dot{\theta} + \frac{\theta\partial_{\underset{\sim}{g}}\tilde{e}(\theta,\underset{\sim}{q}) - Z^T(\theta)\underset{\sim}{q}}{\theta^2}\cdot\dot{\underset{\sim}{q}} - \frac{\underset{\sim}{q}\cdot \underset{\sim}{K}(\theta)^{-1}\underset{\sim}{q}}{\theta^2}\Big)d\xi\,, \tag{36}$$

in which θ, $\underset{\sim}{q}$, $\dot{\theta}$, and $\dot{\underset{\sim}{q}}$ stand for $\theta(\xi)$, $\underset{\sim}{q}(\xi)$, $\dot{\theta}(\xi)$ and $\dot{\underset{\sim}{q}}(\xi)$. Therefore, $s(P_t,\sigma)$ can be written as a sum,

$$s(P_t,\sigma) = J_1 + J_2\,, \tag{37}$$

of two terms, the first of which,

$$J_1 = J_1(c) \doteq \int_c \Big(\frac{\partial_\theta \tilde{e}(\theta,\underset{\sim}{q})}{\theta}\,d\theta + \frac{\theta\partial_{\underset{\sim}{g}}\tilde{e}(\theta,\underset{\sim}{q}) - Z(\theta)^T\underset{\sim}{q}}{\theta^2}\cdot d\underset{\sim}{q}\Big)\,, \tag{38}$$

is a line integral independent of the parameterization of c, and the second,

$$J_2 = -\int_0^t \frac{\underset{\sim}{q}\cdot \underset{\sim}{K}(\theta)^{-1}\underset{\sim}{q}}{\theta^2}\,d\xi\,, \tag{39}$$

does depend on the parameterization, but has the bound

$$|J_2| \leq Mt\,, \tag{40}$$

where

$$M = \sup_{c} |\underset{\sim}{g} \cdot \underset{\sim}{K}(\theta)^{-1} \underset{\sim}{g}/\theta^2| \tag{41}$$

is finite because $\underset{\sim}{K}$ is continuous on $\underline{I} \subset (0,\infty)$, and c is a compact subset of $\Sigma = \underline{I} \times A$.

Now, let c be not only an oriented p-C^1 curve in Σ but also a closed curve. For each p-C^1 parameterization of c, the corresponding pair (P_t, σ) is cyclic, and hence (35) and (37) yield

$$J_1(c) + J_2 \leq 0 . \tag{42}$$

In view of the bound (40) on J_2, the relation (42) can hold for <u>all</u> p-C^1 parameterizations of c only if

$$J_1(c) \leq 0 . \tag{43}$$

For the curve -c that differs from c only in orientation, equation (38) yields $J_1(-c) = -J_1(c)$, but the argument that gave (43) yields also $J_1(-c) \leq 0$, and, therefore,

$$J_1(c) = 0 . \tag{44}$$

As c is an arbitrary p-C^1 closed curve in Σ, and Σ is an open connected subset of $R \oplus V^{(3)}$, (44) and a familiar theorem about the existence of potentials for vector fields yield the existence of a continuously differentiable real-valued function $\tilde{\eta}$ on Σ such that

$$\theta \partial_\theta \tilde{\eta}(\theta, \underset{\sim}{g}) = \partial_\theta \tilde{e}(\theta, \underset{\sim}{g}) , \tag{45}$$

$$\theta^2 \partial_{\underset{\sim}{g}} \tilde{\eta}(\theta, \underset{\sim}{g}) = \theta \partial_{\underset{\sim}{g}} \tilde{e}(\theta, \underset{\sim}{g}) - \underset{\sim}{Z}(\theta)^T \underset{\sim}{g} , \tag{46}$$

and, for any oriented p-C^1 curve c in Σ, closed or not,

$$J_1(c) = \tilde{\eta}(\sigma_2) - \tilde{\eta}(\sigma_1) , \tag{47}$$

with σ_1 the initial and σ_2 the final point of c .

Equations (37) and (36) imply that for each pair (P_t, σ) with σ in $D(P_t)$,

$$s(P_t,\sigma) = \eta(\rho_{P_t}\sigma) - \eta(\sigma) + J_2 , \tag{48}$$

where J_2 is as in equation (39).

If for a given state $\sigma_0 = (\theta_0, q_0)$ and positive number t, we let $P_t^\circ = (\xi^\circ, \underset{\sim}{g}^\circ)$ be the process defined on $[0,t)$ by

$$\left.\begin{aligned} \xi^\circ &= 0 , \\ \underset{\sim}{g}^\circ &\equiv -\underset{\sim}{K}(\theta_0)^{-1} \underset{\sim}{q}_0 , \end{aligned}\right\} \tag{49}$$

the, the corresponding solution of (19) is the constant

$$\theta \equiv \theta_0 , \qquad q \equiv q_0 , \tag{50}$$

then pair (P_t°, σ_0) is cyclic, and (35), (48), and (39) yield

$$0 \leq s(P_t^\circ, \sigma_0) = -t\underset{\sim}{q}_0 \cdot \underset{\sim}{K}(\theta_0)^{-1} \underset{\sim}{q}_0/\theta_0^2 . \tag{51}$$

Thus, the second law implies that for each θ in $\underline{I}$,

$$\underset{\sim}{g} \cdot \underset{\sim}{K}(\theta)^{-1} \underset{\sim}{g} \geq 0 , \tag{52}$$

for all $\underset{\sim}{g}$ in $\underline{A}$ and, as $\underset{\sim}{K}(\theta)^{-1}$ is an invertible tensor and $\underline{A}$ contains a spherical neighborhood of $\underset{\sim}{0}$, this implies that $\underset{\sim}{K}(\theta)$ <u>is positive definite for each</u> θ <u>in</u> $\underline{I}$. Furthermore, whether or not (P_t, σ) is cyclic, (51) and (39) together imply

$$J_2 \leq 0 , \tag{53}$$

and hence (48) yields

$$s(P_t,\sigma) \leq \tilde{\eta}(\rho_{P_t}\sigma) - \tilde{\eta}(\sigma) . \tag{54}$$

As this relation holds for all pairs (P_t,σ) with P_t in Π and σ in $D(P_t)$, we may assert that $\tilde{\eta}$ is an <u>entropy function</u>.* Of course, the

---

* It follows from (54) and the continuity of $\tilde{\eta}$ that $\tilde{\eta}$ is also an "upper potential" for s in the sense of Definition 3.2 of [20].

existence of this entropy function implies that $s(P_t,\sigma)$ is not positive when the pair (P_t,σ) is cyclic. Employing that continuity of $\tilde{\eta}$, one may show further that (54) implies that s has the Clausius property at each state in Σ, and hence we can assert

Remark 3. The response functions $\underset{\sim}{T}$, $\underset{\sim}{K}$, and $\tilde{e}$ are compatible with the second law of thermodynamics if, and only if, there is a continuously differentiable function $\tilde{\eta} : \Sigma \to R$ obeying (54) for all pairs (P_t,σ) with P_t in Π and σ in $D(P_t)$.

Suppose now that we have, in addition to the function $\tilde{\eta}$ of equation (47), another function $\bar{\eta} : \Sigma \to R$ that is an entropy function for the same material element, i.e., that obeys the relation

$$\bar{\eta}(\rho_{P_t}\sigma) - \bar{\eta}(\sigma) \geq s(P_t,\sigma) \tag{55}$$

for each pair (P_t,σ) with σ in $D(P_t)$. For each pair of states (σ_1,σ_2), there is an oriented p-C^1 curve c that has σ_1 as its initial point and σ_2 as its final point; for each p-C^1 parameterization of c, (55) and (37) yield

$$\bar{\eta}(\sigma_2) - \bar{\eta}(\sigma_1) \geq J_1(c) + J_2 \tag{56}$$

and hence, by (40),

$$\bar{\eta}(\sigma_2) - \bar{\eta}(\sigma_1) \geq J_1(c) \ . \tag{57}$$

If we now interchange σ_1 and σ_2, and replace c by the curve $-c$ that differs from c only in orientation, the same argument gives

$$\bar{\eta}(\sigma_1) - \bar{\eta}(\sigma_2) \geq J_1(-c) = -J_1(c) \ , \tag{58}$$

which is compatible with (57) only if $\bar{\eta}(\sigma_2) - \bar{\eta}(\sigma_1) = J_1(c)$, and, in view (47), we may conclude that

$$\bar{\eta}(\sigma_2) - \bar{\eta}(\sigma_1) = \tilde{\eta}(\sigma_2) - \tilde{\eta}(\sigma_1) \ ; \tag{59}$$

that is, $\tilde{\tilde{\eta}}$ can differ from $\tilde{\eta}$ by only a constant.*

The existence of an entropy function, $\tilde{\tilde{\eta}}$, i.e., a function from Σ to R obeying (55), implies that validity of (35), from which, as we have shown, there follows the existence of a continuously differential entropy function $\tilde{\eta}$ that can differ from $\tilde{\tilde{\eta}}$ by at most a constant. Thus we have

Remark 4. If a material element has an entropy function, it is continuously differentiable and unique to within a constant.

The value η of the entropy function $\tilde{\eta}$ is called, of course, the entropy; here $\tilde{\eta}$ is unique if we assign the value 0 to the entropy in a "standard state" σ^0 . If $\underline{I}$ has the form $(0,\alpha)$, and the function η_0 , defined on $\underline{I}$ by

$$\eta_0(\theta) = \tilde{\eta}(\theta,0) , \tag{60}$$

has a limit as $\theta \to 0$, then a natural normalization of $\tilde{\eta}$ is obtained by putting

$$\lim_{\theta\to 0} \eta_0(\theta) = 0 . \tag{61}$$

The value ψ of the function $\tilde{\psi}$ defined on Σ by

$$\tilde{\psi}(\theta,\underset{\sim}{g}) = \tilde{e}(\theta,\underset{\sim}{g}) - \theta\tilde{\eta}(\theta,\underset{\sim}{g}) \tag{62}$$

is the Helmholtz free energy. The assumed smoothness of $\tilde{e}$ and the derived smoothness of $\tilde{\eta}$ imply that $\tilde{\psi}$ is continuously differentiable, and, in view of equation (45), we have, throughout Σ ,

$$\partial_\theta\tilde{\psi}(\theta,\underset{\sim}{g}) = -\tilde{\eta}(\theta,\underset{\sim}{g}) , \tag{63}$$

and equation (46) yields

$$\theta\partial_{\underset{\sim}{g}}\tilde{\psi}(\theta,\underset{\sim}{g}) = \underset{\sim}{Z}(\theta)^T\underset{\sim}{g} . \tag{64}$$

This last relation tells us that $\tilde{\psi}$ must have the form shown in (7c), and once

---

* Our proofs of the existence, differentiability, and uniqueness of $\tilde{\eta}$ parallel proofs we gave in [19] and rest on arguments introduced in the discussion of the thermodynamics of elastic elements with heat conduction in [20].

that is known, the relation (63) implies that $\tilde{\eta}$ must be as shown in (7b). Clearly, (7b), (7c), and (62) imply that $\tilde{e}$ must be as in (7a).

Of course, (63) and (45) are the same as the relations (8). From (64) we conclude that, for each θ in $\underline{I}$, the function $\underset{\sim}{q} \to \tilde{\psi}(\theta,\underset{\sim}{q})$ has a gradient of order two given by

$$\theta \partial^2_{\underset{\sim}{q}} \tilde{\psi}(\theta,\underset{\sim}{q}) = \underset{\sim}{Z}(\theta)^T , \tag{65}$$

which implies that $\underset{\sim}{Z}(\theta)$ <u>is symmetric for each</u> θ in $\underline{I}$, i.e., (6) holds.

We summarize in the following

Theorem.* <u>The second law implies that for each value of</u> θ:

(i) <u>the tensor</u> $\underset{\sim}{K}(\theta)$ <u>is positive definite;</u>

(ii) <u>the tensor</u> $\underset{\sim}{Z}(\theta) = \underset{\sim}{K}(\theta)^{-1}\underset{\sim}{T}(\theta)$ <u>is symmetric;</u>

(iii) e, η, <u>and</u> ψ <u>are not independent of</u> $\underset{\sim}{q}$, <u>but are instead given by functions</u> $\tilde{e}$, $\tilde{\eta}$ <u>and</u> $\tilde{\psi}$ <u>that are related as shown in</u> (8) <u>and have the forms shown in</u> (7).

<u>Remark</u> 5. It is a consequence of the relations (45) and (46) and the derived smoothness of the entropy function that when θ and $\underset{\sim}{q}$ are continuous so also is $\dot{\eta}$, and

$$\begin{aligned}\dot{\eta} &= \partial_\theta \tilde{\eta}(\theta,\underset{\sim}{q})\dot{\theta} + \partial_{\underset{\sim}{q}} \tilde{\eta}(\theta,\underset{\sim}{q})\cdot\dot{\underset{\sim}{q}} \\ &= [\theta \partial_\theta \tilde{e}(\theta,\underset{\sim}{q})\dot{\theta} + \theta \partial_{\underset{\sim}{q}} \tilde{e}(\theta,\underset{\sim}{q})\cdot\dot{\underset{\sim}{q}} - \underset{\sim}{q}\cdot\underset{\sim}{Z}(\theta)\dot{\underset{\sim}{q}}]/\theta^2 \\ &= \dot{e}/\theta - \underset{\sim}{q}\cdot\underset{\sim}{Z}(\theta)\dot{\underset{\sim}{q}}/\theta^2 .\end{aligned} \tag{66}$$

The quantity

$$\gamma = \dot{\eta} + \mathrm{div}(q/\theta) - r/\theta \tag{67}$$

is called that <u>rate of production of entropy</u>; it follows from (66), (27), (28), and (4) that here

---

* Cf. [19], Theorem 4.1.

$$\gamma = \underset{\sim}{g} \cdot \underset{\sim}{K}(\theta)^{-1} \underset{\sim}{g} / \theta^2 . \tag{68}$$

The positive-definiteness of $\underset{\sim}{K}$ implies that γ is not negative and vanishes only if $\underset{\sim}{g} = \underset{\sim}{0}$. Thus the Clausius-Duhem inequality holds in the present theory.

Singular Surfaces

Suppose that at each point $\underset{\sim}{x}$ of a region $\underline{R}$ of a Euclidean point space, the constitutive relations,

$$\underset{\sim}{T}(\theta)\dot{\underset{\sim}{q}} + \underset{\sim}{q} = -\underset{\sim}{K}(\theta)\underset{\sim}{g} , \tag{69}$$

$$e = \tilde{e}(\theta,\underset{\sim}{g}) , \tag{70}$$

hold with $\underset{\sim}{T}$, $\underset{\sim}{K}$, and $\tilde{e}$ the continuously differentiable functions of (18). Suppose further that these functions are compatible with thermodynamics and hence obey the conclusions (i), (ii), and (iii) of the theorem of the previous section, so that, in particular, $\tilde{e}$ has the form (7a).

We are here interested in cases in which, for some $t^* > 0$, the time-dependent fields θ , $\underset{\sim}{q}$, and r are continuous on $\underline{R}X(0,t^*)$, but $\underline{R}x(0,t^*)$ contains a smooth hypersurface $\underline{S}$ across which $\dot{\theta}$, $\underset{\sim}{g}$, $\dot{\underset{\sim}{q}}$, and $\text{grad}_{\underset{\sim}{x}}\, \underset{\sim}{q}$ may have jumps although they are continuous on the complement of $\underline{S}$. Let $(\underset{\sim}{n},-U)$, with $|\underset{\sim}{n}| = 1$ and $U \gtrless 0$, be the normal to $\underline{S}$ at a point $(\underset{\sim}{x}_0,t_0)$ in the interior of $\underline{S}$; $\underset{\sim}{n}$ is the <u>direction of propagation</u> and U the <u>speed</u> of $\underline{S}$ at (x_0,t_0) . The <u>jump</u> $[f]$ experienced by a field f (such as $\dot{\theta}$, $\underset{\sim}{g}$, etc.) as $\underline{S}$ "passes through the place $\underset{\sim}{x}_0$ at time t_0" is

$$[f] = \lim_{t\to t_0^+} f(\underset{\sim}{x}_0,t) - \lim_{t\to t_0^-} f(\underset{\sim}{x}_0,t) . \tag{71}$$

We assume that $[\dot{\theta}] \neq 0$, and we call the hypersurface $\underline{S}$ a <u>temperature-rate wave</u>.*

---

* The term was introduced by Gurtin and Pipkin [15], and our treatment of the subject in [19] drew on observations made by them. See also the recent papers of Morro [16] [25] and Cattaneo's now classical study [2] of waves of order two, i.e., surfaces across which θ and $\underset{\sim}{q}$ and their first derivatives are continuous, but their second derivatives suffer jumps. Our discussion of temperature-rate waves in [19] was based on constitutive assumptions of greater generality than those employed here.

In [19] we showed that a temperature-rate wave cannot be purely transverse, i.e., cannot be such that $[\dot{q}]\cdot n = 0$.* We also showed there that U must obey a quadratic equation** which in the present context takes the form

$$U^2 \partial_\theta \tilde{e}(\theta, q) + U n\cdot Z(\theta)^{-1} \partial_q e(\theta, q) - n\cdot Z(\theta)^{-1} n = 0 . \tag{72}$$

Here $\theta = \theta(x_0, t_0)$ and $q = q(x_0, t_0)$ with (x_0, t_0) a point on S at which the wave speed is U and the direction of propagation is n. By (13),

$$\partial_\theta \tilde{e}(\theta, q) = c_0(\theta) + q\cdot B(\theta) q , \tag{73}$$

and

$$\partial_q \tilde{e}(\theta, q) = 2A(\theta) q = \frac{2}{\theta} Z(\theta) q - \frac{d}{d\theta} Z(\theta) q , \tag{74}$$

with c_0, B, A, and Z as in (10), (15), (12), and (5).

Let us assume now, in accord with experience, that for each θ in I : (I) $Z(\theta)$ is positive definite, and (II) $c_0(\theta)$ is positive. As we have shown that $K(\theta)$ is positive-definite, for crystals of high enough symmetry (e.g., cubic crystals) (I) is implied by the physical observation that $T(\theta)$, the tensor of relaxation times, is positive definite. The assumption (II) that the heat capacity is positive when $q = 0$, is obviously in accord with observation and statistical mechanical models; as $\partial_\theta \tilde{e}$ is continuous, (II) implies that, for each θ , there is a neighborhood N_θ of the origin in $V^{(3)}$ such that $\partial_\theta e(\theta, q)$ is positive for each q in N_θ .

From the relations (72)-(74) we read off

Remark 6. Suppose θ and q are such that $\partial_\theta \tilde{e}(\theta, q) > 0$, and define $U_0(\theta, q, n)$ by the relation

---

* See Remark 5.1 of [19]. Cattaneo [2] obtained an analogous result for waves of order two in materials that obey his theory (in which the dependence of e on q is not taken into account).

** Equation (5.13) of [19]. Analogues of this equation occur also in the papers by Gurtin and Pipkin [15] and Morro [16][25].

$$U_0(\theta,\underset{\sim}{q},\underset{\sim}{n}) = \sqrt{\frac{\underset{\sim}{n} \cdot \underset{\sim}{Z}(\theta)^{-1}\underset{\sim}{n}}{\partial_\theta \tilde{e}(\theta,\underset{\sim}{q})}} \quad . \tag{75}$$

When $\underset{\sim}{q}(\underset{\sim}{x}_0,t_0) = \underset{\sim}{0}$, i.e., when the temperature-rate wave is propagating into a region in which $\underset{\sim}{q} = \underset{\sim}{0}$, the speed U of the wave is

$$U(\theta,0,\underset{\sim}{n}) = U_0(\theta,\underset{\sim}{0},\underset{\sim}{n}) = \sqrt{\underset{\sim}{n}\cdot\underset{\sim}{Z}(\theta)^{-1}\underset{\sim}{n}/c_0(\theta)} \ . \tag{76}$$

In general, the equation (72) for U (with $\partial_\theta e(\theta,\underset{\sim}{q}) > 0$) has a unique positive solution $U(\theta,\underset{\sim}{q},\underset{\sim}{n})$ that can be written in the form

$$U(\theta,\underset{\sim}{q},\underset{\sim}{n}) = U_0(\theta,\underset{\sim}{q},\underset{\sim}{n}) \left[\sqrt{1 + (\underset{\sim}{\nu}\cdot\underset{\sim}{n})^2} - \underset{\sim}{\nu}\cdot\underset{\sim}{n}\right], \tag{77}$$

with $\underset{\sim}{\nu}$ in $V^{(3)}$ given by

$$\begin{aligned}\underset{\sim}{\nu} = \underset{\sim}{\nu}(\theta,\underset{\sim}{q},\underset{\sim}{n}) &= \underset{\sim}{Z}(\theta)^{-1}\ \partial_{\underset{\sim}{q}}\tilde{e}(\theta,\underset{\sim}{q})/2\ U_0(\theta,\underset{\sim}{q},\underset{\sim}{n})\ \partial_\theta\tilde{e}(\theta,\underset{\sim}{q}) \\ &= \underset{\sim}{Z}(\theta)^{-1}\ \underset{\sim}{A}(\theta)\underset{\sim}{q}/U_0(\theta,\underset{\sim}{q},n)\partial_\theta\tilde{e}(\theta,\underset{\sim}{q}) \\ &= \frac{2\underset{\sim}{q} - \theta\underset{\sim}{Z}(\theta)^{-1}\ (d\underset{\sim}{Z}(\theta)/d\theta)\underset{\sim}{q}}{2\theta\ U_0(\theta,\underset{\sim}{q},n)\ \partial_\theta\tilde{e}(\theta,\underset{\sim}{q})} \ .\end{aligned} \tag{78}$$

Therefore, when, as is expected for dielectric crystals, $\underset{\sim}{Z}(\theta)^{-1}\underset{\sim}{A}(\theta)$ is positive definite and hence $\underset{\sim}{q} \neq \underset{\sim}{0}$ implies $\underset{\sim}{\nu}\cdot\underset{\sim}{q} > 0$, a temperature-rate wave propagating in the direction of the heat flux vector travels more slowly than one propagating in the opposite direction:*

$$\underset{\sim}{q} \neq \underset{\sim}{0}\ ,\ \underset{\sim}{n} = \underset{\sim}{q}/|\underset{\sim}{q}| \quad \text{implies} \quad U(\theta,\underset{\sim}{q},\underset{\sim}{n}) < U(\theta,\underset{\sim}{q},-\underset{\sim}{n}) \ . \tag{79}$$

---

* Toward the end of their discussion of waves, Gurtin and Pipkin [15] make an assumption that leads them to a conclusion opposite to the present.

REFERENCES

[1] B.D. Coleman and W. Noll, The thermodynamics of elastic materials with heat conduction and viscosity, Arch. Rational Mech. Anal. **13**, 167-178 (1963).

[2] C. Cattaneo, Sulla conduzione del calore, Atti Sem. Mat. Fis. Univ. Modena **3**, 83-101 (1948).

[3] C. Truesdell and R.G. Muncaster, Fundamentals of Maxwell's Kinetic Theory of a Simple Monatomic Gas, New York, etc.: Academic Press (1980).

[4] M. Chester, Second sound in solids, Phys. Rev. **131**, 2013-2015 (1963).

[5] R.A. Guyer and J.A. Krumhansl, Solution of the linearized phonon Boltzmann equation, Phys. Rev. **148**, 766-778 (1966).

[6] J.C. Ward and J. Wilks, Second sound and the thermo-mechanical effect at very low temperatures, Phil. Mag. **43**, 48-50 (1952).

[7] R.B. Dingle, The velocity of second sound in various media, Proc. Roy. Soc. (London) **A65**, 1044-1050 (1952).

[8] J.A. Sussman and A. Thellung, Thermal conductivity of perfect dielectric crystals in the absence of umklapp processes, Proc. Phys. Soc. (London) **81**, 1122-1130 (1963).

[9] A. Griffin, On the detection of second sound in crystals by light scattering, Phys. Letters **17**, 208-210 (1965).

[10] E.W. Prohofsky and J.A. Krumhansl, Second-sound propagation in dielectric solids, Phys. Rev. **133**, A1403-A1410 (1964).

[11] R.A. Guyer and J.A. Krumhansl, Dispersion relation for second sound in solids, Phys. Rev. **133**, A1411-A1417 (1964).

[12] C.P. Enz, One-particle densities, thermal propagation, and second sound in dielectric crystals, Ann. Phys. (N.Y.) **46**, 114-173 (1968).

[13] P.C. Kwok, Dispersion and damping of second sound in non-isotropic solids, Physics **3**, 221-229 (1967).

[14] R.J. Hardy, Phonon Boltzmann equation and second sound in solids, Phys. Rev. B **2**, 1193-1207 (1970).

[15] M.E. Grutin and A.C. Pipkin, A general theory of heat conduction with finite wave speeds, Arch. Rational Mech. Anal. **31**, 113-126 (1968).

[16] A. Morro, Wave propagation in thermo-viscous materials with hidden variables, Arch. Mech. (Warszawa) **32**, 145-161 (1980).

[17] Y.-H. Pao and D.K. Banerjee, Thermal pulses in dielectric crystals, Lett. Appl. Eng. Sci. **1**, 35-41, (1973).

[18] D.K. Banerjee and Y.-H. Pao, Thermoelastic waves in anisotropic solids, J. Acoustic. Soc. Am. **56**, 1444-1454 (1974).

[19] B.D. Coleman, M. Fabrizio, and D.R. Owen, On the thermodynamics of second sound in dielectric crystals, Arch. Rational Mech. Anal. **80**, 135-158 (1982).

[20] B.D. Coleman and D.R. Owen, A mathematical foundation for thermodynamics, Arch. Rational Mech. Anal. **54**, 1-104 (1974).

[21] B.D. Coleman and D.R. Owen, On thermodynamics and elastic-plastic materials, Arch. Rational Mech. Anal. **59**, 25-51 (1975); erratum, ibid., **62**, 396. On the thermodynamics of elastic-plastic materials with temperature-dependent moduli and yield stresses, ibid. **70**, 340-354 (1979).

[22] B.D. Coleman and V.J. Mizel, Thermodynamics and departure from Fourier's law of heat conduction, Arch. Rational Mech. Anal. **13**, 245-261 (1963).

[23] B.D. Coleman and V.J. Mizel, Existence of caloric equations of state in thermodynamics, J. Chem. Phys. **40**, 1116-1125 (1964).

[24] B.D. Coleman and M.E. Gurtin, Thermodynamics with internal state variables, J. Chem. Phys. **47**, 597-613 (1967).

[25] A. Morro, Acceleration waves in thermo-viscous fluids, Rend. Sem. Mat. Univ. Padova **63**, 169-184 (1980).

DISSIPATION, STABILIZATION AND THE SECOND LAW OF THERMODYNAMICS

C. M. Dafermos
Division of Applied Mathematics
Brown University
Providence, R. I., 02912, USA

1. Introduction

The main objective of these lectures is to explore the implications of the second law of thermodynamics on the analytical structure of thermomechanical processes. Using as model the hierarchy of material classes: thermoviscoelasticity, thermoelasticity with heat conduction and adiabatic thermoelasticity, we attempt to demonstrate how internal dissipation affects the smoothness of processes. We show that when internal dissipation is very weak (or non existent) the development of shock waves is inevitable and discuss the dual role of the second law as an admissibility criterion on processes with discontinuities and as a guarantor of stability.

The bibliography at the end of each section provides only a sample of the relevant literature and is far from comprehensive.

This work was supported in part by NSF grants #MCS-79-05774-05, CME 80-23824 and in part by the U.S. Army under contract #ARO-DAAG-29-79-C-0161.

2. Bodies and Motions

In continuum physics, the mathematical model of an n-dimensional body (n = 1,2 or 3) is a manifold characterized by a reference configuration, that is, an open subset $\mathcal{B}$ of the reference space R^n. The typical point $\underset{\sim}{X}$ in $\mathcal{B}$ is called a material particle. In the applications the standard procedure for constructing a reference configuration of a body is to identify "molecules" with the point in space that they happen to occupy at a certain fixed time instant. In general, however, a reference configuration need not be an actual configuration of the body but only an abstrct representation of it. In fact it is instructive to keep constantly in mind that the reference space is not necessarily the physical space itself but just a copy of it.

A configuration of the body $\mathcal{B}$ is a Lipschitz homeomorphism $\underset{\sim}{x} = \underset{\sim}{x}(\underset{\sim}{X})$ from $\mathcal{B}$ to the physical space R^n. A motion of $\mathcal{B}$ is an evolution $\underset{\sim}{x} = \underset{\sim}{x}(\underset{\sim}{X},t)$ of configurations. Thus, $\underset{\sim}{x}(\underset{\sim}{X},t)$ is the position of particle $\underset{\sim}{X}$ at time t; the curve $\underset{\sim}{x}(\underset{\sim}{X},\cdot)$ is the trajectory of particle $\underset{\sim}{X}$; finally, $\underset{\sim}{x}(\cdot,t)$ is the configuration of $\mathcal{B}$ at time t.

In continuum physics one seeks to determine the time evolution of the fields of various physical quantities, such as density, stress, temperature, entropy, etc., defined over the moving body. Since every configuration is homeomorphic to the reference configuration, these field quantities can be represented equally well as functions of $(\underset{\sim}{X},t)$ (referential description) or of $(\underset{\sim}{x},t)$ (spatial description). With the exception of Section 13, we shall be employing here the referential description which is more convenient for the purposes of the present lectures.

As regards notation, we shall be using the same letter to denote a field as well as its value, writing, for example, $\theta = \theta(\underset{\sim}{X},t)$ for temperature. Cartesian components of the particle $\underset{\sim}{X}$ will be designated by Greek subscripts, such as X_α, X_β, X_γ, while Cartesian components of $\underset{\sim}{x}$ will be designated by Latin subscripts, e.g., x_i, x_j, x_k, etc. The partial derivatives $\frac{\partial}{\partial X_\alpha}$ and $\frac{\partial}{\partial t}$ will be denoted by $,\alpha$ and a dot $(\cdot)$. We adopt throughout the usual repeated index summation convention.

Two kinematical fields that play a key role in Continuum Physics are deformation gradient $\underset{\sim}{F}(\underset{\sim}{X},t)$ and velocity $\underset{\sim}{v}(\underset{\sim}{X},t)$, namely the space and time derivatives of the motion. In components form:

$$F_{i\alpha} = x_{i,\alpha}, \qquad v_i = \dot{x}_i. \tag{2.1}$$

Since $\underset{\sim}{x}(\underset{\sim}{X},t)$ is a homeomorphism, we assume

$$\det \underset{\sim}{F}(\underset{\sim}{X},t) \neq 0. \tag{2.2}$$

Bibliography

GURTIN, M.E., An Introduction to Continuum Mechanics. Academic Press, New York 1981.

TRUESDELL, C. A., A First Course in Rational Continuum Mechanics. Vol. 1. Academic Press, New York 1977.

TRUESDELL, C. A., and TOUPIN, R. A., The Classical Field Theories. Handbuch der Physik III/1. Springer-Verlag, Berlin 1960.

3. Balance Laws

Every continuum physical theory rests upon a set of balance laws. The referential description of the typical balance law for a moving body $\mathcal{B}$ is an equation of the form

$$(3.1)\quad \int_{\Omega} G(\underset{\sim}{X},t_2)d\underset{\sim}{X} - \int_{\Omega} G(\underset{\sim}{X},t_1)d\underset{\sim}{X} = \int_{t_1}^{t_2}\int_{\partial\Omega} \underset{\sim}{P}(\underset{\sim}{X},t)\cdot\underset{\sim}{N}(\underset{\sim}{X})dS\,dt + \int_{t_1}^{t_2}\int_{\Omega} H(\underset{\sim}{X},t)d\underset{\sim}{X}\,dt$$

that holds for any time interval (t_1,t_2) and every subdomain Ω of $\mathcal{B}$ with sufficiently smooth boundary $\partial\Omega$. $\underset{\sim}{N}(\underset{\sim}{X})$ denotes the unit normal on $\partial\Omega$ at $\underset{\sim}{X}$ in $\partial\Omega$. Equation (3.1) states that the set function with density G is conserved in the sense that changes in the quantity stored in Ω are generated solely by flux through $\partial\Omega$ and/or production inside Ω.

In dealing with balance laws it is convenient to use the theory of functions of bounded variation (BV) in the sense of Tonelli and Cesari. A bounded measurable function $\chi(\underset{\sim}{X},t)$ on $\mathcal{B}\times(t_1,t_2)$ is of class BV when its partial derivatives $\chi_{,\alpha}$, $\dot{\chi}$ are Borel measures. The domain $\mathcal{B}\times(t_1,t_2)$ of χ in BV can be decomposed into the union of three pairwise disjoint sets $\mathcal{C}$, $\mathcal{J}$ and $\mathcal{R}$ with the following properties:

(a) $\mathcal{C}$ is the set of points of approximate continuity (in the sense of Lebesgue) of χ.

(b) $\mathcal{J}$ is the set of points of approximate jump discontinuity of χ in the following sense: Through each point $(\underset{\sim}{X},t)$ in $\mathcal{J}$ there is a "tangential" n-dimensional hyperplane from either side of which χ attains Lebesgue approximate limits $\chi^-(\underset{\sim}{X},t)$ and $\chi^+(\underset{\sim}{X},t)$ at $(\underset{\sim}{X},t)$. The difference $\chi^+(\underset{\sim}{X},t) - \chi^-(\underset{\sim}{X},t)$ is the jump of χ across $\mathcal{J}$ at $(\underset{\sim}{X},t)$ and is denoted by $[\chi(\underset{\sim}{X},t)]$.

(c) $\mathcal{R}$ is the residual set.

$\mathcal{R}$ is "small" in that its n-dimensional Hausdorff measure is zero. $\mathcal{J}$ is the countable union of sets each of which is the image of a Lipschitz map of a bounded subset of R^n. Thus $\mathcal{J}$ can be visualized as the countable union of n-dimensional "hypersurfaces" embedded in $\mathcal{B}\times(t_1,t_2)$.

It turns out that BV is the broadest function class in which the Gauss-Green theorem is applicable (over any set of "finite perimeter" whose characteristic function is of class BV).

As we shall see in the course of these lectures it is internal dissipation in the continuous medium that regulates the smoothness of the motion and the relevant fields. For the present purposes we assume that the fields $G(\underset{\sim}{X},t)$ and $\underset{\sim}{P}(\underset{\sim}{X},t)$ in the balance law (3.1) are of class BV while $H(\underset{\sim}{X},t)$ is in L^1. Then, by virtue of the Gauss-Green theorem, (3.1) can be reduced to the equivalent local form

$$\dot{G} = \mathrm{Div}\, \underset{\sim}{P} + H \tag{3.2}$$

the equality being understood in the sense of measures. In particular, (3.2) applied to the set $\mathcal{J}$ of points of approximate jump discontinuity of G and $\underset{\sim}{P}$ yields

$$s[G] + [\underset{\sim}{P}]\cdot\underset{\sim}{N} = 0 \tag{3.3}$$

where $(\underset{\sim}{N},-s)$ is any $(n+1)$-vector orthogonal to the tangential hyperplane to $\mathcal{J}$. For convenience, we normalize this vector so that $|\underset{\sim}{N}| = 1$. Then, if we visualize $\mathcal{J}$ as a family of $(n-1)$-dimensional hypersurfaces moving in $\mathcal{B}$, $\underset{\sim}{N}$ is the direction and s is the speed of propagation. Thus BV is the natural function class for the description of motions with shock waves. In that context $\mathcal{J}$ is the shock set, (3.3) are the <u>Rankine-Hugoniot</u> jump conditions and $\mathcal{R}$ contains essentially points of shock interactions.

Bibliography

DAFERMOS, C.M., Quasilinear hyperbolic systems that result from conservation laws. In "Nonlinear Waves" S. Leibovich and A. R. Seebass, eds., Cornell U. Press, Ithaca, N. Y., 1974.

TRUESDELL, C.A., and TOUPIN, R.A., The Classical Field Theories. Handbuch der Physik III/1. Springer-Verlag, Berlin 1960.

VOLPERT, A.I., The spaces BV and quasilinear equations. Math. USSR Sbornik 2(1967), 225-267.

DIPERNA, R.J., Singularities of solutions of nonlinear hyperbolic systems of conservation laws. Arch. Rational Mech. Anal. 60(1975), 75-100.

4. The Balance Laws of Continuum Thermomechanics

Continuum thermomechanics rests upon the balance laws of mass, linear and angular momentum, and energy as well as the Clausius-Duhem inequality.

In its referential form the balance law of mass simply states that the reference mass density ρ depends solely upon $\underset{\sim}{X}$. For simplicity let us assume that ρ is an a priori assigned positive constant.

The field equations of the balance laws of linear and angular momentum read

$$\rho\dot{\underset{\sim}{v}} = \operatorname{Div} \underset{\sim}{T} + \rho\underset{\sim}{f} \tag{4.1}$$

$$\rho\overline{\underset{\sim}{v}\times\underset{\sim}{x}}^{\bullet} = \operatorname{Div}(\underset{\sim}{T}\times\underset{\sim}{x}) + \rho\underset{\sim}{f}\times\underset{\sim}{x} \tag{4.2}$$

where $\underset{\sim}{T}$ is the Piola-Kirchhoff stress tensor and $\underset{\sim}{f}$ the body force per unit mass. Since $\underset{\sim}{x}$ is Lipschitz continuous, it is readily seen that, whenever (4.1) holds, (4.2) is equivalent to

$$\underset{\sim}{T}\underset{\sim}{F}^T = \underset{\sim}{F}\underset{\sim}{T}^T. \tag{4.3}$$

The field equation of the balance law of energy takes the form

$$\rho\overline{(\varepsilon + \frac{1}{2}\,\underset{\sim}{v}\cdot\underset{\sim}{v})}^{\bullet} = \operatorname{Div}(\underset{\sim}{T}\cdot\underset{\sim}{v} + \underset{\sim}{Q}) + \rho\underset{\sim}{f}\cdot\underset{\sim}{v} + \rho r \tag{4.4}$$

where ε denotes internal energy, $\underset{\sim}{Q}$ is heat flux and r is energy supply per unit mass. The standard chain rule applies to the product of a Lipschitz continuous function and a BV function but not to the product of two BV functions. Thus when the velocity field $\underset{\sim}{v}(\underset{\sim}{X},t)$ is Lipschitz continuous one may use (4.1) to reduce (4.4) into the simpler form

$$\rho\dot{\varepsilon} = \operatorname{tr}(\underset{\sim}{T}\dot{\underset{\sim}{F}}^T) + \operatorname{Div}\underset{\sim}{Q} + \rho r \tag{4.5}$$

but this reduction is not allowed when $\underset{\sim}{v}(\underset{\sim}{X},t)$ is merely of class BV.

Within the present framework the second law of Thermodynamics is expressed by the Clausius-Duhem inequality which states that the increase in the entropy stored in any part Ω of the body $\mathscr{B}$ exceeds the sum of entropy flux through $\partial\Omega$ and entropy supply inside Ω. Under the classical assumption that entropy flux is heat flux over temperature and entropy supply is heat supply over temperature, the local form of the Clausius-Duhem inequality reads

$$(4.6) \qquad \rho\dot{\eta} - \mathrm{Div}\,\frac{\underset{\sim}{Q}}{\theta} - \rho\,\frac{r}{\theta} \geq 0$$

where η denotes the (specific) entropy and θ is absolute temperature. When the fields $\underset{\sim}{v}(\underset{\sim}{X},t)$ and $\theta(\underset{\sim}{X},t)$ are Lipschitz continuous one may combine (4.6) with (4.5) to deduce

$$(4.7) \qquad \rho\dot{\varepsilon} - \rho\theta\dot{\eta} - \mathrm{tr}(\underset{\sim}{T}\dot{\underset{\sim}{F}}^T) - \frac{1}{\theta}\,\underset{\sim}{Q}\cdot\mathrm{Grad}\,\theta \leq 0.$$

Under the same smoothness assumptions one may rewrite (4.7) into the equivalent form

$$(4.8) \qquad \rho\dot{\psi} + \rho\eta\dot{\theta} - \mathrm{tr}(\underset{\sim}{T}\dot{\underset{\sim}{F}}^T) - \frac{1}{\theta}\,\underset{\sim}{Q}\cdot\mathrm{Grad}\,\theta \leq 0,$$

which involves the Helmholtz free energy function $\psi = \varepsilon - \theta\eta$. However, (4.7) and (4.8) are not necessarily valid when $\underset{\sim}{v}(\underset{\sim}{X},t)$ and/or $\theta(\underset{\sim}{X},t)$ are merely of class BV.

We close this section with remarks on the function of the balance laws recorded above in the edifice of continuum thermomechanics. The balance laws of linear momentum (4.1) and energy (4.4) are being regarded as restrictions on motions, that is as evolution equations which will determine, in conjunction with constitutive relations, the thermomechanical process of the body. In contrast, the balance law of angular momentum (4.3) is viewed as a compatibility restriction which will have to be satisfied a priori by the constitutive relation for the stress. Finally, the Clausius-Duhem inequality plays a dual role: In the framework of smooth processes it is to be satisfied automatically (just like the balance law of angular momentum) and in this capacity it induces a priori restrictions on constitutive relations. On the other hand, in the context of processes with shock waves it is a restriction on motions (like the balance laws of linear momentum and energy). We will return to these ideas in the following sections.

Bibliography

TRUESDELL, C.A., and NOLL, W., The Nonlinear Field Theories of Mechanics. Handbuch der Physik III/3, Springer-Verlag, Berlin 1965.

TRUESDELL, C.A., and TOUPIN, R.A., The Classical Field Theories. Handbuch der Physik III/1. Springer-Verlag, Berlin 1960.

5. Constitutive Relations

Constitutive equations, which express how the fields appearing in the balance laws are related to the motion, serve to identify the material of the body. Out of the wealth of types of constitutive relations discussed in the standard treatises on continuum mechanics, we will describe here those that will serve as models in our further discussion.

A material is homogeneous thermoviscoelastic, of the differential type, when free energy, stress, entropy and heat flux at any point $(\underset{\sim}{X},t)$ are solely determined by deformation gradient, velocity gradient, temperature and temperature gradient at $(\underset{\sim}{X},t)$. As stated in Section 4, the constitutive relations must comply with the requirement that every smooth process, compatible with the balance laws of momentum and energy, must automatically satisfy the Clausius-Duhem inequality (4.6) and thereby also its reduced form (4.8). This leads to

$$(5.1)\qquad \begin{cases} \psi = \hat{\psi}(\underset{\sim}{F},\theta) \\ \underset{\sim}{T} = \rho \dfrac{\partial\hat{\psi}(\underset{\sim}{F},\theta)}{\partial \underset{\sim}{F}} + \hat{\underset{\sim}{S}}(\underset{\sim}{F},\dot{\underset{\sim}{F}},\theta,\mathrm{Grad}\,\theta) \\ \eta = -\dfrac{\partial\hat{\psi}(\underset{\sim}{F},\theta)}{\partial\theta} \\ \underset{\sim}{Q} = \hat{\underset{\sim}{Q}}(\underset{\sim}{F},\dot{\underset{\sim}{F}},\theta,\mathrm{Grad}\,\theta) \end{cases}$$

where $\hat{\underset{\sim}{S}}$ and $\hat{Q}$ must satisfy

$$(5.2)\qquad \mathrm{tr}[\hat{\underset{\sim}{S}}(\underset{\sim}{F},\underset{\sim}{W},\theta,\underset{\sim}{G})\underset{\sim}{W}^T] + \frac{1}{\theta}\hat{\underset{\sim}{Q}}(\underset{\sim}{F},\underset{\sim}{W},\theta,\underset{\sim}{G})\cdot\underset{\sim}{G} \geq 0.$$

The limiting case of a thermoviscoelastic material with stress and heat flux independent of velocity gradient yields the thermoelastic material in which (5.1), (5.2) take the form

(5.3)
$$\begin{cases} \psi = \hat{\psi}(\underset{\sim}{F},\theta) \\ \underset{\sim}{T} = \rho \dfrac{\partial\hat{\psi}(\underset{\sim}{F},\theta)}{\partial \underset{\sim}{F}} \\ \eta = -\dfrac{\partial\hat{\psi}(\underset{\sim}{F},\theta)}{\partial\theta} \\ \underset{\sim}{Q} = \hat{\underset{\sim}{Q}}(\underset{\sim}{F},\theta,\text{Grad }\theta) \end{cases}$$

(5.4)
$$\hat{\underset{\sim}{Q}}(\underset{\sim}{F},\theta,\underset{\sim}{G})\cdot\underset{\sim}{G} \geq 0.$$

When variations in temperature are neglected (for instance energy supply is controlled so that $\theta(\underset{\sim}{X},t) \equiv$ constant) (5.3) reduces to <u>isothermal thermoelasticity</u>

(5.5)
$$\begin{cases} \psi = \hat{\psi}(\underset{\sim}{F}) \\ \underset{\sim}{T} = \rho \dfrac{\partial\hat{\psi}(\underset{\sim}{F})}{\partial \underset{\sim}{F}} \,. \end{cases}$$

Isothermal motions are to be determined solely from the balance law of linear momentum (4.1). Thus isothermal thermoelasticity appears as a purely mechanistic theory, isomorphic to <u>hyperelasticity</u>. However, this theory inherits from thermodynamics the Second Law in the following way. Combining (4.4) with (4.6) and using that $\theta \equiv$ const. we obtain

(5.6)
$$\rho\overline{(\psi + \tfrac{1}{2}\underset{\sim}{v}\cdot\underset{\sim}{v})}^{\bullet} - \text{Div}(\underset{\sim}{T}\cdot\underset{\sim}{v}) - \rho\underset{\sim}{f}\cdot\underset{\sim}{v} \leq 0.$$

Of course, when the fields $\underset{\sim}{F}(\underset{\sim}{X},t)$ and $\underset{\sim}{v}(\underset{\sim}{X},t)$ are Lipschitz continuous, (5.6) is automatically satisfied, as an equality, by virtue of (4.1) and (5.5). However, in the framework of fields of class BV (i.e., motions with shock waves) (5.6) is an extraneous condition which plays the role of the Clausius-Duhem inequality. The measure on the left-hand side of (5.6) is concentrated on the set of points of approximate jump discontinuity of $\underset{\sim}{F}(\underset{\sim}{X},t)$ and $\underset{\sim}{v}(\underset{\sim}{X},t)$ (shock set).

A thermoelastic material with $\hat{\underset{\sim}{Q}} \equiv \underset{\sim}{0}$ is a <u>thermoelastic nonconductor of heat</u>. It will prove convenient to write the constitutive relations for thermoelastic nonconductors in an alternative, albeit equivalent, form to (5.3) by reversing the roles of θ and η and thus postulating that internal energy, stress and temperature at any point $(\underset{\sim}{X},t)$ are solely determined by deformation gradient and entropy at $(\underset{\sim}{X},t)$. The most general constitutive relations of this form which are compatible with the requirement that every smooth process satisfies automatically the reduced form (4.7) of the Clausius-Duhem inequality read

$$(5.7) \qquad \begin{cases} \varepsilon = \hat{\varepsilon}(\underset{\sim}{F},\eta) \\ \underset{\sim}{T} = \rho \dfrac{\partial\hat{\varepsilon}(\underset{\sim}{F},\eta)}{\partial \underset{\sim}{F}} \\ \theta = \dfrac{\partial\hat{\varepsilon}(\underset{\sim}{F},\eta)}{\partial\eta} \; . \end{cases}$$

In thermoelastic nonconductors of heat, when the fields $\underset{\sim}{F}(\underset{\sim}{X},t)$ and $\eta(\underset{\sim}{X},t)$ are Lipschitz continuous, (4.5) together with (5.7) yield

$$(5.8) \qquad \dot{\eta} - \frac{r}{\theta} = 0,$$

that is, the Clausius-Duhem inequality is satisfied identically, as an equality, in the framework of smooth processes. More generally, it can be shown that when $\underset{\sim}{F}(\underset{\sim}{X},t)$ and $\eta(\underset{\sim}{X},t)$ are of class BV the measure $\dot{\eta} - r/\theta$ is concentrated on the set of points of approximate jump discontinuity (shock set).

All constitutive relations recorded above must satisfy additional restrictions induced by the balance of angular momentum (4.3) and the principle of material frame indifference. For the thermoelastic nonconductor these restrictions are expressed by

$$(5.9) \qquad \hat{\varepsilon}(\underset{\sim}{O}\underset{\sim}{F},\eta) = \hat{\varepsilon}(\underset{\sim}{F},\eta), \quad \text{for all orthogonal } \underset{\sim}{O}\,.$$

Bibliography

COLEMAN, B.D., and MIZEL, V.J., Existence of caloric equations of state in thermodynamics. J. Chem. Phys. 40(1964), 1116-1125.

COLEMAN, B.D., and NOLL, W., The thermodynamics of elastic materials with heat conduction and viscosity. Arch. Rational Mech. Anal. 13(1963), 167-178.

TRUESDELL, C.A., and NOLL, W., The Nonlinear Field Theories of Mechanics. Handbuch der Physik III/3. Springer-Verlag, Berlin 1965.

6. Internal Dissipation

The Clausius-Duhem inequality (4.6) indicates that all thermomechanical processes are dissipative. The degree of dissipativeness varies, however, from material class to material class. One end of the spectrum is occupied by thermoelastic nonconductors, in which no dissipation is present in the context of smooth processes (cf. (5.8)) and $\dot{\eta} - r/\theta$ is concentrated on the shock set. On the other end lie thermoviscoelastic materials in which, both, viscosity and heat diffusion induce internal dissipation evidenced by (5.2). Thermoelastic conductors fall somewhere in the middle with heat diffusion being the only source of internal dissipation in the context of smooth processes.

Dissipation generally affects the smoothness as well as the asymptotic behavior of thermomechanical processes. With regard to smoothening effects it is natural to classify materials into one of the following categories:

I. Dissipation is so overpowering that it can even smoothen out "rough" initial states, always yielding smooth processes.

II. Dissipation is sufficiently powerful to preserve the smoothness of smooth initial states but incapable of smoothening out rough initial states.

III. Dissipation preserves the smoothness of smooth initial states "near" equilibrium but cannot prevent the breaking of acceleration waves of large amplitude so smooth initial states "far" from equilibrium generally generate processes that develop shock waves (of class BV).

IV. Dissipation is so weak that even acceleration waves of small amplitude break. Thus even smooth initial states near equilibrium generate processes that generally develop shock waves (of class BV).

Continuous media in Category I are rather rare and too tame to be of much interest (typical representatives are rigid conductors of heat.) We will thus focus our attention, in these lectures, upon material classes that fall into one of the remaining categories II, III, or IV.

Bibliography

DAFERMOS, C.M., Conservation laws with dissipation. In "Nonlinear Phenomena in Mathematical Sciences" V. Lakshmikantham, Ed., Academic Press, New York (to appear).

7. Development of Singularities in Adiabatic Processes in Thermoelasticity

A thermomechanical process of a thermoelastic nonconductor is identified by fields $\{\underset{\sim}{F}(\underset{\sim}{X},t),\ \underset{\sim}{v}(\underset{\sim}{X},t),\eta(\underset{\sim}{X},t)\}$, generally of class BV, which satisfy, in the sense of measures, the system of evolution equations

$$
(7.1)\qquad \begin{cases} \dot{\underset{\sim}{F}} = \operatorname{Grad} \underset{\sim}{v} \\ \rho\dot{\underset{\sim}{v}} = \operatorname{Div} \underset{\sim}{T} + \rho\underset{\sim}{f} \\ \rho\overline{(\varepsilon+\frac{1}{2}\underset{\sim}{v}\cdot\underset{\sim}{v})}^{\bullet} = \operatorname{Div}(\underset{\sim}{T}\cdot\underset{\sim}{v}) + \rho\underset{\sim}{f}\cdot\underset{\sim}{v} + \rho r \end{cases}
$$

together with the inequality

$$
(7.2)\qquad \dot{\eta} - \frac{r}{\theta} \geq 0,
$$

where $\varepsilon,\underset{\sim}{T},\theta$ are given by constitutive relations (5.7) and the fields $\underset{\sim}{f}(\underset{\sim}{X},t)$, $r(\underset{\sim}{X},t)$ are assigned a priori.

Uniformly Lipschitz continuous processes $\{\overline{\underset{\sim}{F}}(\underset{\sim}{X},t),\overline{\underset{\sim}{v}}(\underset{\sim}{X},t),\overline{\eta}(\underset{\sim}{X},t)\}$ are called smooth and satisfy the reduced form of (7.1):

$$
(7.3)\qquad \begin{cases} \dot{\overline{\underset{\sim}{F}}} = \operatorname{Grad} \overline{\underset{\sim}{v}} \\ \rho\dot{\overline{\underset{\sim}{v}}} = \operatorname{Div} \overline{\underset{\sim}{T}} + \rho\overline{\underset{\sim}{f}} \\ \rho\dot{\overline{\varepsilon}} = \operatorname{tr}(\overline{\underset{\sim}{T}}\dot{\overline{\underset{\sim}{F}}}^{T}) + \rho\overline{r} \end{cases}
$$

and thereby, as we saw in Section 5, (7.2) as an equality, i.e.,

$$
(7.4)\qquad \dot{\overline{\eta}} - \frac{\overline{r}}{\overline{\theta}} = 0.
$$

The central problem is to determine the process in $\mathscr{B}$ which is generated by an initial state $\{\underset{\sim}{F}^{o},\underset{\sim}{v}^{o},\eta^{o}\}$ and is appropriately controlled in $\partial\mathcal{B}$. In mathematical terms we have to solve the system (7.1), (7.2) under prescribed initial and boundary conditions. In discussing the solvability of this problem, the first step is to impose restrictions upon the function $\hat{\varepsilon}(F,\eta)$ guaranteeing that the response of the material is reasonable.

With every fixed state $(\bar{F},\bar{\eta})$ and any unit n-vector N we associate the symmetric acoustic tensor $Q(\bar{F},\bar{\eta};N)$,

$$Q_{ij}(\bar{F},\bar{\eta};N) = \frac{\partial^2\hat{\varepsilon}(\bar{F},\bar{\eta})}{\partial F_{i\alpha}\partial F_{j\beta}} N_\alpha N_\beta . \tag{7.5}$$

We say that the strong ellipticity condition holds at $(\bar{F},\bar{\eta})$ if $Q(\bar{F},\bar{\eta};N)$ is positive definite for all N. Strong ellipticity implies that the system (7.3) is hyperbolic. The characteristic speeds in the direction N are 0, with multiplicity $n(n-1)+1$, and $\pm\lambda_k^{1/2}$, $k = 1,\dots,n$, where $\lambda_1,\dots,\lambda_n$ are the eigenvalues of $Q(\bar{F},\bar{\eta};N)$. Note that strong ellipticity is not incompatible with the principle of material frame indifference (5.9).

Under the hyperbolicity assumption and certain other reasonable hypotheses, it can be shown that the initial-boundary value problem of place for (7.3) admits a unique smooth solution on a certain time interval $[0,t_o)$. It turns out, however, that the smooth solution eventually develops singularities, unless, of course, $\hat{\varepsilon}(F,\eta)$ happens to be quadratic in F, a condition incompatible with (5.9). One can see this phenomenon by considering the one-dimensional version of (7.3) and showing that first derivatives of the solution generally blow up in a finite time. Here, however, we will exhibit the development of singularities by monitoring the evolution of the amplitude of acceleration waves.

Let us consider a plane acceleration wave which propagates in the direction N with speed s into a region in which the material is at rest at a homogeneous state $(\bar{F},\bar{\eta})$. This means that there is an n-dimensional plane $\mathcal{S}$, embedded in space-time, with normal $(N,-s)$ and a smooth (i.e., Lipschitz continuous) solution $\{F(X,t),v(X,t),\eta(X,t)\}$ which is equal to $\{\bar{F},0,\bar{\eta}\}$ on the side of $\mathcal{S}$ into which $(N,-s)$ points and is twice continuously differentiable on the opposite side of $\mathcal{S}$. Thus $\{F(X,t),$ $v(X,t),\eta(X,t)\}$ is continuous but its first and second derivatives may experience jump discontinuities across $\mathcal{S}$.

By virtue of (7.4), first derivatives of $\eta(\underset{\sim}{X},t)$ are continuous across $\mathscr{S}$. First derivatives of $\underset{\sim}{F}(\underset{\sim}{X},t)$ and $\underset{\sim}{v}(\underset{\sim}{X},t)$ may jump but the jumps are restricted by the compatibility conditions

$$(7.6)\qquad [\dot{\underset{\sim}{F}}] = -\frac{1}{s}[\dot{\underset{\sim}{v}}]\otimes\underset{\sim}{N},\quad [\text{Grad }\underset{\sim}{F}] = \frac{1}{s^2}[\dot{\underset{\sim}{v}}]\otimes\underset{\sim}{N}\otimes\underset{\sim}{N}.$$

The jumps of second derivatives satisfy more complicated, iterated, compatibility conditions.

Taking the jump of $(7.3)_2$ across $\mathscr{S}$ and using (5.7), (7.6) and (7.5) we obtain

$$(7.7)\qquad (\underset{\sim}{Q}(\overline{\underset{\sim}{F}},\overline{\eta};\underset{\sim}{N}) - s^2\underset{\sim}{I})[\dot{\underset{\sim}{v}}] = \underset{\sim}{0}$$

which shows that s^2 is an eigenvalue of $\underset{\sim}{Q}(\overline{\underset{\sim}{F}},\eta;\underset{\sim}{N})$ and

$$(7.8)\qquad [\dot{\underset{\sim}{v}}] = \alpha\underset{\sim}{r}$$

where $\underset{\sim}{r}$ is a unit eigenvector associated with s^2. The magnitude α of the jump of the acceleration is an appropriate measure of the <u>amplitude</u> of the wave.

In order to see how α evolves with time, we apply the <u>displacement derivative</u> operator

$$(7.9)\qquad \frac{\delta}{\delta t} = \frac{\partial}{\partial t} + s\underset{\sim}{N}\cdot\frac{\partial}{\partial\underset{\sim}{X}}$$

on (7.3), we take the inner product with $\underset{\sim}{r}$ and form the jump across $\mathscr{S}$ of the resulting equation. This yields

$$(7.10)\qquad \frac{\delta\alpha}{\delta t} + \frac{1}{2s^3}\frac{\partial^3\hat{\varepsilon}(\overline{\underset{\sim}{F}},\overline{\eta})}{\partial F_{i\alpha}\partial F_{j\beta}\partial F_{k\gamma}} N_\alpha N_\beta N_\gamma r_i r_j r_k \alpha^2 = 0.$$

We say that the field $\underset{\sim}{r}$ is genuinely nonlinear at $(\overline{\underset{\sim}{F}},\overline{\eta})$ if

$$(7.11)\qquad \frac{\partial^3\hat{\varepsilon}(\overline{\underset{\sim}{F}},\overline{\eta})}{\partial F_{i\alpha}\partial F_{j\beta}\partial F_{k\gamma}} N_\alpha N_\beta N_\gamma r_i r_j r_k \neq 0.$$

Under the genuine nonlinearity assumption, it follows from (7.10) that when α has the opposite sign than the expression in (7.11) then it will blow up in finite time. Generally, a shock wave is generated at any point where the amplitude of an acceleration wave explodes.

Bibliography

CHEN, P.J., Growth and Decay of Waves in Solids. Handbuch der Physik IIIa/3. Springer-Verlag, Berlin 1973.

LAX, P.D., Development of singularities of nonlinear hyperbolic partial differential equations. J. Math. Physics 5 (1964), 611-613.

TRUESDELL, C.A., and NOLL, W., The Nonlinear Field Theories of Mechanics. Handbuch der Physik III/3. Springer-Verlag, Berlin 1965.

TRUESDELL, C.A., and TOUPIN, R.A., The Classical Field Theories. Handbuch der Physik III/1. Springer-Verlag, Berlin 1960.

8. Entropy Admissibility Criterion

As shown in the previous section, the class of smooth processes is too narrow to support global existence theorems. The general expectation is that BV is the proper function class for processes that are defined globally in time but this conjecture has been verified thus far only in one dimension and for processes that start out in a neighborhood of a rest state.

A new difficulty arises in the theory of processes of class BV. By means of simple, one dimensional, examples it can be shown that the balance laws of momentum and energy (7.1) do not determine uniquely the thermomechanical process of a thermoelastic nonconductor, generated by a specified initial state, within the function class BV. In other words, the initial (or initial-boundary) value problem for (7.1) may have more than one solution of class BV. Therefore, (7.1) must be supplemented with admissibility criteria which serve to identify the physically relevant process. The Clausius-Duhem inequality (7.2) induces such a criterion which we call the entropy criterion. The obvious questions are:

(a) Is the entropy criterion capable of ruling out all physically undesirable processes? If not, what supplementary criteria have to be adopted and what is their physical motivation?

(b) Are processes compatible with the entropy criterion endowed with "stability"? Such an occurrence would satisfy those who speculate that the second law of thermodynamics is essentially a statement of stability.

As a partial answer to question (b), above, we now proceed to show that the entropy admissibility criterion at least induces uniqueness and stability of smooth thermomechanical processes within the broader class of BV processes.

THEOREM 8.1. Let $\mathcal{B}$ be a bounded body with smooth boundary $\partial\mathcal{B}$. Assume that $\{\bar{F}(X,t), \bar{v}(X,t), \bar{\eta}(X,t)\}$ is a smooth process, defined on $\mathcal{B} \times [0,t_o]$, with supply terms $\{\bar{f}(X,t), \bar{r}(X,t)\}$ in $L^\infty(\mathcal{B} \times [0,t_o])$.

Suppose that there is a positive constant ν such that

$$(8.1)\qquad \frac{\partial^2\hat{\varepsilon}(\bar{\underset{\sim}{F}},\bar{\eta})}{\partial F_{i\alpha}\partial F_{j\beta}} N_\alpha N_\beta \xi_i \xi_j + 2\frac{\partial^2\hat{\varepsilon}(\bar{\underset{\sim}{F}},\bar{\eta})}{\partial F_{i\alpha}\partial \eta} N_\alpha \xi_i \zeta + \frac{\partial^2\hat{\varepsilon}(\bar{\underset{\sim}{F}},\bar{\eta})}{\partial \eta^2}\zeta^2$$

$$\geq \nu(|\underset{\sim}{\xi}|^2|\underset{\sim}{N}|^2 + \zeta^2),$$

for any state $(\bar{F},\bar{\eta})$ in the range of the given process, every $\underset{\sim}{\xi}$ and $\underset{\sim}{N}$ in R^n and all ζ in R. Then there are positive constants δ,α,β,K,M with the following property. For any process $\{\underset{\sim}{F}(\underset{\sim}{X},t),\underset{\sim}{v}(\underset{\sim}{X},t),\eta(\underset{\sim}{X},t)\}$ of class BV, with supply terms $\{\underset{\sim}{f}(\underset{\sim}{X},t),r(\underset{\sim}{X},t)\}$ in $L^1([0,t_o];L^2(\mathscr{B}))$, which satisfies the entropy admissibility criterion together with

$$(8.2)\qquad |\underset{\sim}{F}(\underset{\sim}{X},t) - \bar{\underset{\sim}{F}}(\underset{\sim}{X},t)| + |\eta(\underset{\sim}{X},t) - \bar{\eta}(\underset{\sim}{X},t)| < \delta, \quad (\underset{\sim}{X},t)\in\mathscr{B}\times[0,t_o],$$

$$(8.3)\qquad \underset{\sim}{x}(\underset{\sim}{X},t) = \bar{\underset{\sim}{x}}(\underset{\sim}{X},t), \qquad (\underset{\sim}{X},t)\in\partial\mathscr{B}\times[0,t_o],$$

we have

$$(8.4)\qquad ||\{\underset{\sim}{F}-\bar{\underset{\sim}{F}},\underset{\sim}{v}-\bar{\underset{\sim}{v}},\eta-\bar{\eta}\}(\cdot,\tau)||_{L^2(\mathscr{B})}$$

$$\leq M\exp(\alpha\tau+\beta\tau^2)||\{\underset{\sim}{F}-\bar{\underset{\sim}{F}},\underset{\sim}{v}-\bar{\underset{\sim}{v}},\eta-\bar{\eta}\}(\cdot,0)||_{L^2(\mathscr{B})}$$

$$+ N\exp(\alpha\tau+\beta\tau^2)\int_0^\tau ||\{\underset{\sim}{f}-\bar{\underset{\sim}{f}},r-\bar{r}\}(\cdot,t)||_{L^2(\mathscr{B})}dt.$$

Before proceeding to the proof of the theorem, the following remarks are in order. Assumption (8.1) on material response combines the strong ellipticity condition, introduced in Section 7, with the standard thermodynamic assumptions of Gibb's stability. The estimate (8.4) establishes continuous dependence of the process upon initial data and supply terms and thus, in particular, uniqueness. We wish to reemphasize that even though $\{\bar{\underset{\sim}{F}}(\underset{\sim}{X},t),\bar{\underset{\sim}{v}}(\underset{\sim}{X},t),\bar{\eta}(\underset{\sim}{X},t)\}$ has been assumed smooth, the comparison process $\{\underset{\sim}{F}(\underset{\sim}{X},t),\underset{\sim}{v}(\underset{\sim}{X},t),\eta(\underset{\sim}{X},t)\}$ is merely of class BV.

We now sketch the proof of Theorem 8.1. We define

(8.5) $$\hat{H}(\underset{\sim}{F},\underset{\sim}{v},\eta;\overline{\underset{\sim}{F}},\overline{\underset{\sim}{v}},\overline{\eta}) = \tfrac{1}{2}\rho(\underset{\sim}{v}-\overline{\underset{\sim}{v}})\cdot(\underset{\sim}{v}-\overline{\underset{\sim}{v}}) + \rho\hat{\varepsilon}(\underset{\sim}{F},\eta) - \rho\hat{\varepsilon}(\overline{\underset{\sim}{F}},\overline{\eta})$$

$$- \operatorname{tr}[\hat{T}(\overline{\underset{\sim}{F}},\overline{\eta})(\underset{\sim}{F}-\overline{\underset{\sim}{F}})^T] - \rho\hat{\theta}(\overline{\underset{\sim}{F}},\overline{\eta})(\eta-\overline{\eta}),$$

(8.6) $$\hat{\underset{\sim}{G}}(\underset{\sim}{F},\underset{\sim}{v},\eta;\overline{\underset{\sim}{F}},\overline{\underset{\sim}{v}},\overline{\eta}) = -(\hat{\underset{\sim}{T}}(\underset{\sim}{F},\eta) - \hat{\underset{\sim}{T}}(\overline{\underset{\sim}{F}},\overline{\eta}))\cdot(\underset{\sim}{v}-\overline{\underset{\sim}{v}})$$

and let

(8.7) $$H(\underset{\sim}{X},t) = \hat{H}(\underset{\sim}{F}(\underset{\sim}{X},t),\underset{\sim}{v}(\underset{\sim}{X},t),\eta(\underset{\sim}{X},t);\ \overline{\underset{\sim}{F}}(\underset{\sim}{X},t),\overline{\underset{\sim}{v}}(\underset{\sim}{X},t),\overline{\eta}(\underset{\sim}{X},t)),$$

(8.8) $$\underset{\sim}{G}(\underset{\sim}{X},t) = \underset{\sim}{G}(\underset{\sim}{F}(\underset{\sim}{X},t),\underset{\sim}{v}(\underset{\sim}{X},t),\eta(\underset{\sim}{X},t);\ \overline{\underset{\sim}{F}}(\underset{\sim}{X},t),\overline{\underset{\sim}{v}}(\underset{\sim}{X},t),\overline{\eta}(\underset{\sim}{X},t)).$$

We proceed to estimate $\dot{H}$ + Div $\underset{\sim}{G}$, using that $\{\underset{\sim}{F}(\underset{\sim}{X},t),\underset{\sim}{v}(\underset{\sim}{X},t),\eta(\underset{\sim}{X},t)\}$ satisfies (7.1), (7.2) while $\underset{\sim}{F}(\underset{\sim}{X},t),\underset{\sim}{v}(\underset{\sim}{X},t),\eta(\underset{\sim}{X},t)$ satisfies (7.3), (7.4). The result of the long and tedious calculation is

(8.9) $$\dot{H} + G_{\alpha,\alpha} \le \rho(v_i-\overline{v}_i)(f_i-\overline{f}_i) + \frac{\rho}{\theta}(\theta-\overline{\theta})(r-\overline{r}) - \frac{\rho\overline{r}}{\theta\overline{\theta}}(\theta-\overline{\theta})^2$$

$$+ \dot{\overline{F}}_{i\alpha}\{T_{i\alpha} - \overline{T}_{i\alpha} - \frac{\overline{\partial\hat{T}_{i\alpha}}}{F_{j\beta}}(F_{j\beta}-\overline{F}_{j\beta}) - \frac{\overline{\partial\hat{T}_i}}{\partial\eta}(\eta-\overline{\eta})\}$$

$$+ \frac{\rho\overline{r}}{\overline{\theta}}\{\theta-\overline{\theta} - \frac{\partial\hat{\theta}}{\partial F_{j\beta}}(F_{j\beta}-\overline{F}_{j\beta}) - \frac{\partial\hat{\theta}}{\partial\eta}(\eta-\overline{\eta})\}.$$

We now integrate (8.9) over $\mathscr{B}\times[0,\tau]$. The integral of Div G vanishes, by virtue of the Gauss-Green theorem and (8.3). To estimate the integral of H we use (8.1), (8.2) and go through the procedure that is followed in the theory of elliptic partial differential equations in order to establish Gårding's inequality thus obtaining

(8.10) $$\int_{\mathscr{B}} H(\underset{\sim}{X},\tau)\,d\underset{\sim}{X} \ge \lambda\,||\{\underset{\sim}{F}-\overline{\underset{\sim}{F}},\underset{\sim}{v}-\overline{\underset{\sim}{v}},\eta-\overline{\eta}\}(\cdot,\tau)||^2_{L^2(\mathscr{B})}$$

$$- \kappa\int_{\mathscr{B}} |\underset{\sim}{x}(\underset{\sim}{X},\tau) - \overline{\underset{\sim}{x}}(\underset{\sim}{X},\tau)|^2\,d\underset{\sim}{X}.$$

The last term on the right-hand side of (8.10) can be estimated from above,

$$(8.11)\quad \int_{\mathcal{B}} |\underset{\sim}{x}(\underset{\sim}{X},\tau) - \overline{\underset{\sim}{x}}(\underset{\sim}{X},\tau)|^2 d\underset{\sim}{X} \leq c \int_{\mathcal{B}} |\underset{\sim}{F}(\underset{\sim}{X},0) - \overline{\underset{\sim}{F}}(\underset{\sim}{X},0)|^2 \, d\underset{\sim}{X}$$

$$+ c \int_0^\tau \int_{\mathcal{B}} |\underset{\sim}{v}(\underset{\sim}{X},t) - \overline{\underset{\sim}{v}}(\underset{\sim}{X},t)|^2 d\underset{\sim}{X}\, dt.$$

Finally, we observe that the right-hand side of (8.9) is of quadratic order in $(\underset{\sim}{F}-\overline{\underset{\sim}{F}}, \underset{\sim}{v}-\overline{\underset{\sim}{v}}, \eta-\overline{\eta}, \underset{\sim}{f}-\overline{\underset{\sim}{f}}, r-\overline{r})$. Combining the above ingredients and using Gronwall type inequalities we arrive at (8.4).

Bibliography

DAFERMOS, C.M., The second law of thermodynamics and stability. Arch. Rational Mech. Anal. 70(1979), 167-197.

DAFERMOS, C.M., Stability of motions of thermoelastic fluids. J. Thermal Stresses 2(1979), 127-134.

DIPERNA, R.J., Uniqueness of solutions to hyperbolic conservation laws. Indiana U. Math. J. 28(1979), 137-188.

KOSINSKI, W., On weak solutions, stability and uniqueness in dynamics of dissipative bodies. Arch. Mech. 33(1981), 319-323.

LIU, T.-P., Initial-boundary value problems for gas dynamics. Arch. Rational Mech. Anal. 64(1977), 137-168.

9. The Lax Shock Admissibility Criterion

We take up here the question of admissibility of processes with shock waves. The results of the previous section together with the fact that the measure $\dot{\eta} - r/\theta$ is concentrated on the set of points of approximate jump discontinuity indicate strongly that the issue of admissibility arises only on the shock set and may therefore be settled by means of local shock admissibility criteria. In particular, the entropy admissibility criterion is equivalent to the jump condition

$$s[\eta] \le 0 \tag{9.1}$$

which is obtained by applying the measure (7.2) to the shock set.

The Rankine-Hugoniot jump conditions corresponding to (7.1) read

$$\begin{cases} s[\underset{\sim}{F}] + [\underset{\sim}{v}] \otimes \underset{\sim}{N} = \underset{\sim}{0} \\ \rho s[\underset{\sim}{v}] + \underset{\sim}{N}\cdot[\underset{\sim}{T}] = \underset{\sim}{0} \\ \rho s[\varepsilon + \frac{1}{2}\underset{\sim}{v}\cdot\underset{\sim}{v}] + \underset{\sim}{N}\cdot[\underset{\sim}{T}\cdot\underset{\sim}{v}] = 0. \end{cases} \tag{9.2}$$

We assume that the strong ellipticity condition, introduced in Section 7, holds and discuss briefly the structure of shocks of small amplitude. We recall that the characteristic speeds of (7.1) in the direction $\underset{\sim}{N}$ are 0, with multiplicity $n(n-1) + 1$, and $\pm\lambda_k^{1/2}$, $k = 1,\dots,n$, where $\lambda_1,\dots,\lambda_n$ are the eigenvalues of the acoustic tensor. The propagation speed of a shock of small amplitude has to be close to one of the characteristic speeds.

We observe that (9.2) admits one family of stationary shocks (i.e., $s = 0$) with $[\underset{\sim}{v}] = \underset{\sim}{0}$ and $\underset{\sim}{N}\cdot[\underset{\sim}{T}] = \underset{\sim}{0}$. These are admissible discontinuities that incur no dissipation and need not concern us any further.

We now turn to shocks with nonzero speed. In state space we seek all states $(\underset{\sim}{F},\underset{\sim}{v},\eta)$ in a small neighborhood of a fixed one $(\overline{\underset{\sim}{F}},\overline{\underset{\sim}{v}},\overline{\eta})$ which can be joined to $(\overline{\underset{\sim}{F}},\overline{\underset{\sim}{v}},\overline{\eta})$ by a shock that propagates in a specified direction $\underset{\sim}{N}$ and with speed close to $\lambda_k^{1/2}(\overline{\underset{\sim}{F}},\overline{\eta};\underset{\sim}{N})$ (or $-\lambda_k^{1/2}(\overline{\underset{\sim}{F}},\overline{\eta};\underset{\sim}{N})$). It can be shown by standard bifurcation theory that all these states lie on a smooth curve through $(\bar{F},\bar{v},\bar{\eta})$. To be precise, there are two smooth maps from a small interval $(-\delta,\delta)$ to R and R^{n^2+n+1} which carry τ into $s(\tau)$ and $(F,v,\eta)(\tau)$ with the property $s(0) = \lambda_k^{1/2}(\bar{F},\bar{\eta};N)$ (or $s(0) = -\lambda_k^{1/2}(\bar{F},\bar{\eta};N))$, $(F,v,\eta)(0) = (\bar{F},\bar{v},\bar{\eta})$ and

$$
(9.3)\quad \begin{cases} s(\tau)(F(\tau)-\bar{F}) + (v(\tau)-\bar{v}) \otimes N = 0 \\ \rho s(\tau)(v(\tau)-\bar{v}) + N\cdot(T(\tau)-\bar{T}) = 0 \\ \rho s(\tau)(\varepsilon(\tau)-\bar{\varepsilon}+\frac{1}{2}v(\tau)\cdot v(\tau)-\frac{1}{2}\bar{v}\cdot\bar{v})+N\cdot(T(\tau)\cdot v(\tau)-\bar{T}\cdot\bar{v}) = 0. \end{cases}
$$

Furthermore, $(F,v,\eta)(\tau)$, $-\delta < \tau < \delta$, are the only states in some neighborhood of $(\bar{F},\bar{v},\bar{\eta})$ that can be joined to $(\bar{F},\bar{v},\bar{\eta})$ by a shock with speed close to $\lambda_k^{1/2}(\bar{F},\bar{\eta};N)$ (or $-\lambda_k^{1/2}(\bar{F},\bar{\eta};N))$.

The above shocks will satisfy the entropy admissibility criterion if

$$
(9.4)\qquad s(\tau)(\eta(\tau) - \bar{\eta}) \le 0.
$$

We intend to investigate the implications of (9.4) upon shock stability.

Differentiating (9.3) with respect to τ,

$$
(9.5)\quad \begin{cases} sF' + v'\otimes N = -s'(F-\bar{F}) \\ \rho sv' + N\cdot T' = -\rho s'(v-\bar{v}) \\ \rho s\varepsilon'+\rho sv\cdot v'+N\cdot T'\cdot v+N\cdot T\cdot v' = -\rho s'(\varepsilon-\bar{\varepsilon}+\frac{1}{2}v\cdot v-\frac{1}{2}\bar{v}\cdot\bar{v}) \end{cases}
$$

Using (5.7), (9.5) yields

$$
(9.6)\qquad \rho s\theta\eta' = -s'\{\rho(\varepsilon-\bar{\varepsilon} + \tfrac{1}{2}v\cdot v - \tfrac{1}{2}\bar{v}\cdot\bar{v}) - \mathrm{tr}\{T(F-\bar{F})^T\} - \rho v\cdot(v-\bar{v})\}.
$$

Combining (9.6) with (9.3) and after a long calculation,

$$
(9.7)\qquad s\theta\eta' = s'(v-\bar{v})\cdot(v-\bar{v}).
$$

To estimate $s'(\tau)$, we will compute $s'(0)$. We set $\tau = 0$ in (9.5) and use (5.7), (7.5) to get

$$(9.8)\qquad (\underset{\sim}{Q}(\underset{\sim}{\overline{F}},\overline{\eta};\underset{\sim}{N}) - s^2(0)\underset{\sim}{I})\underset{\sim}{v}'(0) = \underset{\sim}{0}$$

which shows that $\underset{\sim}{v}'(0)$ is an eigenvector of the acoustic tensor associated with the eigenvalue $\lambda_k(\underset{\sim}{\overline{F}},\overline{\eta};\underset{\sim}{N})$. Next we differentiate the first two equations in (9.5) with respect to τ and then set $\tau = 0$ thus obtaining

$$(9.9)\qquad \begin{cases} sF''_{i\alpha} + 2s'F'_{i\alpha} + v''_i N_\alpha = 0 \\[2ex] 2s'v'_i + sv''_i + \dfrac{\partial^2\hat{\varepsilon}}{\partial F_{i\alpha}\partial F_{j\beta}} F''_{j\beta}N_\alpha + \dfrac{\partial^3\hat{\varepsilon}}{\partial F_{i\alpha}\partial F_{j\beta}\partial F_{k\gamma}} F'_{j\beta}F'_{k\gamma}N_\alpha = 0. \end{cases}$$

$(9.5)_1$ with $\tau = 0$ gives $sF'_{i\alpha} = -v'_i N_\alpha$ and this together with (9.9) easily imply

$$(9.10)\qquad 2s'(Q_{ij}v'_j + s^2 v'_i) - s(Q_{ij}v''_j - s^2 v''_i) + \frac{\partial^3\hat{\varepsilon}}{\partial F_{i\alpha}\partial F_{j\beta}\partial F_{k\gamma}} N_\alpha N_\beta N_\gamma v'_j v'_k = 0.$$

Multiplying (9.10) by v'_i and using (9.8) we obtain

$$(9.11)\qquad 4s^2 s' v'_i v'_i = -\frac{\partial^3\hat{\varepsilon}}{\partial F_{i\alpha}\partial F_{j\beta}\partial F_{k\gamma}} N_\alpha N_\beta N_\gamma v'_i v'_j v'_k.$$

Thus whenever the eigenvector $\underset{\sim}{r}$ associated with the eigenvalue $\lambda_k(\underset{\sim}{\overline{F}},\overline{\eta};\underset{\sim}{N})$ satisfies the genuine nonlinearity condition (7.11), (9.11) yields $s'(0) \neq 0$ and thereby $s'(\tau) \neq 0$ for τ small. It then follows from (9.7) that $\eta'(\tau) \neq 0$ and that $s(\tau)\eta'(\tau)$ and $s'(\tau)$ have the same sign. In particular, (9.4) is equivalent to $s(\tau) < s(0) = \lambda_k^{1/2}(\underset{\sim}{\overline{F}},\overline{\eta};\underset{\sim}{N})$ (or $-\lambda_k^{1/2}(\underset{\sim}{\overline{F}},\overline{\eta};\underset{\sim}{N})$). By reversing the roles of left and right state, we also deduce $s(\tau) > \lambda_k^{1/2}(\underset{\sim}{F},\eta;\underset{\sim}{N})$ (or $-\lambda_k^{1/2}(\underset{\sim}{F},\eta;\underset{\sim}{N})$. So, finally, we conclude that under the genuine nonlinearity assumption (7.11) and for shocks of small amplitude the entropy admissibility criterion is equivalent to

$$(9.12)\qquad \lambda_k^{1/2}(\underset{\sim}{F}^-,\eta^-;\underset{\sim}{N}) > s > \lambda_k^{1/2}(\underset{\sim}{F}^+,\eta^+;\underset{\sim}{N})$$

or

$$(9.12)^*\qquad -\lambda_k^{1/2}(\underset{\sim}{F}^-,\eta^-;\underset{\sim}{N}) > s > -\lambda_k^{1/2}(\underset{\sim}{F}^+,\eta^+;\underset{\sim}{N}).$$

The condition (9.11) or (9.11)* constitutes the <u>Lax shock admissibility criterion</u>. It has been shown that when the initial data are smooth except on a smooth surface across which $\{\tilde{F}(\tilde{X},0), \tilde{v}(\tilde{X},0), \eta(\tilde{X},0)\}$ experience jump discontinuities that satisfy, for appropriate s, the Rankine-Hugoniot conditions (9.2) together with Lax's condition (9.11) (or (9.11)*), then this surface will evolve, at least for a short time, as a shock wave. In other words, Lax's shock admissibility criterion and thereby also the entropy admissibility criterion may be interpreted as statements of stability.

Bibliography

DAFERMOS, C.M., The equations of elasticity are special. In "Trends in Applications of Pure Mathematics to Mechanics" Vol. III. R. J. Knops, Ed. Pitman, London 1981.

LAX, P.D., Hyperbolic systems of conservation laws, II, Comm. Pure Appl. Math. 10(1957), 537-566.

LAX, P.D., Shock waves and entropy. In "Contributions to Nonlinear Functional Analysis". E. A. Zarantonello, Ed., Academic Press, New York 1971.

MAJDA, A., The existence of multi-dimensional shock fronts. (To appear).

MALEK-MADANI, R., Energy criteria for finite hyperelasticity. Arch. Rational Mech. Anal. 77(1981), 177-188.

10. Viscosity and Entropy Rate Admissibility Criterion.

In Section 9 we saw that, under the assumption of genuine nonlinearity, the entropy admissibility criterion and the Lax shock admissibility criterion are equivalent, provided shocks have small amplitude. Moreover, the analytical study of the initial value problem for the one-dimensional version of (7.1) seems to indicate that the entropy criterion is capable of singling out a unique solution with prescribed initial data of small variation in the range of genuine nonlinearity. Nevertheless, when genuine nonlinearity fails and/or we are dealing with shocks of large amplitude, it is easy to see, by means of simple examples, that the entropy criterion and Lax's criterion are no longer equivalent and that neither of these criteria (or even both in cooperation) is capable of singling out a unique process. It is clear that a more selective criterion has to be devised.

One point of view is to develop criteria which impose restrictions upon the speed of propagation of admissible shocks, in the spirit of, but more stringent than, the Lax shock admissibility criterion. Successful criteria of this type have indeed been devised but they will not be discussed here since they don't seem to admit any physical interpretation.

A different approach is to attempt to strengthen the entropy criterion. The Clausius-Duhem inequality (4.6) requires that the measure $\rho\dot{\eta} - \mathrm{Div}(\underset{\sim}{Q}/\theta) - \rho r/\theta$ be nonnegative. When this is fulfilled by several processes satisfying the same initial and boundary conditions, it is tempting to single out as admissible the one for which $\rho\dot{\eta} - \mathrm{Div}(\underset{\sim}{Q}/\theta) - \rho r/\theta$ is appropriately "maximized". This leads to an <u>entropy rate admissibility criterion</u>.

Still another type of admissibility criterion arises when the material under investigation happens to lie on the "boundary" of a more elaborate material class in which admissibility of processes is not at issue. For instance, a thermoelastic nonconductor may be visualized as a limiting case of a thermoviscoelastic conductor. According to this philosophy a process of the thermoelastic nonconductor is admissible if it is the limit of processes of a sequence of thermoviscoelastic conductors with viscosity and heat conductivity tending to zero. This is the <u>viscosity criterion.</u>

For illustration purposes we will compare and contrast the various admissibility criteria in the context of one-dimensional isothermal thermoelasticity. The constitutive relations (5.5) can be written in the form

$$(10.1)\qquad \begin{cases} T = \hat{T}(F) \\ \rho\psi = \int_1^F \hat{T}(\xi)d\xi, \end{cases}$$

with $\hat{T}'(F) > 0$, while the balance laws read

$$(10.2)\qquad \begin{cases} F_t = v_X \\ \rho v_t = T_X + \rho f. \end{cases}$$

As explained in Section 5, the second law of thermodynamics is here expressed by the inequality

$$(10.3)\qquad \rho(\psi + \frac{1}{2} v^2)_t - (Tv)_X - \rho f v \leq 0.$$

The Rankine-Hugoniot jump conditions induced by (10.2) are

$$(10.4)\qquad \begin{cases} s[F] + [v] = 0 \\ \rho s[v] + [T] = 0. \end{cases}$$

In particular, the shock speed of propagation is

$$(10.5)\qquad s = \pm\{\frac{[T]}{\rho[F]}\}^{1/2}.$$

The characteristic speeds of (10.2) are $\pm\{\hat{T}'(F)/\rho\}^{1/2}$ so the Lax shock admissibility criterion reduces to

$$(10.6)\qquad \pm\hat{T}'(F^-) > \pm\frac{\hat{T}(F^+)-\hat{T}(F^-)}{F^+-F^-} > \pm\hat{T}'(F^+)$$

in which the plus sign applies for forward shocks and the minus sign for backward shocks.

Inequality (10.3) expresses here the entropy criterion. Its local form is

$$\rho s[\psi + \tfrac{1}{2}v^2] + [Tv] \geq 0. \tag{10.7}$$

Combining (10.1), (10.4) and (10.5) we deduce after a short computation

$$\rho s[\psi+\tfrac{1}{2}v^2] + [Tv] = s\{\int_{F^-}^{F^+} \hat{T}(\xi)d\xi-\tfrac{1}{2}(\hat{T}(F^+)+\hat{T}(F^-))(F^+-F^-)\}. \tag{10.8}$$

We observe that the right-hand side of (10.8) is s times the signed area comprised between the arc of the graph of $\hat{T}(F)$ that joins the points $(F^-,\hat{T}(F^-))$, $(F^+,\hat{T}(F^+))$ and the chord that connects the same points. It follows that Lax's criterion and the entropy criterion are generally unrelated. However, observe that these two criteria are indeed equivalent when $\hat{T}''(F) \neq 0$ which is exactly the present form of the genuine nonlinearity condition.

According to the <u>entropy rate</u> admissibility criterion, a process $\{F(X,t),v(X,t)\}$ on $[a,b] \times [0,t_1]$ is admissible if it minimizes, for every $0 \leq \tau < t_o$, the rate of decrease of mechanical energy

$$\frac{d^+}{dt}\int_a^b \rho(\psi + \tfrac{1}{2}v^2)(X,t)dX\Big|_{t=\tau} \tag{10.9}$$

over the set of processes which satisfy the balance laws, the boundary conditions and coincide with $\{F(X,t),v(X,t)\}$ on $[a,b] \times [0,\tau]$.

It can be shown that the entropy rate criterion is equivalent to the following local statement: Among all possible resolutions of a jump discontinuity into fans of waves admissible is the one that maximizes

$$\Sigma\{\rho s[\psi + \tfrac{1}{2} v^2] + [Tv]\} \tag{10.10}$$

where the summation extends over all shocks in the fan.

We finally consider the <u>viscosity criterion</u>. We visualize our material (10.1) as the $\mu \to 0^+$ limit of the following family of isothermal viscoelastic materials (compare with (5.1)):

$$(10.11) \qquad \begin{cases} T = \hat{T}(F) + \hat{S}(F,\mu\dot{F}) \\ \rho\psi = \int_1^F \hat{T}(\xi)d\xi \end{cases}$$

where

$$(10.12) \qquad \hat{S}(F,W)W > 0, \quad W \neq 0.$$

A BV process $\{F(X,t),v(X,t)\}$ for the elastic material is admissible if it is the a.e. limit of processes $\{F_\mu(X,t),v_\mu(X,t)\}$ of the viscoelastic material, as $\mu \to 0+$.

We proceed to motivate a local version of the viscosity criterion. Let us assume that $\{F_\mu(X,t),v_\mu(X,t)\}$ approximates the profiles of the shock waves of $\{F(X,t),v(X,t)\}$. More precisely, if $(\bar{X},\bar{t})$ is any fixed point on a shock wave, we assume that the derivatives of $\{F_\mu(X,t), v_\mu(X,t)\}$ at $(\bar{X},\bar{t})$ in the tangential direction to the shock wave remain bounded, as $\mu \to 0+$, in contrast with the derivatives in the normal direction which have to blow up, since $\{F_\mu(X,t),v_\mu(X,t)\}$ changes rapidly from near (F^-,v^-) to near (F^+,v^-), across the shock. Let s be the shock speed at the point $(\bar{X},\bar{t})$. We replace (X,t) with new variables

$$(10.13) \qquad \zeta = \frac{1}{\mu}\{s(X-\bar{X}) + t - \bar{t}\}, \quad \xi = \frac{1}{\mu}\{X - \bar{X} - s(t-\bar{t})\},$$

tangential and normal to the shock wave and appropriately scaled. We may visualize $\{F_\mu,v_\mu\}$ as functions of (ζ,ξ) and, as $\mu \to 0$, these functions should converge to $\{F_o(\xi),v_o(\xi)\}$, functions of ξ alone. We recall that $\{F_\mu,v_\mu\}$ satisfy (10.2) with T given by (10.11). Thus, writing these equations in terms of the new variables (ζ,ξ) and letting $\mu \to 0+$ we conclude that $\{F_o(\xi),v_o(\xi)\}$ must satisfy the system of ordinary differential equations

$$(10.14) \qquad \begin{cases} s\dot{F}_o(\xi) + \dot{v}_o(\xi) = 0 \\ \rho s\dot{v}_o(\xi) + \dot{\hat{T}}(F_o(\xi)) + \dot{\hat{S}}(F_o(\xi),\dot{v}_o(\xi)) = 0 \end{cases}$$

together with the boundary conditions

$$(10.15)\quad \begin{cases} F_o(-\infty) = F^-, & v_o(-\infty) = v^- \\ F_o(\infty) = F^+, & v_o(\infty) = v^+ . \end{cases}$$

Eliminating $\dot{v}_o(\xi)$ between the two equations in (10.14) and integrating the resulting equation once we obtain

$$(10.16)\quad \hat{S}(F_o(\xi), -s\dot{F}_o(\xi)) = \rho s^2(F_o(\xi)-F^-) - (\hat{T}(F_o(\xi)) - \hat{T}(F^-)).$$

By virtue of (10.5), F^+ is a critical point of (10.16). We require $F_o(\xi) \to F^+$, as $\xi \to \infty$. For that purpose we need

$$(10.17)\quad \frac{\hat{T}(F) - \hat{T}(F^-)}{F - F^-} - \rho s^2 \begin{cases} > 0 & \text{if } s > 0 \\ < 0 & \text{if } s < 0 \end{cases}$$

for every F between F^- and F^+. (10.17) expresses the local form of the viscosity criterion. On account of (10.5), (10.17) means that when $s[F] < 0$ (or $s[F] > 0$) then the arc of the graph of $\hat{T}(F)$ which joins the points $(F^-,\hat{T}(F^-))$, $(F^+,\hat{T}(F^+))$ remains above (or below) the chord that connects the same points. It is thus clear that the viscosity criterion strengthens the entropy criterion (10.7) and also implies a somewhat weakened version of Lax's criterion (10.6), allowing contact discontinuities. Quite surprisingly it turns out that the viscosity criterion is equivalent to the entropy rate criterion. The proof is long and cumbersome and will not be given here. A similar investigation was undertaken for polytropic nonconductors and showed that the entropy rate criterion is equivalent to the viscosity criterion only when the gas is monatomic.

Bibliography

CHANG, T., and HSIAO, L., The extremal character of the wave curve in the system of conservation laws of aerodynamics. Acta Math. Sinica 1979.

DAFERMOS, C.M., The entropy rate admissibility criterion for solutions of hyperbolic conservation laws. J. Diff. Eqs. 14 (1973), 202-212.

DAFERMOS, C.M., The entropy rate admissibility criterion in thermoelasticity. Rend. Accad. Naz. Lincei, Ser. VIII, Vol. LVII (1974), 113-119.

DAFERMOS, C.M., Structure of solutions of the Riemann problem for hyperbolic systems of conversation laws. Arch. Rational Mech. Anal. 53 (1974), 203-217.

DIPERNA, R.J., Convergence of approximate solutions to conservation laws. Arch. Rational Mech. Anal. (to appear).

HSIAO, L., The entropy rate admissibility criterion in gas dynamics. J. Diff. Eqs., 38(1980), 226-238.

LIU, T.-P. The entropy condition and the admissibility of shocks. J. Math. Anal. Appl. 53(1976), 78-88.

WENDROFF, B., The Riemann problem for materials with nonconvex equations of state. J. Math. Anal. Appl. 38(1972), 454-466.

11. Thermoelasticity

Even though thermoelastic materials that conduct heat are more dissipative than thermoelastic nonconductors, it is my conjecture that in two or three dimensions internal dissipation is too weak to prevent the breaking of acceleration waves of even small amplitude. In other words I believe that two- and three-dimensional thermoelastic materials should be classified in Category IV of Section 6. In contrast, in one-dimensional thermoelasticity the dissipation manages to preserve the smoothness of processes with smooth initial data near equilibrium. Thus one-dimensional thermoelastic materials are classified in Category III. In order to avoid lengthy and cumbersome calculations we will attempt to communicate the essential ingredients of the proof by means of very simple model cases.

Let us first consider the initial value problem

$$u_t + uu_X = 0, \qquad -\infty < X < \infty, \qquad t \geq 0, \tag{11.1}$$

$$u(X,0) = \bar{u}(X), \qquad -\infty < X < \infty, \tag{11.2}$$

and show that it does not admit globally defined smooth solutions, unless $\bar{u}_X(X) \geq 0$. To this end, we monitor the evolution of $u_X(X,t)$ along characteristics

$$\frac{dX}{dt} = u(X,t). \tag{11.3}$$

Differentiating (11.1) with respect to X and setting $u_X = w$ we obtain

$$\frac{\delta w}{\delta t} + w^2 = 0, \tag{11.4}$$

where

$$\frac{\delta}{\delta t} = \frac{\partial}{\partial t} + u\frac{\partial}{\partial X} \tag{11.5}$$

is the directional derivative along a characteristic. Thus $w = u_X$ will blow up in a finite time, unless $w(0) = \bar{u}_X(\cdot) \geq 0$.

The reader may have noticed already the similarity between the above proof and the proof that the amplitude of acceleration waves in thermoelastic nonconductors blows up in a finite time, presented in Section 7. In fact, (11.1) may be viewed as a very simple model of (7.1).

Let us now add to (11.1) a "frictional" term which will induce weak internal dissipation. Consider the equation

$$u_t + uu_X + u = 0, \qquad -\infty < X < \infty, \quad t \geq 0, \tag{11.6}$$

with initial conditions (11.2). Upon differentiating (11.6) with respect to X and setting $u_X = w$ we obtain, in the place of (11.4),

$$\frac{\delta w}{\delta t} + w^2 + w = 0 \tag{11.7}$$

which shows that $w = u_X$ will remain globally bounded if $w(0) = \bar{u}_X(\cdot) \geq -1$ but it will blow up in finite time if $w(0) < -1$.

Thus (11.1) should be classified in Category IV but (11.6) falls in Category III. We now proceed to establish another version of this last result which is less precise but whose proof is particularly versatile.

<u>THEOREM 11.1</u>. Consider the initial value problem (11.6), (11.2) where $\bar{u}_X(\cdot)$ and $\bar{u}_{XX}(\cdot)$ are both in $L^2(-\infty,\infty)$. Then if

$$\|\bar{u}_X\|_{L^2} \; \|\bar{u}_{XX}\|_{L^2} < \frac{2}{25}, \tag{11.8}$$

there exists a global C^1-smooth solution $u(X,t)$ such that $u_X(\cdot,t)$, $u_{XX}(\cdot,\ )$ are in $L^2(-\infty,\infty)$, for any $t \geq 0$.

We first give the idea of the proof. Assuming that a sufficiently smooth solution $u(X,t)$ exists on $(-\infty,\infty) \times [0,t_o)$, we differentiate (11.6) with respect to X, we multiply by $2u_X$, we integrate over $(-\infty,\infty) \times [0,\tau)$, $0 < \tau < t_o$, and integrate by parts thus arriving at

$$\int_{-\infty}^{\infty} u_X^2(X,\tau)\,dX + \int_0^{\tau}\int_{-\infty}^{\infty} (2+u_X)u_X^2\,dxdt = \int_{-\infty}^{\infty} \bar{u}_X^2(X)\,dX, \tag{11.9}$$

from which we could get an L^2 bound on u_X, uniform in time, if we had

$|u_X(X,t)| < 2$. This appears, of course, useless since pointwise bounds are locally stronger than L^2 bounds so one would have to assume more to get less. Nevertheless, one may attempt to obtain pointwise bounds on u_X by establishing first L^2 bounds on u_{XX}. To this end we differentiate (11.6) twice with respect to X, we muptiply by $2u_{XX}$, we integrate over $(-\infty,\infty) \times [0,\tau)$ and integrate by parts to derive the analog of (11.9) for second derivatives. The anticipated difficulty is that we now may have to assume pointwise bounds on u_{XX} in order to obtain L^2 bounds on u_{XX}. This danger, however, does not materialize! The derived estimate reads

$$\text{(11.10)} \qquad \int_{-\infty}^{\infty} u_{XX}^2(X,\tau)\,dX + \int_0^{\tau}\int_{-\infty}^{\infty}(2+5u_X)\,u_{XX}^2\,dX\,dt = \int_{-\infty}^{\infty} \bar{u}_{XX}^2(X)\,dX.$$

The miracle is that only a pointwise bound $5|u_X(X,t)| < 2$ is needed in order to get a uniform L^2 bound on u_{XX}. This is not a coincidence but rather a consequence of the algebraic structure of the operator of differentiation.

It is now easy to synthesize our proof. Let $[0,t_o)$, $t_o \le \infty$, be the maximal interval with the property that there is a C^1 smooth solution $u(X,t)$ on $(-\infty,\infty)\times[0,t_o)$ such that $u_X(\cdot,t)$, $u_{XX}(\cdot,t)$ are in $L^{\infty}_{loc}([0,t_o);L^2(-\infty,\infty))$ and

$$\text{(11.11)} \qquad \|u_X(\cdot,t)\|_{L^2}\,\|u_{XX}(\cdot,t)\|_{L^2} < \frac{2}{25}, \quad 0 \le t < t_o.$$

For $0 < \tau < t_o$ we have the estimates (11.9), (11.10). By account of (11.11) and

$$\text{(11.12)} \qquad u_X^2(X,t) = \int_{-\infty}^{X} (u_X^2)_X\,dX \le 2\|u_X(\cdot,t)\|_{L^2}\,\|u_{XX}(\cdot,t)\|_{L^2},$$

we get

$$\text{(11.13)} \qquad |u_X(X,t)| < \frac{2}{5}, \quad -\infty < X < \infty, \quad 0 \le t < t_o,$$

which, in conjunction with (11.9) and (11.10), implies that $\|u_X(\cdot,\tau)\|_{L^2}$ and $\|u_{XX}(\cdot,\tau)\|_{L^2}$ are nonincreasing on $[0,t_o)$.

Thus, if $t_o < \infty$, we may extend u(X,t) up to $t = t_o$ and (11.11) will now hold for $t = t_o$. But then, by the local existence theorem, u(X,t) can be extended onto a short time interval beyond t_o, still satisfying (11.11), and this is a contradiction since $[0,t_o)$ was assumed maximal. Therefore, $t_o = \infty$ and the solution is global.

To review, the strategy in the above proof was to assume at the outset a pointwise bound for the solution (cf. (11.13)) which then induces a chain of L^2-estimates for derivatives of the solution (cf. (11.9), (11.10) which, in turn, yield back the originally assumed pointwise bound thus closing the cycle. This approach has wide applicability and it has been employed frequently in the program of classification of material models with weak dissipation, induced by fading memory, heat diffusion, etc. In particular, for the case of one-dimensional thermoelasticity the analysis shows that initial data {F(X,0), v(X,0), θ(X,0)} such that F(X,0), $F_X(X,0)$, $F_{XX}(X,0)$, v(X,0), $v_X(X,0)$, $v_{XX}(X,0)$, $v_{XXX}(X,0)$, θ(X,0), $\theta_X(X,0)$, $\theta_{XX}(X,0)$, $\theta_{XXX}(X,0)$, $\theta_{XXXX}(X,0)$ are all in L^2 and their L^2 norms are sufficiently small generate globally defined smooth processes.

Bibliography

DAFERMOS, C.M., Can dissipation prevent the breaking of waves? Trans. 26th Conf. Army Math. (1981), pp. 187-198.

DAFERMOS, C.M., and NOHEL, J.A., A nonlinear hyperbolic Volterra equation in viscoelasticity. Am. J. Math. 1981 (suppl. dedicated to P. Hartman), pp. 87-116.

HRUSA, W.J., A nonlinear functional differential equation in Banach space with applications to materials with fading memory. Arch. Rational Mech. Anal. (to appear).

MACCAMY, R.C., A model for one-dimensional, nonlinear viscoelasticity. Q. Appl. Math. 35(1977), 21-33.

MATSUMURA, A., Global existence and asymptotics of the solution of the second order quasilinear hyperbolic equations with first order dissipation. Publ. Res. Inst. Math. Sci. Kyoto U., Ser. A 13 (1977), 349-379.

SLEMROD, M., Global existence, uniqueness and asymptotic stability of classical smooth solutions in one-dimensional non-linear thermo-elasticity. Arch. Rational Mech. Anal. 76 (1981), 97-133.

12. Thermoviscoelasticity

In thermoviscoelasticity internal dissipation is induced by, both, thermal diffusion and viscosity. It is thus reasonable to expect that processes of thermoviscoelastic bodies are smoother than the processes of thermoelastic ones. Indeed, it can be shown by the method outlined in Section 11 that two or three-dimensional thermoviscoelastic materials (5.1) belong to Category III. In one dimension dissipation is even more effective (as we saw in Section 11, this is also the case in thermoelasticity). In this section we show that one-dimensional viscoelastic materials should be classified in Category II.

Let us consider a one-dimensional homogeneous thermoviscoelastic body with reference configureation the interval [0,1], reference density $\rho = 1$ and constitutive relations

$$(12.1)\qquad \begin{cases} \varepsilon = \hat{\varepsilon}(F,\theta) \\ T = -\hat{p}(F,\theta) + \hat{\mu}(F)v_X \\ Q = -\hat{\kappa}(F,\theta)\theta_X \end{cases}$$

which satisfy the following assumptions:

$$(12.2)\qquad \hat{\varepsilon}_F(F,\theta) = -\hat{p}(F,\theta) + \theta\hat{p}_\theta(F,\theta),$$

$$(12.3)\qquad \hat{\mu}(F)F \geq \mu_o > 0, \quad 0 < F < \infty,$$

$$(12.4)\qquad \hat{\varepsilon}(F,\theta) \geq 0, \quad 0 < \nu \leq \hat{\varepsilon}_\theta(F,\theta) \leq N(1+\theta^{1/3}),$$

$$(12.5)\qquad \begin{cases} \hat{p}(F,\theta) \geq 0, & 0 < F < F_1, \quad 0 \leq \theta < \infty \\ \hat{p}(F,\theta) \leq 0, & F_2 < F < \infty, \quad 0 \leq \theta < \infty, \end{cases}$$

$$(12.6)\qquad |\hat{p}_F(F,\theta)| \leq N(1+\theta^{4/3}), \quad |\hat{p}_\theta(F,\theta)| \leq N(1+\theta^{1/3}),$$

$$(12.7)\qquad 0 < \kappa_o \leq \hat{\kappa}(F,\theta) \leq N, \ |\hat{\kappa}_F(F,\theta)| \leq N, \ |\hat{\kappa}_\theta(F,\theta)| \leq N.$$

where N is some positive constant. Assumption (12.2) is equivalent to

$(5.1)_2$. (12.3) states that viscosity is uniformly positive. Similarly, according to (12.4) and (12.7), the specific heat and heat conductivity are uniformly positive. Finally, by (12.5) the elastic part of the stress is compressive at high density and tensile at low density, i.e., the material is solid-like. The remaining growth assumptions are technical.

The balance equations of momentum and energy read

$$(12.8)\qquad \begin{cases} F_t - v_X = 0 \\ v_t - T_X = 0 \\ (\varepsilon + \frac{1}{2}v^2)_t - (Tv)_X + Q_X = 0. \end{cases}$$

We assume that the endpoints of the body are traction-free and thermally insulated, that is,

$$(12.9)\qquad \begin{cases} T(0,t) = T(1,t) = 0, & t \geq 0 \\ Q(0,t) = Q(1,t) = 0, & t \geq 0. \end{cases}$$

We also prescribe initial conditions

$$(12.10)\qquad F(X,0) = F_o(X), \quad v(X,0) = v_o(X), \quad \theta(X,0) = \theta_o(X), \; 0 \leq X \leq 1.$$

THEOREM 12.1. Assume that $F_o(X)$, $F_{oX}(X)$, $v_o(X)$, $v_{oX}(X)$, $v_{oXX}(X)$, $\theta_o(X)$, $\theta_{oX}(X)$, $\theta_{oXX}(X)$ are all in $C^\alpha[0,1]$ and let $F_o(X) > 0$, $\theta_o(X) > 0$, $0 \leq X \leq 1$. Then there exists a unique solution $\{F(X,t), v(X,t), \theta(X,t)\}$ of (12.8), (12.9), (12.10) on $[0,1] \times [0,\infty)$ such that F, F_X, F_t, F_{Xt}, v, v_X, v_t, v_{XX}, θ, θ_X, θ_t, θ_{XX} are all in $C^{\alpha,\alpha/2}$. Moreover, $\theta(X,t) > 0$, $c < F(X,t) < C$, for $0 \leq X \leq 1$, $0 \leq t < \infty$, where c and C are positive constants depending on the initial data.

The theorem thus establishes that the thermoviscoelastic material introduced above should be classified in Category II. The proof is based on an application of the Leray-Schauder fixed point theorem. The key element in the proof is the derivation of a priori estimates. The main obstacle is posed by coupling terms, such as stress power, that have superlinear growth and may thus induce blow up of solutions in a finite time. One may control the destabilizing effect of such terms by means of interpolation inequalities, provided that a basic set of a priori bounds is already available. The bounds on total momentum and total energy and the Clausius-Duhem inequality can be employed for that purpose. The analysis is too long to be reproduced here.

Bibliography

DAFERMOS, C.M., Global smooth solutions to the initial-boundary value problem for the equations of one-dimensional nonlinear thermo-viscoelasticity. SIAM J. Math. Analysis 13(1982), 397-408.

DAFERMOS, C.M., and HSIAO, L., Global smooth thermomechanical processes in one-dimensional nonlinear thermoviscoelasticity. J. Nonlinear Analysis (to appear).

KAZHIKOV, A.M., and SHELUKHIN, V. V., Unique global solution with respect to time of initial-boundary value problems for one dimensional equations of a viscous gas. Appl. Math. Mech. 41 (1977), 273-282.

MATSUMURA, A. and NISHIDA, T., The initial value problem for the equations of motion of viscous and heat-conductive gases. J. Math. Kyoto Univ. 20(1980), 67-104.

13. Adiabatic Shearing of Incompressible Fluids with Temperature Dependent Viscosity

In Section 11 we investigated thermoelastic materials with thermal diffusion but no viscosity. Here we look at the dual situation where viscosity is present but there is no thermal diffusion. A convenient test case arises in the problem of adiabatic shearing of an incompressible Newtonian fluid with viscosity that varies with temperature, $\mu = \mu(\theta)$, between two parallel plates occupying the planes $x = 0$ and $x = 1$. The flow is in the direction of the y axis. We let v denote the y-component of velocity. Then the shearing stress is

$$\sigma = \mu(\theta) v_x. \tag{13.1}$$

Assuming that the specific heat of the fluid is constant, say $c > 0$, the balance equations of momentum and energy take the form

$$\begin{cases} \rho v_t = [\mu(\theta) v_x]_x \\ c\theta_t = \mu(\theta) v_x^2 \end{cases} \qquad 0 \le x \le 1, \quad t \ge 0, \tag{13.2}$$

while the corresponding boundary and initial conditions read

$$v(0,t) = 0, \quad v(1,t) = 1, \quad t \ge 0, \tag{13.3}$$

$$v(x,0) = v_o(x), \ \theta(x,0) = \theta_o(x), \quad 0 \le x \le 1. \tag{13.4}$$

The relevant question here is whether the solution to (13.2), (13.3), (13.4) exists for all t and approaches asymptotically the uniform shearing flow:

$$v_x(x,t) = 1, \quad v_t(x,t) = 0, \quad \theta(x,t) = \Theta(t) \tag{13.5}$$

where

$$\int_{\theta_o}^{\Theta(t)} \frac{d\xi}{\mu(\xi)} = ct,$$

or else whether velocity gradient localizations may occur.

In (13.2) there is competition between the destabilizing effect of the stress power in the energy balance equation and the stabilizing effect of viscosity in the balance equation of momentum.

The existence and asymptotic behavior of solutions to (13.2), (13.3), (13.4) have been investigated under a variety of assumptions on $\mu(\theta)$. In order to convey the flavor of the conclusion, we state below the result in a representative case.

THEOREM 13.1. Let $\mu(\theta) = \beta\theta^{-\gamma}$, $\beta > 0$, $0 < \gamma < 1$. Assume $v_o(x) \in W^{2,2}(0,1)$, $\theta_o(x) \in W^{1,2}(0,1)$, $v_o(0) = 0$, $v_o(1) = 1$, $\theta_o(x) > 0$, $0 < x < 1$. Then there is a unique solution $v(x,t)$ of (13.2), (13.3), (13.4) on $[0,1] \times [0,\infty)$ and, as $t \to \infty$,

$$v_x(x,t) = 1 + O(t^{-\frac{1-\gamma}{1+\gamma}}), \tag{13.7}$$

$$v_t(x,t) = O(t^{-1}), \tag{13.8}$$

$$\int_{\theta_o(x)}^{\theta(x,t)} \frac{d\xi}{\mu(\xi)} = ct + O(t^{\frac{2\gamma}{\gamma+1}}). \tag{13.9}$$

The proof of the theorem rests upon a number of a priori estimates. As an illustration we present below a few of these estimates. The letter K will denote throughout a positive generic constant.

Multiplying $(13.2)_1$ by v_t and integrating over $[0,1] \times [0,t]$ we obtain, after two integrations by parts,

$$\rho\int_0^t\int_0^1 v_t^2 dx d\tau + \frac{\beta}{2}\int_0^1 \theta^{-\gamma}(x,t) v_x^2(x,t)dx + \frac{\beta^2\gamma}{2}\int_0^t\int_0^1 \theta^{-2\gamma-1} v_x^4 dx d\tau = \text{const.} \tag{13.10}$$

whence

$$\int_0^t\int_0^1 v_t^2 \, dx d\tau \le K. \tag{13.11}$$

We next multiply $(13.2)_2$ by $\mu(\theta) = \beta\theta^{-\gamma}$ to get

$$(13.12)\qquad \frac{c\beta}{1-\gamma}[\theta^{1-\gamma}(x,t)]_t = \sigma^2(x,t)$$

$$= \int_0^1 \sigma^2(y,t)\,dy + 2\int_0^1\int_y^x \sigma(\xi,t)v_t(\xi,t)\,d\xi\,dy.$$

Hence, setting

$$(13.13)\qquad \phi(t) = 1 + \int_0^t\int_0^1 \sigma^2\,dx\,d\tau,$$

we obtain easily from (13.12)

$$(13.14)\qquad \frac{1}{K}\,\phi^{\frac{1}{1-\gamma}}(t) \le \theta(x,t) \le K\phi^{\frac{1}{1-\gamma}}(t).$$

By virtue of (13.14),

$$(13.15)\qquad \int_0^1 \sigma^2(x,t)\,dx = \int_0^1 \beta^2\theta^{-2\gamma}(x,t)v_x^2(x,t)\,dx \ge \frac{1}{K}\,\phi^{-\frac{2\gamma}{1-\gamma}}(t).$$

On the other hand, again by (13.14),

$$(13.16)\qquad \int_0^1 \sigma^2(x,t)\,dx \le K\phi^{-\frac{\gamma}{1-\gamma}}(t)\int_0^1 \theta^{-\gamma}(x,t)v_x^2(x,t)\,dx.$$

In order to estimate the right-hand side of (13.16), we multiply $(13.2)_1$ by tv_t, we integrate over $[0,1]\times[0,t]$ and integrate by parts thus obtaining

$$(13.17)\qquad \rho\int_0^t\int_0^1 \tau v_t^2\,dx\,d\tau + \frac{\beta t}{2}\int_0^1 \theta^{-\gamma}(x,t)v_x^2(x,t)\,dx$$

$$+ \frac{\beta^2\gamma}{2}\int_0^t\int_0^1 \tau\theta^{-2\gamma-1}v_x^4\,dx\,d\tau = \frac{\beta}{2}\int_0^t\int_0^1 \theta^{-\gamma}v_x^2\,dx\,d\tau.$$

By account of $(13.2)_2$,

$$(13.18)\qquad \beta\int_0^t\int_0^1 \theta^{-\gamma}v_x^2\,dx\,d\tau = c\int_0^1 \theta(x,t)\,dx - c\int_0^1 \theta_o(x)\,dx.$$

Combining (13.16), (13.17) and (13.18),

$$(13.19) \qquad \int_0^1 \sigma^2(x,t)dx \le K\phi^{-\frac{1+\gamma}{1-\gamma}}(t)\frac{1}{t}\{\int_0^1 \theta(x,t)dx\}^2.$$

We now multiply $(13.2)_1$ by v and integrate over $[0,1] \times [0,t]$ to get, with the help of (13.18),

$$(13.20) \qquad \frac{\rho}{2}\int_0^1 v^2(x,t)dx + \int_0^1 \theta(x,t)dx = \int_0^t \sigma(1,\tau)d\tau + \text{const.}$$

Therefore, by Schwarz's inequality,

$$(13.21) \qquad [\int_0^1 \theta(x,t)dx]^2 \le t\int_0^t \sigma^2(1,\tau)d\tau + \text{const.}$$

$$\le 2t\int_0^t\int_0^1 \sigma^2 dx d\tau + t\int_0^t\int_0^1 \sigma_x^2 dx d\tau + \text{const.}$$

Hence, using (13.11), (13.13) we get, for lage t,

$$(13.22) \qquad [\int_0^1 \theta(x,t)dx]^2 \le Kt\phi(t).$$

Combining (13.19) with (13.22) and recalling (13.15),

$$(13.23) \qquad \frac{1}{K}\phi^{-\frac{2\gamma}{1-\gamma}}(t) \le \int_0^1 \sigma^2(x,t)dx \le K\phi^{-\frac{2\gamma}{1-\gamma}}(t).$$

Since

$$(13.24) \qquad \frac{d\phi}{dt} = \int_0^1 \sigma^2(x,t)dx,$$

(13.23) yields

$$(13.25) \qquad \frac{1}{K}t^{\frac{1-\gamma}{1+\gamma}} \le \phi(t) \le Kt^{\frac{1-\gamma}{1+\gamma}}$$

and so (13.14) implies

$$(13.26) \qquad \frac{1}{K}t^{\frac{1}{1+\gamma}} \le \theta(x,t) \le Kt^{\frac{1}{1+\gamma}}$$

which is at least the right order for (13.9).

The remaining estimates proceed at the same pattern and eventually yield (13.7), (13.8) and (13.9).

Bibliography

DAFERMOS, C.M., and HSIAO, L., Adiabatic shearing of incompressible fluids with temperature dependent viscosity. Quart. Appl. Math. (to appear).

Rational Thermodynamics of Mixtures of Fluids

by

Ingo Müller

FB 9 - Hermann-Föttinger-Institut
TU Berlin

Course given at Centro Internazionale Matematico Estivo on "Thermodynamics and Constitutive Equations" in Noto, Italy June 23 - to July 2, 1982

Contents

3. Simple Mixtures and Mixtures of Ideal Gases

3.1. Simple Mixtures

3.1.1. Definition and Interpretation

3.1.2. Chemical Potentials and Free Energy of a Simple Mixture

3.2. Mixtures of Ideal Gases

3.2.1. Dalton's Law

3.2.2. Gibbs Paradox

3.3. Law of Mass Action in Mixtures of Ideal Gases

3.3.1. A Reduced Form of the Chemical Potentials

3.3.2. Law of Mass Action

3.3.3. Mass Conservation for a Single Reaction

3.3.4. Typical Example for the Application of the Law of Mass Action; Haber Bosch Synthesis

3.3.5. le Chatelier's Principle

4. Diffusion and Wave Propagation in Simple Mixtures

4.1. Transport Equations for Diffusion and Heat Conduction

4.1.1. Phenomenological Equations of Linear Irreversible Thermodynamics

4.1.2. Fick's Law of Diffusion as a Mutilated Form of the Equations of Balance of Partial Momenta. Fourier's Law

4.1.3. Symmetry Relations for Diffusion Coefficients

4.1.4. Equality of Coefficients of Thermal Diffusion and Diffusion-Thermo-Coefficients

4.2. Wave Propagation in a Binary Mixture of Non-Reacting Constituents

4.2.1. Linearized Equations of Balance for a Binary Non-Reacting Mixture

4.2.2. First and Second Sound

4.2.3. Plane Harmonic Waves

4.2.4. Plane Harmonic Waves in a Binary Mixture of Non-Reacting Constituents

1. THERMODYNAMIC PROCESSES

1.1. Objective of Thermodynamics of Mixtures of Fluids

The objective of thermodynamics of mixtures of ν fluids is the determination of the fields of

$$\begin{aligned} &\rho_\alpha(x_n,t) \text{ - densities,} \\ &v_i^\alpha(x_n,t) \text{ - velocities,} \\ &T(x_n,t) \text{ - temperature} \end{aligned} \tag{1.1}$$

in all points of the body and at all times.

In order to achieve this objective, we need field equations and in thermodynamics it is customary to derive these from the equations of balance of mechanics and thermodynamics.

1.2. Equations of Balance

1.2.1. General

If Ψ is any additive property of the body we may write the balance of Ψ in a material volume V consisting of regular parts R^+ and R^- and of a singular surface s in the form (e.g. see [1])

$$\int_{R^++R^-} \frac{\partial \Psi^v}{\partial t}\,dv + \int_{S^++S^-} (\Psi^v v_i + \Phi_i^v)\,da_i - \int_s [\Psi^v] u_i\,da_i - \int_{R^++R^-} (\sigma^v + s^v)\,dv =$$

$$= \frac{d}{dt}\int_s \Psi^s da - \int_{\partial s} \Phi_i^s \tau_i\,ds + \int_s (\sigma^s + s^s)\,da \,. \tag{1.2}$$

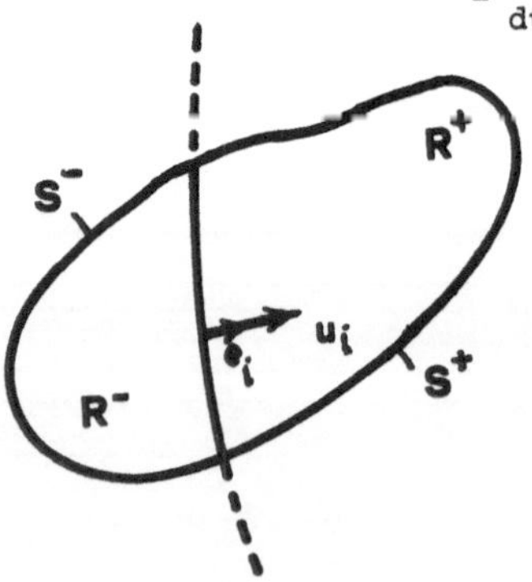

Figure 1

Ψ^v and Ψ^s are the volume density and surface density of Ψ respectively and similarly σ^v, s^v and σ^s, s^s are the densities of production and supply in the volume and on the surface. Φ_i^v and Φ_i^s are the densities of the non-convective flux of Ψ through S^+, S^- and ∂s respectively. u_i is the normal velocity of the singular surface and τ_i is a unit tangent vector on s which is normal to the line ∂s. The square bracket $[\Psi]$ stands for $\Psi^+ - \Psi^-$ as usual.

Singular surfaces play an important role in thermodynamics of mixtures, since they represent walls separating different bodies. As far as possible we shall neglect the surface densities and in particular we shall always consider Ψ^s and s^s as zero.

In regular points of the body the equation (1.2) implies the differential equation

$$\frac{\partial \Psi^v}{\partial t} + \frac{\partial}{\partial x_i}(\Psi^v v_i + \Psi_i^v) = \sigma^v + s^v , \tag{1.3}$$

whereas on the singular surface it implies the jump condition

$$[\Psi^v (v_i - u_i) + \Phi_i^v] e_i = \lim_{A\to 0} \frac{1}{A} \left(- \int_{\partial A} \Phi_i^s \tau_i ds + \int_A \sigma^s da\right) , \tag{1.4}$$

where A is an infinitesimal area on s.

1.2.2. Equation of Balance for Mass, Momentum and Energy

The conservation laws of mass, momentum and energy are special cases of the general equation of balance with properties and definitions that are summarized in the table

Ψ	Ψ^v	Φ_j^v	σ^v	s^v	Φ_j^s	σ^s
Mass	ρ	0	0	0	0	0
Momentum	ρv_i	$-t_{ij}$	0	ρf_i	$-t_{ij}^s$	0
Energy	$\rho(\varepsilon + \frac{v^2}{2})$	$-t_{ij} v_i + q_j$	0	$\rho(f_i v_i + r)$	$-t_{ij}^s v_i + q_j^s$	0

(1.5)

All productions are zero, since mass, momentum and energy are conserved in "closed bodies". Moreover the non-convective mass fluxes Φ_j^v, Φ_j^s as well as the mass-supply s^v are zero, because there is no flux of mass through a material surface, nor is there any supply of mass.

It follows that the equation of mass in regular points and on singular surfaces reads

$$\frac{\partial\rho}{\partial t} + \frac{\partial\rho v_i}{\partial x_i} = 0 ,$$

$$[\rho(v_i - u_i)e_i] = 0 . \tag{1.6}$$

Similarly, simply by insertion, we obtain the equation of balance of momentum in regular points and on singular surfaces

$$\frac{\partial\rho v_i}{\partial t} + \frac{\partial}{\partial x_j}(\rho v_i v_j - t_{ij}) = \rho f_i ,$$

$$[\rho v_i(v_j - u_j) - t_{ij}]e_j = \lim_{A\to 0}\frac{1}{A}\left(\int_{\partial A} \overset{s}{t}_{ij}\tau_j ds\right) . \tag{1.7}$$

t_{ij} is the stress tensor in the body and $\overset{s}{t}_{ij}\tau_j$ is the force per line element on the singular surface. f_i is the specific body force, usually gravitational. In a non-inertial frame f_i contains the specific inertial forces

$$i_i = 2\,\Omega_{ik}(v_k - \dot{b}_k) - \Omega_{ik}^2(x_k - b_k) + \dot{\Omega}_{ik}(x_k - b_k) + \ddot{b}_i \tag{1.8}$$

consisting of Coriolis force, centrifugal force, Euler force and force of relative translation. Ω_{ik} is the matrix of angular velocity of the non-inertal frame with respect to an inertial one and b_i is the radius vector between the origins of the two frames.

Finally we write the balance of energy in regular and singular points

$$\frac{\partial\rho(\varepsilon + \frac{v^2}{2})}{\partial t} + \frac{\partial}{\partial x_j}(\rho(\varepsilon + \frac{1}{2}v^2)v_j - t_{ij}v_i + q_j) = \rho(f_i v_i + r) ,$$

$$[\rho(\varepsilon + \frac{1}{2}v^2)(v_j - u_j) - t_{ij}v_i + q_j]e_i = \lim_{A\to 0}\frac{1}{A}\int_{\partial A}(\overset{s}{t}_{ij}v_i\tau_j - \overset{s}{q}_j\tau_j)ds . \tag{1.9}$$

ε and q_j are the specific value and the flux of internal energy respectively and ρr is the density of its supply due to radiation. $\overset{s}{q}_j$ is the flux of internal energy along the singular surface which we shall always neglect.

The reason for splitting the energy into two parts, viz. internal energy $\rho\varepsilon$ and kinetic energy $\frac{\rho}{2}v^2$ is that these contributions have different transformation properties as we shall see. The same is true for the decomposition of the fluxes and of the supply of energy.

1.2.3. Equations of Balance of Moment of Mementum, Kinetic Energy and Internal Energy

Multiplication of the balance of momentum by $\varepsilon_{k\ell i} x_i$ leads to the equation

$$\frac{\partial \rho(\underset{\sim}{x}\times\underset{\sim}{v})_k}{\partial t} + \frac{\partial}{\partial x_j}(\rho(\underset{\sim}{x}\times\underset{\sim}{v})_k v_j - \varepsilon_{k\ell i} x_\ell t_{ij}) = \varepsilon_{kij} t_{ij} + (\underset{\sim}{x}\times\rho\underset{\sim}{f})_k \tag{1.10}$$

which has the form of an equation of balance for moment of momentum whose density is $\rho(\underset{\sim}{x}\times\underset{\sim}{v})_k$. Therefore $\varepsilon_{kij} t_{ij}$ must be interpreted as the density of production of moment of momentum and if we assume that moment of momentum is conserved which is a valid assumption in non-polar fluids we conclude that the stress must be symmetric:

$$t_{ij} = t_{ji} \ . \tag{1.11}$$

Multiplication of the balance of momentum by v_i leads to the equation

$$\frac{\partial \frac{\rho}{2} v^2}{\partial t} + \frac{\partial}{\partial x_j}(\frac{\rho}{2} v^2 v_j - t_{ij} v_i) = -t_{ij}\frac{\partial v_i}{\partial x_j} + \rho f_i v_i \ , \tag{1.12}$$

which is the balance of kinetic energy, whose density is $\frac{\rho}{2} v^2$. The term $-t_{ij}\frac{\partial v_i}{\partial x_j}$ is the density of production of kinetic energy.

Subtraction of the balance of kinetic energy from the balance of energy (1.9) leads to the balance of internal energy, also called the first law of thermodynamics.

$$\frac{\partial \rho\varepsilon}{\partial t} + \frac{\partial}{\partial x_j}(\rho\varepsilon v_j + q_j) = t_{ij}\frac{\partial v_i}{\partial x_j} + \rho r \ . \tag{1.13}$$

The production density is $t_{ij}\frac{\partial v_i}{\partial x_j}$.

The balance of internal energy on a singular surface results from combining the jump conditions $(1.9)_2$ and $(1.7)_2$. It is necessary for this to assume that the velocity on the singular surface element A is the same everywhere and is equal to u_i, the normal velocity of the surface. In that case $(1.9)_2$ can be written as

$$[\rho(\varepsilon + \frac{1}{2} v^2)(v_j - u_j) - t_{ij} v_i + q_j] e_j = \lim_{A\to 0} \frac{1}{A}(u_i \int_{\partial A} t^s_{ij} \tau_j ds) \tag{1.14}$$

and now the integral may be eliminated between (1.14) and (1.7) with the result

$$[q_j - t_{ij}(v_i - u_i)]e_j + [\varepsilon + \frac{1}{2}(v_k - u_k)(v_k - u_k)]\rho(v_j - u_j)e_j = 0 , \qquad (1.15)$$

where the mass balance on the surface has also been used. This is the balance of internal energy on the singular surface.

In particular, for an impermeable surface and if the tangential velocities vanish on both sides, (1.15) reduces to the simple statement that the normal component of the flux of internal energy has no jump on the singular surface.

1.2.4. Equations of Balance for the Constituents

The following treatment on the equations of balance of constituents and the equations of balance of the mixture follows closely Truesdell's considerations on these subjects. See [2], [3].

The equations of balance of mass, momentum and energy of the constituents of a mixture have the same forms as those for a single body, except that the productions are not zero, because the partial quantities need not be conserved. We have

$$\frac{\partial \rho_\alpha}{\partial t} + \frac{\partial}{\partial x_j}(\rho_\alpha \overset{\alpha}{v}_j) = \tau_\alpha , \qquad (1.16)_1$$

$$\frac{\partial \rho_\alpha \overset{\alpha}{v}_i}{\partial t} + \frac{\partial}{\partial x_j}(\rho_\alpha \overset{\alpha}{v}_i \overset{\alpha}{v}_j - \overset{\alpha}{t}_{ij}) = \overset{\alpha}{m}_i + \rho_\alpha \overset{\alpha}{f}_i , \qquad (1.16)_2$$

$$\frac{\partial \rho_\alpha (\varepsilon_\alpha + \frac{\overset{\alpha}{v}{}^2}{2})}{\partial t} + \frac{\partial}{\partial x_j}(\rho_\alpha(\varepsilon_\alpha + \frac{\overset{\alpha}{v}{}^2}{2})\overset{\alpha}{v}_j - \overset{\alpha}{t}_{ij}\overset{\alpha}{v}_i + \overset{\alpha}{q}_j) = \ell_\alpha + \rho_\alpha(\overset{\alpha}{f}_i \overset{\alpha}{v}_i + r_\alpha) . \qquad (1.16)_3$$

Here τ_α, $\overset{\alpha}{m}_i$ and ℓ_α are the production densities of the constituents due to chemical reaction and to the exchange of momentum and energy between the constituents. The production densities must obey the conditions

$$\sum_{\alpha=1}^{\nu} \tau_\alpha = 0 , \qquad \sum_{\alpha=1}^{\nu} \overset{\alpha}{m}_i = 0 , \qquad \sum_{\alpha=1}^{\nu} \ell_\alpha = 0 , \qquad (1.17)$$

which express the conservation of mass, momentum and energy of the mixture as a whole. The production densities τ_α of mass are further restricted by stoichiometric requirements, by which only so many are independent as there are independent chemical reactions. Let the index a $(a = 1,2,\ldots,n)$ denote a chemical reaction, of which there are n independent ones and let $\overset{a}{\gamma}_\alpha$ be the stoichiometric coefficient of constituent α with molecular weight M_α in reaction a. Then the conservation of mass in each reaction requires

$$\sum_{\alpha=1}^{\nu} \overset{a}{\gamma}_\alpha M_\alpha \mu = 0 \tag{1.18}$$

where μ is a reference mass which may be taken to be the mass of a hydrogen atom. The number density of chemical reactions of type a occuring per unit time is called the reaction rate density and is denoted by $\overset{a}{\Lambda}$. The mass production of constituent α is then easily seen to be given by

$$\tau_\alpha = \sum_{a=1}^{n} (\overset{a}{\gamma}_\alpha M_\alpha \mu)\overset{a}{\Lambda} \quad . \tag{1.19}$$

Thus, indeed the number of independent mass productions is equal to n, the number of reaction rate densities.

The partial equations of balance of mass, momentum and energy on a singular surface are somewhat problematic, because of the production terms. Indeed, in general there will be production of mass, momentum and energy on those surfaces and in that case the jump conditions provide no useful information, unless we know how big the production is. Therefore we shall avoid using the partial equations of balance on the surface with one exception, the balance of mass.

For many singular surfaces of thermodynamics we can safely say that there is no mass production on them, and in that case the conservation of mass requires

$$[\rho_\alpha(v_i^\alpha - u_i)]e_i = 0 \ , \tag{1.20}$$

which states that the mass of constituent α that flows into the surface on one side, comes out on the other side.

1.2.5. Equations of Balance for the Mixture

Summation of all equations of balance of mass leads to

$$\frac{\partial \sum\limits_{\alpha=1}^{\nu} \rho_\alpha}{\partial t} + \frac{\partial}{\partial x_j}\left(\sum_{\alpha=1}^{\nu} \rho_\alpha v_j^\alpha\right) = 0 \ .$$

With the definitions

$$\rho \equiv \sum_{\alpha=1}^{\nu} \rho_\alpha \qquad \text{and} \qquad v_i \equiv \sum_{\alpha=1}^{\nu} \frac{\rho_\alpha}{\rho} v_i^\alpha \tag{1.21}$$

for the density and the velocity of the mixture we obtain the equation of balance of the mixture

$$\frac{\partial \rho}{\partial t} + \frac{\partial}{\partial x_j}(\rho v_j) = 0 \tag{1.22}$$

which has the same form as the equation of balance of mass for a single body.

We shall now proceed to do the same to the equations of balance of momentum and energy of the mixture, namely make them formally identical to the corresponding equations in a single body. And we shall chose the definitions of stress, internal energy and flux of internal energy accordingly. First of all we define a diffusion velocity $u_i^\alpha = v_i^\alpha - v_i$ which, by (1.21) satisfies the identity

$$\sum_{\alpha=1}^{\nu} \rho_\alpha u_i^\alpha = 0 \ . \tag{1.23}$$

If we define the stress of the mixture by

$$t_{ij} \equiv \sum_{\alpha=1}^{\nu} (t_{ij}^\alpha - \rho_\alpha u_i^\alpha u_j^\alpha) \ , \tag{1.24}$$

we obtain from $(1.16)_2$ by summation over all α the balance of momentum of the mixture, viz.

$$\frac{\partial \rho v_i}{\partial t} + \frac{\partial}{\partial x_j}(\rho v_i v_j - t_{ij}) = \sum_{\alpha=1}^{\nu} \rho_\alpha f_i^\alpha \ . \tag{1.25}$$

Similarly the definition of the specific internal energy ε and of the flux of internal energy q_i by

$$\varepsilon \equiv \sum_{\alpha=1}^{\nu} \frac{\rho_\alpha}{\rho}\left(\varepsilon_\alpha + \frac{u_\alpha^2}{2}\right) \quad \text{and} \quad q_i \equiv \sum_{\alpha=1}^{\nu}\left(q_i^\alpha + \rho_\alpha\left(\varepsilon_\alpha + \frac{u_\alpha^2}{2}\right) u_i^\alpha - t_{ji}^\alpha u_j^\alpha\right) \tag{1.26}$$

and summation over all equations $(1.16)_3$ leads to the balance of energy for the mixture, viz.

$$\frac{\partial\rho(\varepsilon+\frac{v^2}{2})}{\partial t}+\frac{\partial}{\partial x_j}\left(\rho(\varepsilon+\frac{v^2}{2})v_j-t_{ij}v_i+q_j\right)=\sum_{\alpha=1}^{\nu}\rho_\alpha(f_i^\alpha v_i^\alpha+r_\alpha)\,, \tag{1.27}$$

which again has the same form as the energy balance for a single body.

The equations of balance of internal energy can be obtained from (1.25) and (1.27) in the same manner as in a single body and the result is also the same:

$$\frac{\partial\rho\varepsilon}{\partial t}+\frac{\partial}{\partial x_j}(\rho\varepsilon v_j+q_j)=t_{ij}\frac{\partial v_i}{\partial x_j}+\sum_{\alpha=1}^{\nu}\rho_\alpha(f_i^\alpha u_i^\alpha+r_\alpha) \tag{1.28}$$

except that there is a supply of internal energy due to body forces, viz. the power of the body forces on the diffusive motion.

On a singular surface the equations of balance for the mixture have the same form as those for a single body and, in particular, the balance of mass and the balance of internal energy reads.

$$\begin{gathered}[\rho(v_i-u_i)]e_i=0\\ [q_j-t_{ij}(v_i-u_i)]e_j+[\varepsilon+\frac{1}{2}(v_k-u_k)(v_k-u_k)]\rho(v_j-u_j)e_j=0\,.\end{gathered} \tag{1.29}$$

1.3. Constitutive Relations

1.3.1. Relevant Balance Equations

For the purpose of determining the $4\nu+1$ fields (1.1) we shall rely upon the partial equations of balance of masses and momenta and upon the balance of internal energy of the mixture. These equations are summarized here.

$$\begin{gathered}\frac{\partial\rho_\alpha}{\partial t}+\frac{\partial\rho_\alpha v_j^\alpha}{\partial x_j}=\sum_{a=1}^{n}\gamma_\alpha^a M_\alpha\mu\Lambda^a\,,\\ \frac{\partial\rho_\alpha v_i^\alpha}{\partial t}+\frac{\partial}{\partial x_j}(\rho_\alpha v_i^\alpha v_j^\alpha-t_{ij}^\alpha)=m_i^\alpha+\rho_\alpha f_i^\alpha\,,\\ \frac{\partial\rho\varepsilon}{\partial t}+\frac{\partial}{\partial x_j}(\rho\varepsilon v_j+q_j)=t_{ij}\frac{\partial v_i}{\partial x_j}+\sum_{\alpha=1}^{\nu}\rho_\alpha(f_i^\alpha u_i^\alpha+r_\alpha)\,.\end{gathered} \tag{1.30}$$

1.3.2. General Constitutive Relations, Thermodynamic Processes

It is true that the balance laws (1.30) cannot serve as field equations for the fields (1.1) in the present form, even if, as we shall assume f_i^α and r_α are given fields. Indeed, these balance laws contain the additional quantities

$$\Lambda^a, \ t_{ij}^\alpha, \ m_i^\alpha, \ \varepsilon, \ q_i \ . \tag{1.31}$$

These are called constitutive quantities and experience tells us that they are related to the thermodynamic fields $\rho_\alpha(\underset{\sim}{x},t)$, $v_i^\alpha(\underset{\sim}{x},t)$ and $T(\underset{\sim}{x},t)$ in a materially dependent manner by what we call constitutive relations.

In particular, if the constitutive relations have the general form

$$\begin{aligned}
&\Lambda^a = \Lambda^a\,(\rho_\beta, \rho_{\beta,i}, v_i^\beta, T, T_{,i}) && (a = 1,2,\ldots,n)\\
&t_{ij}^\alpha = t_{ij}^\alpha\,(\ldots\ldots\ldots\ldots\ldots\ldots) && \text{with} \ \sum_{\alpha=1}^{\nu} t_{ij}^\alpha = \sum_{\alpha=1}^{\nu} t_{ji}^\alpha\\
&m_i^\alpha = m_i^\alpha\,(\ldots\ldots\ldots\ldots\ldots\ldots) && \text{with} \ \sum_{\alpha=1}^{\nu} m_i^\alpha = 0\\
&\varepsilon = \varepsilon\,(\ldots\ldots\ldots\ldots\ldots\ldots)\\
&q_i = q_i\,(\ldots\ldots\ldots\ldots\ldots\ldots)
\end{aligned} \tag{1.32}$$

we say that they characterize a mixture of inviscid fluids. [In mixtures of viscous fluids we should have to add velocity gradients to the list of variables. This addition offers no serious complication, but it makes for more complicated results].

If the consitutive functions were known explicitly, we could eliminate Λ^a, t_{ij}^α, m_i^α, ε and q_i between the equations of balance (1.30) and the constitutive relations (1.32) and thus obtain a set of explicit field equations. Every solution $\rho_\alpha(\underset{\sim}{x},t)$, $v_i^\alpha(\underset{\sim}{x},t)$, $T(\underset{\sim}{x},t)$ of the field equations is called a thermodynamic process.

If indeed the constitutive functions were known, the determination of a thermodynamic process would be a purely mathematical problem. However, the specific form of the constitutive functions is unknown and it is the task of the thermodynamic constitutive theory to find restrictions on the generality of these functions.

We proceed to exploit the principle of material frame indifference and the entropy principle to derive such restrictions.

1.3.3. The Interaction Force

The principle of material frame indifference can best be exploited, if it is applied to scalar, vectorial and tensorial constitutive quantities, because in this case the representation theorems for isotropic functions can be used.

Among the quantities (1.31) the momentum productions m^α_i are not objective vectors. Instead

$$m^\alpha_i - \tau_\alpha v^\alpha_i \tag{1.33}$$

is an objective vector. Indeed, in an arbitrary Euclidean frame we may rewrite the balance of momenta $(1.30)_2$ in the form

$$\rho_\alpha(\grave{v}^{*\alpha}_i - i^{*\alpha}_i) - \frac{\partial t^{*\alpha}_{ij}}{\partial x^*_j} = (m^{*\alpha}_i - \tau_\alpha v^{*\alpha}_i) + \rho_\alpha f^{*\alpha}_i , \tag{1.34}$$

where

$$\begin{aligned} \grave{v}^{*\alpha}_i &\equiv \frac{\partial v^{*\alpha}_i}{\partial t} + v^{*\alpha}_j \frac{\partial v^{*\alpha}_i}{\partial x^*_j} , \\ i^{*\alpha}_i &\equiv 2\,\Omega^*_{ik}(v^{*\alpha}_k - \dot{b}^*_k) - \Omega^{*2}_{ik}(x^*_k - b^*_k) + \dot{\Omega}^*_{ik}(x^*_k - b^*_k) + \ddot{b}^*_i \end{aligned} \tag{1.35}$$

are the acceleration and inertial acceleration of constituent α respectively. Since $\grave{v}^{*\alpha}_i - i^{*\alpha}_i$ is an objective vector and since $\rho_\alpha f^{*\alpha}_i$ as well as $\frac{\partial t^{*\alpha}_{ij}}{\partial x^*_j}$ are also objective vectors, (1.34) implies that $m^{*\alpha}_i - \tau_\alpha v^{*\alpha}_i$ is objective as well. Since m^α_i and τ_α or Λ^a are constitutive quantities, so is $m^\alpha_i - \tau_\alpha v^\alpha_i$. But note that, while $\sum_{\alpha=1}^{\nu} m^\alpha_i = 0$ and $\sum_{\alpha=1}^{\nu} \tau_\alpha = 0$ hold, the sum over the interaction forces is non-zero in general.

1.3.4. Intrinsic Values of the Internal Energy and of the Flux of Internal Energy

By the definitions (1.26) of ε and q_i, these quantities contain several terms which are explicit in the variables ρ_α and v^α_i. It is useful to separate these parts of ε and q_i and define the intrinsic values of internal energy and flux of internal energy by

$$\varepsilon_I \equiv \varepsilon - \sum_{\alpha=1}^{\nu} \frac{\rho_\alpha}{\rho} u_\alpha^2 \ ,$$
$$q_i^I \equiv q_i - \sum_{\alpha=1}^{\nu} \frac{\rho_\alpha}{2} u_\alpha^2 u_i^\alpha \ . \tag{1.36}$$

Note that ε_I and q_i^I are constitutive quantities since ε and q_i are such quantities. The intrinsic quantities ε_I and q_i^I are related to the partial internal energies ε_α and the partial heat fluxes q_i^α by the equations

$$\varepsilon_I = \sum_{\alpha=1}^{\nu} \frac{\rho_\alpha}{\rho} \varepsilon_\alpha \ ,$$
$$q_i^I = \sum_{\alpha=1}^{\nu} q_i^\alpha + \sum_{\alpha=1}^{\nu} (\rho_\alpha \varepsilon_\alpha \delta_{ij} - t_{ij}^\alpha) u_j^\alpha \tag{1.37}$$

which follow from (1.36) and (1.26).

1.3.5. Principle of Material Frame Indifference

From the considerations of the two previous sections we conclude that the set of constitutive relations

$$\Lambda^a = \Lambda^a (\rho_\beta, \rho_{\beta,i}, v_i^\beta, T, T_{,i}) \qquad (a = 1,2,\ldots,n)$$
$$t_{ij}^\alpha = t_{ij}^\alpha (\ldots\ldots\ldots\ldots\ldots\ldots\ldots) \qquad \text{with} \ \sum_{\alpha=1}^{\nu} t_{ij}^\alpha = \sum_{\alpha=1}^{\nu} t_{ij}^\alpha$$
$$m_i^\alpha - \tau_\alpha v_i^\alpha = M_i^\alpha (\ldots\ldots\ldots\ldots\ldots\ldots\ldots) \qquad \text{with} \ \sum_{\alpha=1}^{\nu} M_i^\alpha = -\sum_{\alpha=1}^{\nu-1} \tau_\alpha (v_i^\alpha - v_i^\nu) \tag{1.38}$$
$$\varepsilon_I = \varepsilon_I (\ldots\ldots\ldots\ldots\ldots\ldots\ldots)$$
$$q_i^I = q_i^I (\ldots\ldots\ldots\ldots\ldots\ldots\ldots)$$

is equivalent to the equations (1.32).

The principle of material frame indifference in the present case states that, if

$$C = \hat{C} (\rho_\beta, \rho_{\beta,i}, v_i^\beta, T, T_{,i}) \tag{1.39}$$

is any one of the constitutive equations (1.38) in an inertial frame, we must have

$$C^* = \hat{C} (\rho_\beta, \rho_{\beta,i}^*, v_i^{*\beta}, T, T_{,i}^*) \tag{1.40}$$

in an arbitrary Euclidean frame. The two frames are related by the equation

$$x_i^* = O_{ij}^* x_j + b_i^* .$$

Note that the constitutive <u>function</u> $\hat{C}$ is the same one in the two frames while its value has changed unless C is a scalar. $v_i^{*\beta}$, $\rho_{\beta,i}^*$ and $T_{,i}$ are the velocities and gradients of densities and temperature in the Euclidean frame

$$\rho_{\beta,i}^* = O_{ij}^* \rho_{\beta,j} , \qquad v_i^{*\beta} = O_{ij}^* v_j^\beta + \dot{b}_i^* , \qquad T_{,i}^* = O_{ij}^* T_{,j} . \tag{1.41}$$

The relation (1.40) for scalar, vectorial and tensorial constitutive quantities reads

$$\begin{aligned} S(\rho_\beta, \rho_{\beta,i}, v_i^\beta, T, T_{,i}) &= S(\rho_\beta, O_{ij}^* \rho_{\beta,j}, O_{ij}^* v_j^\beta + \dot{b}_i^*, T, O_{ij}^* T_{,j}) \\ O_{ij}^* V_j(\ldots\ldots\ldots\ldots\ldots) &= V_i(\ldots\ldots\ldots\ldots\ldots\ldots\ldots\ldots) \\ O_{ij}^* O_{k\ell}^* T_{j\ell}(\ldots\ldots\ldots\ldots\ldots) &= T_{ik}(\ldots\ldots\ldots\ldots\ldots\ldots\ldots\ldots) \end{aligned} \tag{1.42}$$

and these equations must hold for all O_{ij}^* and $\dot{b}_i^*$. In particular they must hold for

$$O_{ij}^* = \delta_{ij} \qquad \text{and} \qquad \dot{b}_i^* = - v_i^\nu \quad .$$

In this case (1.42) can be summarized in the form

$$\hat{C}(\rho_\beta, \rho_{\beta,i}, v_i^\beta, T, T_{,i}) = \hat{C}(\rho_\beta, \rho_{\beta,i}, v_i^\beta - v_i^\nu, T, T_{,i}) ,$$

whence we conclude that the constitutive quantities cannot depend on all ν velocities independently, but only on the $\nu - 1$ relative velocities

$$V_i^\alpha = v_i^\alpha - v_i^\nu , \tag{1.43}$$

which are objective vectors.

The requirements (1.42) of material frame indifference may therefore be written as

$$\begin{aligned} S(\rho_\beta, \rho_{\beta,i}, V_i^\beta, T, T_{,i}) &= S(\rho_\beta, O_{ij}^* \rho_{\beta,j}, O_{ij}^* V_j^\beta, T, O_{ij}^* T_{,i}) \\ O_{ij}^* V_i(\ldots\ldots\ldots\ldots\ldots) &= V_i(\ldots\ldots\ldots\ldots\ldots\ldots\ldots\ldots) \\ O_{ij}^* O_{k\ell}^* T_{j\ell}(\ldots\ldots\ldots\ldots\ldots) &= T_{ik}(\ldots\ldots\ldots\ldots\ldots\ldots\ldots\ldots) \end{aligned} \tag{1.44}$$

and we express this by saying that the constitutive functions must be scalar, vectorial and tensorial isotropic functions.

1.3.6. Linear Representations

The general solution of the functional equations (1.44) are given by representation theorems for isotropic functions. E.g. see [4], [5] for details. It follows that in the present case a scalar can only depend on ρ_β, T and on all scalar products of $\rho_{\beta,i}, v_i^\beta$ and $T_{,i}$.

A vector must have additive terms in the directions of $\rho_{\beta,i}, v_i^\beta$ and $T_{,i}$ with scalar coefficients. And finally a tensor can have an isotropic part plus dyadic products of the form $\rho_{\beta,i} v_j^\alpha$ formed from all available vectors.

We do not use such general representations here, since we shall limit the attention to the study of constitutive functions that are linear in $\rho_{\beta,i}, v_i^\beta$ and $T_{,i}$. In this case the most general form of the constitutive equations, that is compatible with the isotropy requirement, reads

$$
\begin{aligned}
\Lambda^a &= \Lambda^a(\rho_\beta, T) \,, \\
t_{ij}^\alpha &= -p^\alpha(\rho_\beta, T)\,\delta_{ij} \,, \\
m_i^\alpha - \tau_\alpha v_i^\alpha &= \sum_{\beta=1}^{\nu} M_\rho^{\alpha\beta}\rho_{\beta,i} + \sum_{\beta=1}^{\nu-1} M_V^{\alpha\beta} v_i^\beta + M_T^\alpha T_{,i} \,, \\
q_i^I &= \sum_{\beta=1}^{\nu} q_\rho^\beta\, \rho_{\beta,i} + \sum_{\beta=1}^{\nu-1} q_V^\beta v_i^\beta + q_T T_{,i} \,, \\
\varepsilon^I &= \varepsilon^I(\rho_\beta, T) \,.
\end{aligned}
\tag{1.45}
$$

p^α is called the partial pressure and q_T is the heat conductivity. The coefficients $M_\rho^{\alpha\beta}$, $M_V^{\alpha\beta}$, M_T^α as well as q_ρ^β, q_V^β, and q_T may all depend on ρ_β and T.
The values $M_\rho^{\alpha\beta}$, $M_V^{\alpha\beta}$ and M_T^β are not all independent, because of the restriction $(1.38)_3$ on the functions M_i^α. We have

$$\sum_{\alpha=1}^{\nu} M_\rho^{\alpha\beta} = 0 \,, \qquad (p = 1,2,\ldots,\nu) \tag{1.46$_1$}$$

$$\sum_{\alpha=1}^{\nu} M_V^{\alpha\beta} + \tau_\beta = 0 \,, \qquad (p = 1,2,\ldots,\nu-1) \quad \text{and} \tag{1.46$_2$}$$

$$\sum_{\alpha=1}^{\nu} M_T^\alpha = 0 \,. \tag{1.46$_3$}$$

The equations (1.45) are the most general linear constitutive relations that are compatible with the principle of material frame indifference. We conclude that linearity and frame indifference have reduced the $12\nu + n - 2$ unknown functions of $7\nu + 4$ variables each in the equations (1.38) to $2\nu^2 + \nu + n + 1$ functions of only $\nu + 1$ variables each. For small values of ν this is a considerable and very useful reduction.

Note that while the consitutive functions are linear, the theory as a whole is not, because the explicit nonlinear terms in t_{ij}, ε and q_i are taken into account.

2. ENTROPY PRINCIPLE

2.1. Purpose and Statement

2.1.1. Purpose of the Entropy Principle

Drastic as the restrictions of material frame indifference may be, they still leave us with the large number of coefficient-functions in the equations (1.45). The entropy principle is used to acquire some knowledge on these coefficient functions.

For this purpose it is unnecessary to include the external supplies f_i^α and r_α in the analysis, because the restrictions on the constitutive equations are unaffected by the presence or absence of those supplies.

2.1.2. Statement of the Entropy Principle

The entropy principle has been motivated in thermostatics and linear irreversible thermodynamics. It is stated here in a form that was first proposed for mixtures by Müller [6].

The principle consists of four parts:

i.) (Entropy Balance)

The entropy is an additive quantity whose equation of balance in regular points of the body we write as

$$\frac{\partial \rho\eta}{\partial t} + \frac{\partial}{\partial x_j}(\rho\eta v_j + \Phi_j) = \sigma \quad . \tag{2.1}$$

ii.) (Constitutive Property)

The specific entropy η and its flux Φ_i are an objective scalar and vector respectively and both are given by constitutive relations that obey the principle of material frame indifference. In particular in a mixture of inviscid fluids we have

$$\eta = \eta(\rho_\beta, T) \,,$$
$$\Phi_i = \sum_{\beta=1}^{\nu} \varphi_\rho^\beta \rho_{\beta,i} + \sum_{\beta=1}^{\nu-1} \varphi_V^\beta V_i^\beta + \varphi_T T_{,i} \,, \tag{2.2}$$

where φ_ρ^β, φ_V^β and φ_T may be functions of ρ_β, T .

iii.) (Entropy Inequality)

The entropy production σ is non-negative for all thermodynamic processes, so that the inequality

$$\frac{\partial \rho\eta}{\partial t} + \frac{\partial}{\partial x_j} (\rho\eta v_j + \Phi_j) \geq 0 \tag{2.3}$$

holds.

iv.) (Ideal Walls)

An ideal wall has no entropy production and the temperature is assumed continuous across it, so that we have

$$[\Phi_i e_i] + [\eta]\rho(v_i - u_i)e_i = 0 \qquad \text{and} \qquad [T] = 0 \,, \tag{2.4}$$

i.e. the normal component of the entropy flux is continuous and T is continuous.

2.1.3. Lagrange Multipliers

According to iii.) the inequality (2.3) does not hold for all fields $\rho_\alpha(x,t), v_i^\alpha(x,t)$, $T(x,t)$ but only for those which are solutions of the field equations. We may free ourselves of this constraint by the use of Lagrange multipliers. In the present case of mixtures we write the new inequality

$$\frac{\partial \rho\eta}{\partial t} + \frac{\partial}{\partial x_j} (\rho\eta v_j + \Phi_j) - \sum_{\alpha=1}^{\nu} \Lambda^{\rho_\alpha} \left(\frac{\partial \rho_\alpha}{\partial t} + \frac{\partial \rho_\alpha v_j^\alpha}{\partial x_j} - \tau_\alpha\right) -$$
$$- \sum_{\alpha=1}^{\nu} \Lambda^{v_i^\alpha} \left(\frac{\partial \rho_\alpha v_i^\alpha}{\partial t} + \frac{\partial}{\partial x_j} (\rho_\alpha v_i^\alpha v_j^\alpha - t_{ij}^\alpha) - m_i^\alpha\right) - \Lambda^\varepsilon \left(\frac{\partial \rho\varepsilon}{\partial t} + \frac{\partial}{\partial x_j} (\rho\varepsilon v_j + q_j) - t_{ij} \frac{\partial v_i}{\partial x_j}\right) \geq 0 \tag{2.5}$$

Liu [7] has shown that this inequality must hold for all fields ρ_α, v_i^α, T rather than only for thermodynamic processes. The Lagrange multipliers Λ^{ρ_α}, $\Lambda^{v_i^\alpha}$ and Λ^ε may be functions of the variables $\rho_\beta, \rho_{\beta,i}, v_i^\beta, T, T_{,i}$.

As it is our objective to obtain restrictions on the constitutive relations (1.45); the newly introduced quantities η and Φ_i as well as the Lagrange multipliers must be eliminated from the final results.

2.2. Evaluation of the Entropy Principle

2.2.1. Results from the Entropy Inequality

Insertion of the constitutive relations for η, Φ_i and for Λ^a, t_{ij}^α, $m_i^\alpha - \tau_\alpha v_i^\alpha$, ε, and q_i into the inequality (2.5) leads to an inequality whose left hand side is explicitly linear in the derivatives

$$\frac{\partial T}{\partial t}, \; T_{,ik}, \; \frac{\partial \rho_\beta}{\partial t}, \; \rho_{\beta,ik}, \; \frac{\partial v_i^\beta}{\partial t}, \; v_{i,j}^\beta . \tag{2.6}$$

Since the inequality must hold for all fields ρ_α, v_i^α and T, it must hold in particular for arbitrary values of the derivatives (2.6). Therefore we could violate the inequality, if a term with these derivatives were present on its left and side. It follows that the coefficients of the derivatives must vanish and this argument leads to the following conditions

$$\frac{\partial \eta}{\partial T} - \Lambda^\varepsilon \frac{\partial \varepsilon}{\partial T} = 0 , \tag{2.7$_1$}$$

$$\frac{\partial \Phi_i}{\partial T_{,j}} - \Lambda^\varepsilon \frac{\partial q_i}{\partial T_{,j}} = 0 , \tag{2.7$_2$}$$

$$\frac{\partial \eta}{\partial \rho_\beta} - \Lambda^\varepsilon \frac{\partial \varepsilon}{\partial \rho_\beta} + \frac{1}{\rho}(\eta - \Lambda^\varepsilon \varepsilon) - \frac{1}{\rho}\Lambda^{\rho_\beta} - \frac{1}{\rho} v_k^\beta \Lambda^{v_k^\beta} = 0 , \tag{2.7$_3$}$$

$$\frac{\partial \Phi_i}{\partial \rho_{\beta,j}} - \Lambda^\varepsilon \frac{\partial q_i}{\partial \rho_{\beta,j}} = 0 , \tag{2.7$_4$}$$

$$\frac{\partial \eta}{\partial v_i^\beta} - \Lambda^\varepsilon \frac{\partial \varepsilon}{\partial v_i^\beta} - \frac{\rho_\beta}{\rho}\Lambda^{v_i^\beta} = 0 , \qquad (\beta = 1,2,\dots,\nu-1) \tag{2.7$_5$}$$

$$-\sum_{\beta=1}^{\nu-1}\left(\frac{\partial \eta}{\partial v_i^\beta} - \Lambda^\varepsilon \frac{\partial \varepsilon}{\partial v_i^\beta}\right) - \frac{\rho_\nu}{\rho}\Lambda^{v_i^\nu} = 0 , \tag{2.7$_6$}$$

$$(\frac{\partial\eta}{\partial\rho_\alpha} - \Lambda^\varepsilon \frac{\partial\varepsilon}{\partial\rho_\alpha})\,\delta_{ij} - \frac{1}{\rho\rho_\alpha}(\frac{\partial\Phi_j}{\partial v_i^\alpha} - \Lambda^\varepsilon \frac{\partial q_j}{\partial v_i^\alpha}) + \frac{1}{\rho}\Lambda^{v_i^\alpha} u_j^\alpha - \frac{1}{\rho^2}\Lambda^\varepsilon t_{ij} = 0\,, \qquad (\alpha = 1,2,\ldots,\nu-1) \qquad (2.7)_7$$

$$(\frac{\partial\eta}{\partial\rho_\nu} - \Lambda^\varepsilon \frac{\partial\varepsilon}{\partial\rho_\nu})\,\delta_{ij} + \sum_{\beta=1}^{\nu-1} \frac{1}{\rho\rho_\nu}(\frac{\partial\Phi_j}{\partial v_i^\beta} - \Lambda^\varepsilon \frac{\partial q_j}{\partial v_i^\beta}) - \frac{1}{\rho}\Lambda^{v_i^\nu} u_j^\nu - \frac{1}{\rho^2}\Lambda^\varepsilon t_{ij} = 0\,. \qquad (2.7)_8$$

There remains the inequality

$$\sum_{\beta=1}^{\nu} [(\frac{\partial\Phi_j}{\partial\rho_\beta} - \Lambda^\varepsilon \frac{\partial q_j}{\partial\rho_\beta}) - \rho(\frac{\partial\eta}{\partial\rho_\beta} - \Lambda^\varepsilon \frac{\partial\varepsilon}{\partial\rho_\beta})\,u_j^\beta - \sum_{\alpha=1}^{\nu} \Lambda^{v_j^\alpha} \frac{\partial p^\alpha}{\partial\rho_\beta} + \Lambda^\varepsilon t_{ij}\frac{1}{\rho}u_i^\beta]\,\rho_{\beta,j} +$$

$$+ (\frac{\partial\Phi_j}{\partial T} - \Lambda^\varepsilon \frac{\partial q_j}{\partial T} - \sum_{\alpha=1}^{\nu} \Lambda^{v_j^\alpha} \frac{\partial p_\alpha}{\partial T})\,T_{,j} + \sum_{\alpha=1}^{\nu} \Lambda^{\rho_\alpha}\tau_\alpha + \sum_{\alpha=1}^{\nu} \Lambda^{v_i^\alpha} m_i^\alpha \geq 0\,. \qquad (2.8)$$

We shall now proceed to evaluate these conditions, and start with $(2.7)_{5,6}$. We recall that η is independent of v_i^β and that in ε it is only the explicit term $\frac{1}{2}\sum_{\alpha=1}^{\nu}\frac{\rho_\alpha}{\rho}u_\alpha^2$ that depends on v_i^β. By use of the identities

$$u_i^\beta = \sum_{\gamma=1}^{\nu-1} (\delta_{\beta\gamma} - \frac{\rho_\gamma}{\rho})\, v_i^\gamma \qquad \text{and} \qquad v_i^\gamma = \sum_{\beta=1}^{\nu-1} (\delta_{\gamma\beta} + \frac{\rho_\beta}{\rho_\nu})\, u_i^\beta \qquad (2.9)$$

we find $\frac{\partial\varepsilon}{\partial v_i^\beta} = \frac{\rho_\beta}{\rho}u_i^\beta$ and therefore the two equations $(2.7)_{5,6}$ may be written as

$$\Lambda^{v_i^\beta} = -\Lambda^\varepsilon u_i^\beta \qquad (2.10)$$

so that the Lagrange multipliers $\Lambda^{v_i^\beta}$ are now reduced to Λ^ε.

The equations $(2.7)_{1,3}$ may be written in the form

$$\frac{\partial\eta}{\partial T} = \Lambda^\varepsilon \frac{\partial\varepsilon_I}{\partial T}\,,$$

$$\frac{\partial\rho\eta}{\partial\rho_\beta} = \Lambda^\varepsilon \frac{\partial\rho\varepsilon_I}{\partial\rho_\beta} + \Lambda^{\rho_\beta} + \Lambda^\varepsilon \frac{1}{2}u_\beta^2 - \Lambda^\varepsilon v_k^\beta u_k^\beta$$

and it is thus obvious that Λ^ε can only depend on ρ_β and T; also Λ^{ρ_β} can be decomposed into a velocity dependent part and into an intrinsic part that depends only on ρ_β and T :

$$\Lambda^{\rho_\beta} = \Lambda_I^{\rho_\beta}(\rho_\delta, T) + \Lambda^\varepsilon(v_i^\beta u_i^\beta - \frac{1}{2}u_\beta^2)\,.$$

We may thus summarize the two equations for $\frac{\partial\eta}{\partial T}$ and $\frac{\partial\eta}{\partial\rho_\beta}$ in the form

$$d(\rho\eta) = \Lambda^\varepsilon d(\rho\varepsilon^I) + \sum_{\alpha=1}^{\nu} \Lambda_I^{\rho_\alpha} d\rho_\alpha \,. \tag{2.11}$$

If we multiply $(2.7)_7$ by ρ_α and $(2.7)_8$ by ρ_ν and sum all of these equations, we obtain

$$\begin{aligned} \Lambda^\varepsilon \sum_{\alpha=1}^{\nu} p_\alpha &= - \sum_{\alpha=1}^{\nu} \rho\rho_\alpha \left(\frac{\partial\eta}{\partial\rho_\alpha} - \Lambda^\varepsilon \frac{\partial\varepsilon^I}{\partial\rho_\alpha}\right) \\ \Lambda^\varepsilon \sum_{\alpha=1}^{\nu} p_\alpha &= - \sum_{\alpha=1}^{\nu} \rho_\alpha \Lambda_I^{\rho_\alpha} + \rho(\eta - \Lambda^\varepsilon\varepsilon^I) \,. \end{aligned} \tag{2.12}$$

On the other hand, when we subtract $(2.7)_8$ from $(2.7)_7$ we get after a little calculation by use of (2.9)

$$\frac{\partial\Phi_j}{\partial v_i^\beta} - \Lambda^\varepsilon \frac{\partial q_j^I}{\partial v_i^\beta} = \sum_{\alpha=1}^{\nu} \left(\rho_\beta\delta_{\beta\alpha} - \frac{\rho_\alpha\rho_\beta}{\rho}\right)\left(\frac{\partial\rho\eta}{\partial\rho_\alpha} - \Lambda^\varepsilon \frac{\partial\rho\varepsilon^I}{\partial\rho_\alpha}\right) \delta_{ij} \,.$$

By integration of this as well as of $(2.7)_2$ and $(2.7)_4$ we obtain

$$\Phi_j = \Lambda^\varepsilon q_j^I + \sum_{\beta=1}^{\nu-1} \sum_{\alpha=1}^{\nu} \left(\rho_\beta\delta_{\beta\alpha} - \frac{\rho_\alpha\rho_\beta}{\rho}\right)\left(\frac{\partial\rho\eta}{\partial\rho_\alpha} - \Lambda^\varepsilon \frac{\partial\rho\varepsilon^I}{\partial\rho_\alpha}\right) v_j^\beta \,, \tag*{$(2.13)_1$}$$

$$\Phi_j = \Lambda^\varepsilon q_j^I + \sum_{\alpha=1}^{\nu} \rho_\alpha u_j^\alpha \left(\frac{\partial\rho\eta}{\partial\rho_\alpha} - \Lambda^\varepsilon \frac{\partial\rho\varepsilon^I}{\partial\rho_\alpha}\right) \,, \qquad \text{or} \tag*{$(2.13)_2$}$$

$$\Phi_j = \Lambda^\varepsilon q_j^I + \sum_{\alpha=1}^{\nu} \rho_\alpha u_j^\alpha \Lambda_I^{\rho_\alpha} \,, \qquad \text{or} \tag*{$(2.13)_3$}$$

$$\Phi_j = \Lambda^\varepsilon q_j + \sum_{\alpha=1}^{\nu} \rho u_j^\alpha \left(\frac{\partial\rho\eta}{\partial\rho_\alpha} - \Lambda^\varepsilon \frac{\partial\rho\varepsilon}{\partial\rho_\alpha}\right) \,. \tag*{$(2.13)_4$}$$

With the results derived so far the residual inequality (2.8) may be rewritten in the form

$$\begin{aligned} &\sum_{\gamma=1}^{\nu} \left(\frac{\partial\Lambda^\varepsilon}{\partial\rho_\gamma} q_j^I + \sum_{\alpha=1}^{\nu} \frac{\partial\Lambda_I^{\rho_\alpha}}{\partial\rho_\gamma} \rho_\alpha u_j^\alpha\right) \rho_{\gamma,j} + \left(\frac{\partial\Lambda^\varepsilon}{\partial T} q_j^I + \sum_{\alpha=1}^{\nu} \frac{\partial\Lambda_I^{\rho_\alpha}}{\partial T} \rho_\alpha u_j^\alpha\right) T_{,j} - \\ &- \Lambda^\varepsilon \sum_{\alpha=1}^{\nu} [(m_i^\alpha - \tau_\alpha v_i^\alpha) - p_{,i}^\alpha] u_i^\alpha + \sum_{a=1}^{n} \left(\sum_{\alpha=1}^{\nu} (\Lambda_I^{\rho_\alpha} - \frac{1}{2}\Lambda^\varepsilon u_\alpha^2)\gamma_\alpha^a M_\alpha \mu\right)\Lambda^a \geq 0 \,. \end{aligned} \tag{2.14}$$

The equations (2.10) through (2.14) exhaust the consequences of the entropy inequality.

2.2.2. Results for a Single Fluid

For a single fluid the results from the entropy inequality follow from (2.10) through (2.14) by setting $\nu = 1$. We obtain

$$\Lambda^{v_i} = 0 \quad ,$$

$$d\eta = \Lambda^{\varepsilon}(d\varepsilon - \frac{p}{\rho^2}\, d\rho) \quad ,$$

$$\Phi_i = \Lambda^{\varepsilon} q_i \quad , \tag{2.15}$$

$$\frac{\partial \Lambda^{\varepsilon}}{\partial \rho} q_j \rho,_j + \frac{\partial \Lambda^{\varepsilon}}{\partial T} q_j T,_j \geq 0$$

While the equations (2.10) through (2.14) exhaust the consequences of the entropy inequality, they are of little use in themselves, since they only concern the auxiliary quantities η, Φ_i and Λ^{ρ_α}, $\Lambda^{v_i^\alpha}$, Λ^{ε}. We shall now proceed to show that the jump conditions (2.4), which form part of the entropy principle, can serve to identify the Lagrange multipliers Λ^{ε} and $\Lambda_I^{\rho_\alpha}$.

2.2.3. Λ^{ε} in a Single Fluid

We consider an impermeable ideal wall which separates two different single fluids I and II. Equation (2.4) reads here

$$[\Phi_i e_i] = 0 \qquad \text{and} \qquad [T] = 0 \; . \tag{2.16}$$

Φ_i is equal to $\Lambda^{\varepsilon} q_i$ and since the normal component of the flux of internal energy is continuous, we obtain from (2.16)

$$\Lambda_I^{\varepsilon}(\rho^I, T) = \Lambda_{II}^{\varepsilon}(\rho^{II}, T) \; . \tag{2.17}$$

Nov, ρ^I and ρ^{II} are variables that can be changed independently and therefore this relation can only be valid, if Λ^{ε} is independent of density. We conclude that

$$\Lambda_I^{\varepsilon}(T) = \Lambda_{II}^{\varepsilon}(T) \tag{2.18}$$

holds so that $\Lambda^{\varepsilon}(T)$ is the same function of T in all single fluids. Because of this property we say that $\Lambda^{\varepsilon}(T)$ is a universal function.

The determination of the function $\Lambda^{\varepsilon}(T)$ makes use of the equation $(2.15)_2$ which implies

$$\frac{d\ln\Lambda^{\varepsilon}}{dT} = -\frac{\frac{\partial p}{\partial T}}{p - \rho^2 \frac{\partial \varepsilon}{\partial \rho}} \tag{2.19}$$

as an integrability condition. If the right hand side is known for any one material, as it is for an ideal gas, Λ^{ε} can be calculated as a function of T by integration. It turns out that Λ^{ε} is equal to the reciprocal of T to within a positive multiplicative constant which, without loss of generality, we may choose to be equal to 1. Therefore we have

$$\Lambda^{\varepsilon} = \frac{1}{T} \ . \tag{2.20}$$

Thus in effect the Lagrange multiplier Λ^{ε} has been determined, by bringing an arbitrary single fluid into contact with an ideal gas along an ideal wall. As a result we may now write $(2.15)_2$ in the form

$$d\eta = \frac{1}{T}\,(d\varepsilon - \frac{p}{\rho^2}\,d\rho) \ . \tag{2.21}$$

This is the Gibbs equation for a single fluid. It is of supreme importance in thermodynamics because it permits the determination of the internal energy ε and the entropy η of a single fluid to within additive constants by measurements of pressures and specific heats.

2.2.4. Λ^{ε} in a Mixture

We consider an impermeable ideal wall separating a mixture I from a single fluid II. Here again the jump conditions (2.16) hold and we obtain

$$\Lambda^{\varepsilon}_{I}(\rho^{I}_{\alpha},T) = \frac{1}{T} \ , \tag{2.22}$$

since in a single fluid we already know that $\Lambda^{\varepsilon} = \frac{1}{T}$. Thus the universal character of Λ^{ε} extends to mixtures. We have found this out by bringing a mixture into contact with a single fluid along an ideal wall.

So far we have determined the Lagrange multiplier Λ^{ε} in the results (2.10) through (2.15) of the entropy inequality. It remains to determine the multipliers $\Lambda^{\rho_\alpha}_{I}$.

2.2.5. Chemical Potentials

We consider a semipermeable ideal wall separating two mixtures I and II which is permeable for constituent γ (say); that constituent is therefore present on both sides. We exclude the possibility of tangential velocities at the wall and therefore we have at the wall

$$v_i^\alpha = u e_i \quad \text{for} \quad \alpha \neq \gamma, \qquad \text{and} \qquad v_i = u e_i + \frac{\rho_\gamma}{\rho}(v_i^\gamma - u e_i) \ . \tag{2.23}$$

The balance of internal energy at the wall reads

$$[q_i]e_i - [\frac{t_{ij}e_ie_j}{\rho} - \varepsilon - \frac{1}{2}(\underset{\sim}{v} - u\underset{\sim}{e})^2]\rho(v_j - u_j)e_j = 0$$

according to (1.29). If we introduce q_i^I, ε^I and $t_{ij} = -\sum_{\alpha=1}^{\nu}(p_\alpha\delta_{ij} + \rho_\alpha u_i^\alpha u_j^\alpha)$, this jump condition assumes the form

$$[q_i^I e_i] + [\frac{\sum_{\alpha=1}^{\nu} p_\alpha}{\rho} + \varepsilon_I + \frac{1}{2}(\underset{\sim}{v}^\gamma - u\underset{\sim}{e})^2]\rho(v_\ell - u_\ell)e_\ell = 0 \ . \tag{2.24}$$

The balance of entropy on the ideal wall reads

$$[\Phi_i e_i] + [\eta]\rho(v_\ell - u_\ell)e_\ell = 0$$

according to (2.4), or with $(2.13)_3$

$$[q_i^I e_i] + T\,[\eta + \sum_{\beta=1}^{\nu}(\delta_{\gamma\beta} - \frac{\rho_\beta}{\rho})\,\Lambda_I^{\rho_\beta}]\rho(v_\ell - u_\ell)e_\ell = 0 \ , \tag{2.25}$$

where $\Lambda^\varepsilon = \frac{1}{T}$ and the continuity of T at the ideal wall has been used.

Equations (2.24) and (2.25) represent two equations for the jump of $q_i^I e_i$. If this jump is eliminated between (2.24) and (2.25) we obtain

$$[\varepsilon^I - T\eta + \frac{1}{2}(\underset{\sim}{v}^\gamma - u\underset{\sim}{e})^2 + \frac{\sum_{\alpha=1}^{\nu} p_\alpha}{\rho} - T\sum_{\beta=1}^{\nu}(\delta_{\gamma\beta} - \frac{\rho_\beta}{\rho})\,\Lambda_I^{\rho_\beta}] = 0 \ ,$$

or, upon replacing $\sum_{\alpha=1}^{\nu} p_\alpha$ by $-T\sum_{\alpha=1}^{\nu}\rho_\alpha\Lambda_I^{\rho_\alpha} - \rho(\varepsilon - T\eta)$ according to (2.12) and (2.22):

$$[-T\Lambda_I^{\rho_\gamma} + \frac{1}{2}(\underset{\sim}{v}^\gamma - u\underset{\sim}{e})^2] = 0 \ . \tag{2.26}$$

We conclude that the quantity in brackets is continuous at the ideal semipermeable wall. We call this quantity the chemical potential of constituent γ and denote it by μ_γ:

$$\mu_\gamma \equiv - T \Lambda_I^{\rho_\gamma} + \frac{1}{2} (\underset{\sim}{v}^\gamma - u\underset{\sim}{e})^2 . \tag{2.27}$$

μ_γ contains an explicit part which is the kinetic energy of the motion of constituent γ relative to the semipermeable wall. Splitting this velocity dependent part off, we introduce an intrinsic chemical potential

$$\mu_\gamma^I = - T \Lambda_I^{\rho_\gamma} . \tag{2.28}$$

Thus we have related the Lagrange multipliers $\Lambda_I^{\rho_\gamma}$ to the chemical potentials. This is important, as the chemical potentials, because of their continuity, can be measured. The key to the experimental determination of μ_γ^I, and hence of $\Lambda_I^{\rho_\gamma}$, is the equation (2.12) which can now be written in the form

$$\sum_{\alpha=1}^{\nu} \rho_\alpha \mu_\alpha^I = \rho(\varepsilon_I - T\eta) + \sum_{\alpha=1}^{\nu} p_\alpha \tag{2.29}$$

or for a single fluid

$$\mu^I = \varepsilon - T\eta + \frac{p}{\rho} , \tag{2.30}$$

so that in a single fluid the chemical potential is equal to the specific free enthalpy. Ignoring a fine point for the moment, we may say that μ^I is known, because ε, η, and p are known from measurements of pressures and specific heats as has been explained before.

Let us now consider an ideal semipermeable wall - permeable for constituent γ - and let this wall separate the pure constituent γ from a mixture containing γ. With (2.30) we may write the continuity condition for the chemical potential in the form

$$\underbrace{\{\mu_I^\gamma(\rho_1,\rho_2 \ldots \rho_\gamma,T) + \tfrac{1}{2}(\underset{\sim}{v}^\gamma - u\underset{\sim}{e})^2\}}_{\text{mixture}} = \underbrace{\{\varepsilon(\rho,T) - T\eta(\rho,T) + \frac{p(\rho,T)}{\rho} + \frac{1}{2}(\underset{\sim}{v}^\gamma - u\underset{\sim}{e})^2\}}_{\text{single constituent}} . \tag{2.31}$$

The right hand side of this equation is a known function of ρ, T and $\underset{\sim}{v}^\gamma - u\underset{\sim}{e}$. The determination of $\mu_I^\gamma(\rho_1,\ldots,\rho_\gamma,T)$ proceeds as follows: We measure ρ, T, $\underset{\sim}{v}^\gamma - u\underset{\sim}{e}$ of the single constituent and $\rho_1,\ldots,\rho_\nu,T$ and $\underset{\sim}{v}^\gamma - u\underset{\sim}{e}$ of the mixture and calculate $\mu_I^\gamma(\rho_1,\ldots,\rho_\nu,T)$ from (2.31).

Thus μ^γ or the Lagrange multiplier $\Lambda_I^{\rho_\gamma}$ has been determined by bringing an arbitrary mixture into contact with a single fluid along a semipermeable ideal wall.

In the above argument we have ignored a point which qualifies the result somewhat: the energy ε and the entropy η are known to within additive constants only by measurements of pressures and specific heats. Therefore it follows that (2.31) gives $\mu_I^\gamma(\rho_1,\ldots,\rho_\nu,T)$ only to within a linear function of temperature of the form

$$\beta^\gamma - T\,\alpha^\gamma \,, \tag{2.32}$$

where β^γ and α^γ are the constants in the internal energy and entropy of constituent γ respectively.

2.2.6. Summary of Results

After the Lagrange multipliers Λ^ε and $\Lambda_I^{\rho_\gamma}$ have been identified, the results (2.10) through (2.13) may be summarized in the form

$$\begin{aligned}
\Lambda^{v_i^\beta} &= -\frac{1}{T}u_i^\beta \\
d(\rho\eta) &= \frac{1}{T}\Big(d(\rho\varepsilon_I) - \sum_{\alpha=1}^{\nu}\mu_\alpha^I\, d\rho_\alpha\Big) \,, \\
\rho(\varepsilon_I - T\eta) &= \sum_{\alpha=1}^{\nu}\rho_\alpha\mu_I^\alpha - p \,, \\
\Phi_j &= \frac{1}{T}\Big(q_j^I - \sum_{\alpha=1}^{\nu}\mu_I^\alpha\rho_\alpha u_j^\alpha\Big) \,.
\end{aligned} \tag{2.33}$$

Equation $(2.33)_3$ is called the Gibbs-Duhem equation, it may serve for the calculation of the free energy density $\rho(\varepsilon_I - T\eta)$ once the chemical potentials have been measured.

Equation $(2.33)_2$ is the Gibbs equation for mixtures, it implies that the chemical potential μ^α is the derivative of the free energy density with respect to ρ_α and that the entropy density follows by differentiation of $\rho(\varepsilon_I - T\eta)$ with respect to T

$$\mu_\alpha^I = \frac{\partial(\rho(\varepsilon_I - T\eta))}{\partial\rho_\alpha} \,, \qquad \rho\eta = -\frac{\partial\rho(\varepsilon_I - T\eta)}{\partial T} \,. \tag{2.34}$$

Apart from this the Gibbs equation implies the integrability conditions

$$\frac{\partial \rho\varepsilon_I}{\partial \rho_\alpha} = -T^2 \frac{\partial \frac{\mu_I^\alpha}{T}}{\partial T} \qquad \text{and} \qquad \frac{\partial \mu_I^\alpha}{\partial \rho_\beta} = \frac{\partial \mu_I^\beta}{\partial \rho_\alpha} \tag{2.35}$$

which reduce the necessary effort in the experimental determination of the chemical potentials considerably.

Since we now know that Λ^ε is independent of the densities ρ_γ, inspection of the residual inequality shows that its left hand side is explicitely linear in the gradients $\rho_{\gamma,j}$. Therefore, by the same argument that has led to the restrictions (2.7) we conclude that

$$\sum_{\alpha=1}^{\nu} \frac{\partial \Lambda_I^{\rho_\alpha}}{\partial \rho_\gamma} \rho_\alpha u_j^\alpha + \frac{\partial \Lambda^\varepsilon}{\partial T} q_\rho^\gamma T_{,j} - \Lambda^\varepsilon \sum_{\alpha=1}^{\nu} (M_\rho^{\alpha\gamma} - \frac{\partial p^\alpha}{\partial \rho_\gamma}) u_j^\alpha = 0$$

must hold, which implies with (2.20) and (2.28)

$$q_\rho^\gamma = 0 \qquad \text{and} \qquad M_\rho^{\alpha\gamma} = \frac{\partial p^\alpha}{\partial \rho_\gamma} - \rho_\alpha \frac{\partial \mu_I^\alpha}{\partial \rho_\gamma} . \tag{2.36}$$

It follows that the density gradients do not contribute to the flux of internal energy; by (2.36) the flux of internal energy and the interaction force have the reduced forms

$$\begin{aligned} q_i^I &= q_T T_{,i} + \sum_{\beta=1}^{\nu-1} q_v^\beta v_i^\beta , \\ m_i^\alpha - \tau_\alpha v_i^k &= M_T^\alpha T_{,i} + \sum_{\beta=1}^{\nu-1} M_v^{\alpha\beta} v_i^\beta + \sum_{\beta=1}^{\nu} (\frac{\partial p^\alpha}{\partial \rho_\beta} - \rho_\alpha \frac{\partial \mu_I^\alpha}{\partial \rho_\beta}) \rho_{\beta,i} . \end{aligned} \tag{2.37}$$

There is a residual inequality whose consequences we proceed to exploit.

2.2.7. Equilibrium Properties

With the results summarized in the previous section the residual inequality (2.14) may now be rewritten in the form

$$\begin{bmatrix} -\frac{q_T}{T^2} & -\frac{1}{2T}\{\frac{q_v^\alpha}{T}+[(\frac{M_T^\alpha}{\rho_\alpha}-\frac{M_T^\nu}{\rho_\nu}) - (\frac{\partial\, p_\alpha/\rho_\alpha}{\partial T} - \frac{\partial\, p_\nu/\rho_\nu}{\partial T}) + T\frac{\partial \frac{\mu_I^\alpha-\mu_I^\nu}{T}}{\partial T}]\sum_{\gamma=1}^{\nu-1}(\rho_\alpha\delta_{\alpha\gamma}-\frac{\rho_\alpha\rho_\nu}{\rho})\} \\ -\frac{1}{2T}\{\ \} & -\frac{1}{T}M_v^{\alpha\beta}\end{bmatrix}\begin{bmatrix}T_{,j}\\ V_j^\alpha\end{bmatrix}\begin{bmatrix}T_{,j}\\ V_j^\beta\end{bmatrix}$$

$$-\frac{1}{T}\sum_{a=1}^{n}\left(\sum_{\alpha=1}^{\nu-1}\left(\mu_\alpha^I-\mu_\nu^I+\frac{1}{2}V_\alpha^2\right)\right)\gamma_\alpha^a M_\alpha \mu\Lambda^a \geq 0 \,. \qquad (2.38)$$

The left hand side of this inequality is the entropy production and we denote it by σ. It consists of several parts which are due

i.) to chemical reactions,

ii.) to heat conduction; this is the part which is quadratic in $T_{,j}$,

iii.) to diffusion, which is the part quadratic in V_j^α,

iv.) to thermal diffusion, which is represented by the terms with products of $T_{,j}, V_j^\alpha$.

One might say that chemical reactions, heat conduction, diffusion and thermal diffusion are the dissipative, i.e. entropy producing mechanisms in mixtures of the type considered here.

Equilibrium is characterized as a process in which all constituents are at rest in one frame of reference, the temperature is uniform and time independent and the reaction rate densities Λ^a vanish.

The entropy production σ obviously is a function of ρ_α, T, $T_{,j}$, V_i^α and inspection shows that the function assumes its minimum value, namely zero, in equilibrium. In order to be able to exploit this condition properly we write σ as a function of Λ^a $(a = 1,2,\ldots,n)$, $\rho_{n+1},\ldots,\rho_\nu$, T, $T_{,i}$, V_i^α :

$$\sigma = \Sigma\,(\Lambda^a, \rho_{n+1},\ldots,\rho_\nu, T, T_{,i}, V_i^\alpha) \qquad (2.39)$$

assuming invertibility of $\Lambda^a(\rho_1\ldots\rho_\nu,T)$ with respect to n densities. The advantage of this change of variables is that in (2.39) equilibrium means the vanishing of all variables but $\rho_{n+1},\ldots\rho_\nu,T$ so that we have

$$\sigma\big|_E = \Sigma\,(0,\rho_{n+1},\ldots,\rho_\nu,T,0,0) = 0 \,. \qquad (2.40)$$

As necessary conditions for σ to be minimal in equilibrium we have

$$\left.\frac{\partial\sigma}{\partial\Lambda^a}\right|_E = 0\,, \qquad \left.\frac{\partial\sigma}{\partial T_{,i}}\right|_E = 0\,, \qquad \left.\frac{\partial\sigma}{\partial V_i^\alpha}\right|_E = 0 \tag{2.41}$$

and

$$\left[\left.\frac{\partial^2\sigma}{\partial X_A\,\partial X_B}\right|_E\right] \text{ - positive semi definite ,} \tag{2.42}$$

where X_A stands for any one of the quantities Λ^a, $T_{,i}$, V_i^α.

From the first of these conditions we conclude that

$$\left.\sum_{\alpha=1}^{\nu} \gamma_\alpha^a M_\alpha \mu_I^\alpha\right|_E = 0 \qquad (a = 1,2,\ldots,n) \tag{2.43}$$

must hold. This equation represents the law of mass action and it states that in equilibrium all but $\nu - n$ densities are constrained where n, as always, is the number of independent reactions. We shall come back to these conditions later.

The conditions $(2.41)_{2,3}$ are identically satisfied, while (2.42) implies a number of inequalities. Most conclusions from this latter condition are little suggestive, and therefore we exploit (2.42) only for a binary mixture where it has the form

$$\begin{bmatrix} -\dfrac{q_T}{T^2} & -\dfrac{1}{2T}\left\{\dfrac{q_V^1}{T}+\dfrac{\rho_1\rho_2}{\rho}\left[\left(\dfrac{M_T^1}{\rho_1}-\dfrac{M_T^2}{\rho_2}\right)-\dfrac{\partial\left(\dfrac{p_1}{\rho_1}-\dfrac{p_2}{\rho_2}\right)}{\partial T}+T\,\dfrac{\partial\dfrac{\mu_I^1-\mu_I^2}{T}}{\partial T}\right]\right\} \\ -\dfrac{1}{2T}\{\;\} & -\dfrac{1}{T}M_V^{11} \end{bmatrix} \text{ - positive semi definite} \tag{2.44}$$

The implications of (2.44) read explicitely.

$$q_T \le 0,\quad M_V^{11} \le 0,\quad \frac{q_T M_V^{11}}{T} - \frac{1}{4}\left\{\frac{q_V^1}{T}+\frac{\rho_1\rho_2}{\rho}\left[\left(\frac{M_T^1}{\rho_1}-\frac{M_T^2}{\rho_2}\right)-\frac{\partial\left(\dfrac{p_1}{\rho_1}-\dfrac{p_2}{\rho_2}\right)}{\partial T}+\frac{\partial\dfrac{\mu_I^1-\mu_I^2}{}}{\partial T}\right]\right\}^2 \ge 0. \tag{2.45}$$

The inequalities $(2.45)_{1,2}$ mean that a temperature gradient in one direction creates a flux of internal energy in the opposite direction and that a diffusion velocity creates a force opposite to its own direction. The last inequality in (2.45) is difficult to interpret suggestively.

2.2.8. A Remark on the Strategy of Thermodynamics

We recall that the constitutive theory attempts to derive restrictions on the constitutive functions and, in particular, the entropy principle is used to acquire some knowledge about the coefficient functions of the constitutive relations (1.45).

For that purpose Lagrange multipliers were introduced as auxiliary quantities in (2.5). If we look back now, we can see that the identification of the Lagrange multipliers Λ^{ε} and Λ^{ρ_α} in terms of measurable quantities has proceeded as follows:

i.) A single fluid was brought into contact with an ideal gas along an ideal impermeable wall. The known properties of the gas permitted the calculation of Λ^{ε} in a single fluid and with that knowledge the internal energy $\varepsilon(\rho,T)$ and the pressure $p(\rho,T)$ of a single fluid could be related. Subsequently $\varepsilon(\rho,T)$ and the entropy $\eta(\rho,T)$ could be calculated by integration from a reduced number of measurements of pressures and specific heats.

ii.) A mixture of fluids was brought into contact with a single fluid along an ideal impermeable wall. The now known properties of the single fluid permitted the calculation of Λ^{ε} in the mixture.

iii.) A mixture of fluids was brought into contact with a single fluid along an ideal semipermeable wall. The knowledge of Λ^{ε} in the mixture and the previously acquired knowledge of the functions $\varepsilon(\rho,T)$, $\eta(\rho,T)$ in single fluids permitted the determination of the Lagrange multipliers Λ^{ρ_α} in the mixture. Once these - or, equivalently, the chemical potentials - were known, the free energy $\varepsilon(\rho_1,\ldots,\rho_\nu,T) - T\,\eta(\rho_1,\ldots,\rho_\nu,T)$ of the mixture could be calculated and hence followed $\varepsilon(\rho_1,\ldots,\rho_\nu,T)$ and $\eta(\rho_1,\ldots,\rho_\nu,T)$ themselves.

What has just been described is an example of the strategy of thermodynamics which starts out with the knowledge of properties of the simplest materials and proceeds to form conclusions in the unknown properties of complex bodies by bringing the two types of bodies into contact along an ideal wall.

One might be tempted to try this strategy in other circumstances, e.g. for mixtures with a different temperature for each constituent or for single bodies with a number of internal variables. But in no other case does the strategy work so well as for mixtures, because usually there are no walls which permit the passage of the flux of one of those additional variables while other fluxes are held back.

Even for mixtures the existence of a semipermeable wall for a given constituent cannot be taken for granted and the determination of the chemical potentials is an arduous and uncertain undertaking.

3. SIMPLE MIXTURES AND MIXTURES OF IDEAL GASES

3.1. Simple Mixtures

3.1.1. Definition and Interpretation

There is a special case of some interest within the theory presented heretofore. This is the case of simple mixtures which are characterized as mixtures whose constitutive quantities are independent of the gradients of density.

Now, according to (2.37) the entropy principle has shown that the flux of internal energy is independent of density gradients and that

$$\frac{\partial \mu_I^\alpha}{\partial \rho_\beta} = \frac{1}{\rho_\alpha} \frac{\partial p^\alpha}{\partial \rho_\beta} \tag{3.1}$$

must hold, if the interaction force is to be independent of $\rho_{\beta,i}$. None of the other constitutive quantities (1.45) has depended on density gradients to begin with. Thus (3.1) is the mathematical condition for a simple mixture.

Physically or in terms of an atomistic model, a dependence of $m_i^\alpha - \tau_\alpha v_i^\alpha$ on density gradients would reflect the presence of the long-range attractive forces between the molecules. Thus, if one molecule were surrounded by an inhomogeneous distribution of others, it should feel a force attracting it into the denser region. Accordingly in a simple mixture long-range forces should be absent or at least the effect of the interaction energy should be negligeable. This expectation will be confirmed immediately.

3.1.2. Chemical Potentials and Free Energy of a Simple Mixture

Equation (3.1) implies an integrability condition for μ_I^α which reads

$$-\frac{1}{\rho_\alpha^2} \delta_{\alpha\gamma} \frac{\partial p^\alpha}{\partial \rho_\beta} = -\frac{1}{\rho_\alpha^2} \delta_{\beta\gamma} \frac{\partial p^\alpha}{\partial \rho_\gamma} .$$

For $\beta = \gamma \neq \alpha$ we obtain

$$\frac{\partial p^{\alpha}}{\partial \rho_{\beta}} = 0 \qquad \text{when} \qquad \alpha \neq \beta \tag{3.2}$$

and we conclude that p_α depends only on the density of constituent α, and on temperature of course. By (3.1) this is also true for μ_I^α so that we may write $\mu_I^\alpha = \mu_I^\alpha(\rho_\alpha, T)$.

Therefore it follows from $(2.34)_1$ that

$$\frac{\partial^2 \rho(\varepsilon_I - T\eta)}{\partial \rho_\alpha \partial \rho_\beta} = 0 \qquad \text{when} \qquad \alpha \neq \beta \tag{3.3}$$

and the solution of this differential equation restricts the free energy density $\rho(\varepsilon_I - T\eta)$ to be a sum of functions which depend only on one density and temperature. Equation $(2.34)_2$ shows that the entropy density possesses the same property and so does therefore the density of internal energy. Thus we may write

$$\rho\varepsilon_I = \sum_{\alpha=1}^{\nu} \rho_\alpha \varepsilon_\alpha(\rho_\alpha, T) \qquad \text{and} \qquad \rho\eta = \sum_{\alpha=1}^{\nu} \rho_\alpha \eta_\alpha(\rho_\alpha, T) \; . \tag{3.4}$$

This decomposition of internal energy into a sum of partial internal energies confirms the above expectation that interaction energies may be neglected in a simple mixture. With (3.4) and $p_\alpha = p_\alpha(\rho_\alpha, T)$, $\mu_\alpha^I = \mu_\alpha(\rho_\alpha, T)$ it follows that (2.29) can be decomposed into parts in simple mixtures, viz.

$$\mu_\alpha^I = \varepsilon_\alpha - T\eta_\alpha + \frac{p_\alpha}{\rho_\alpha} \tag{3.5}$$

so that in simple mixtures the chemical potential μ_α^I is equal to the partial free enthalpy.

3.2. Mixtures of Ideal Gases

3.2.1. Dalton's Law

A particular case of a simple mixture is a mixture of ideal gases, whose partial pressures, internal energies and entropies are related to the partial densities and temperature just like those for a single gas, viz.

$$p_\alpha = \rho_\alpha \frac{R}{M_\alpha} T, \qquad \varepsilon_\alpha = z_\alpha \frac{R}{M_\alpha} T + \beta_\alpha, \qquad \eta_\alpha = z_\alpha \frac{R}{M_\alpha} \ln T - \frac{R}{M_\alpha} \ln \rho_\alpha + \alpha_\alpha . \tag{3.6}$$

With respect to the pressures this statement is known as Dalton's Law. α_α and β_α are constants of integration and z_α takes the values 3/2, 5/2 and 3 for one-, two- and more-atomic gases.

The density of free energy of a mixture of ideal gases is therefore given by

$$\rho(\varepsilon_I - T\eta) = \sum_{\alpha=1}^{\nu} (\rho_\alpha z_\alpha \frac{R}{M_\alpha} T(1 - \ln T) + T\rho_\alpha \frac{R}{M_\alpha} \ln \rho_\alpha + \rho_\alpha(\beta_\alpha - T\alpha_\alpha)) . \tag{3.7}$$

The chemical potential μ_I^α follows by differentiation with respect to ρ_α:

$$\mu_I^\alpha(\rho_\alpha, T) = (z_\alpha + 1) \frac{R}{M_\alpha} T - z_\alpha \frac{R}{M_\alpha} T \ln T + \frac{R}{M_\alpha} T \ln \rho_\alpha + (\beta_\alpha - T\alpha_\alpha) . \tag{3.8}$$

It follows from (3.8) that in a mixture of ideal gases not only the chemical potential μ_α^I is continuous across an ideal semipermeable wall, that is permeable for constituent α, but also the density ρ_α and hence the pressure p_α.

3.2.2. Gibbs Paradox

We compare two equilibria of ideal gases in a volume V at temperature T :

i.) A mixture under the pressure p and with partial pressures $p_\alpha = \rho_\alpha \frac{R}{M_\alpha} T$. The entropy of the mixture is given by

$$H_1 = \sum_{\alpha=1}^{\nu} \rho_\alpha V (z_\alpha \frac{R}{M_\alpha} \ln T - \frac{R}{M_\alpha} \ln \rho_\alpha + \alpha_\alpha) \tag{3.9}$$

according to (3.6).

ii.) Each one of the components is compressed into a compartment of volume $V_\alpha = \frac{p_\alpha}{p} V$ inside V. Obviously $\sum_{\alpha=1}^{\nu} V_\alpha = V$ and the pressure in each compartment is equal to p and the density is $\rho_\alpha(p,T)$ such that $\rho_\alpha(p,T)V_\alpha = \rho_\alpha V$. The entropy is

$$H_2 = \sum_{\alpha=1}^{\nu} \rho_\alpha(p,T) V_\alpha (z_\alpha \frac{R}{M_\alpha} \ln T - \frac{R}{M_\alpha} \ln \rho_\alpha(p,T) + \alpha_\alpha) . \tag{3.10}$$

The difference of the entropies in the two situations is

$$H_1 - H_2 = \sum_{\alpha=1}^{\nu} \frac{R}{M_\alpha} \rho_\alpha V \ln \frac{\rho_\alpha(p,T)}{\rho_\alpha} ,$$

$$H_1 - H_2 = \sum_{\alpha=1}^{\nu} \frac{R}{M_\alpha} \rho_\alpha V \ln \frac{V}{V_\alpha} . \tag{3.11}$$

$\frac{R}{M_\alpha} \rho_\alpha V$ is equal to $R\mu N_\alpha$, where μ is the mass of hydrogen atom and N_α is the number of molecules of constituent α. But for ideal gases the number of molecules in a given volume V_α depends only on pressure and temperature and is independent of the gas. Therefore the entropy difference (3.11) is independent of the nature of the ideal gases involved.

In particular, the entropy difference should also have the value (3.11), if all compartments are filled with the <u>same</u> ideal gas. This, however, is paradoxical, since in that case the situations i.) and ii.) are identical.

The situation described above is known as the Gibbs paradox. It is resolved in modern statistical mechanics which for quantum mechanical reasons recognizes that the number of states of a system of N identical particles is considerably less than that of systems of equally many non-identical particles.

3.3. Law of Mass Action in Mixtures of Ideal Gases

3.3.1. A Reduced Form of the Chemical Potentials

Because in a mixture of ideal gases we know the constitutive relations for ε, p, η and hence μ_α explicitly, it is possible to derive a number of interesting practical results for such mixtures, in particular concerning chemical equilibrium. This section conveys an idea of what can be done. More extensive accounts can be found in many books of chemical thermodynamics, e.g. see [8]. In such books the form μ_α^I given below in (3.13) for a mixture of ideal gases is also applied to many solutions with very good success.

In the equation (3.8) for the chemical potential of constituent α in a mixture of ideal gases we introduce the pressure p and the particle densities n_α of the constituents as variables instead of ρ_α. Thus, replacing

$$\rho_\alpha \quad \text{by} \quad \frac{n_\alpha}{\sum_{\beta=1}^{\nu} n_\beta} \frac{p}{\frac{R}{M_\alpha} T}$$

we obtain from (3.8)

$$\mu^I_\alpha(p,T,n_\gamma) = (z_\alpha \frac{R}{M_\alpha}T + \beta_\alpha) - T(z_\alpha \frac{R}{M_\alpha} \ln T - \frac{R}{M_\alpha} \ln \frac{p}{\frac{R}{M_\alpha}T} + \alpha_\alpha) + $$
$$+ \frac{R}{M_\alpha} T + \frac{R}{M_\alpha} T \ln \frac{n_\alpha}{\sum_{\gamma=1}^{\nu} \gamma} \quad . \tag{3.12}$$

This may be written as

$$\mu^I_\alpha(p,T,n_\gamma) = g_\alpha(p,T) + \frac{R}{M_\alpha} T \ln \frac{n_\alpha}{\sum_{\gamma=1}^{\nu} \gamma} \quad , \tag{3.13}$$

where $g_\alpha(p,T) = \varepsilon_a(p,T) - T\eta_\alpha(p,T) + \frac{p}{\rho_\alpha(p,T)}$ is the free enthalpy of the ideal gas α.

3.3.2. Law of Mass Action

The law of mass action in its general form has been given in equation (2.43) where it was derived as an equilibrium condition from the residual inequality. For a mixture of ideal gases we may eliminate μ^I_α between (2.43) and (3.13) and obtain after a little calculation

$$\prod_{\alpha=1}^{\nu} \left(\frac{n_\alpha}{\sum_{\gamma=1}^{\nu} n_\gamma} \right)^{\gamma^a_\alpha} = e^{-\frac{\sum_{\beta=1}^{\nu} g_\beta(T,p)\gamma^a_\alpha M_\alpha}{RT}} \qquad (a = 1,2,\dots,n) \tag{3.14}$$

where, as before $g_\beta(T,p)$ is the free enthalpy of the pure constituent α. It is important to note that the right hand side of (3.14) is independent of the number densities of the constituents; it depends only on p and T, two variables that are often prescribed in a chemical laboratory. Thus for given values of p and T the equations (3.14) represent n restrictions - as many as there are independent reactions - upon the particle densities of constituents in chemical equilibrium.

We may replace n_β in (3.14) by $\frac{p_\beta}{kT} = \frac{p_\beta}{R/L\ T}$, where k and L are the Boltzmann constant and the Avogadro number respectively. Thus we obtain

$$\prod_{\alpha=1}^{\nu} p_\alpha^{\gamma_\alpha^a} = p^{\sum_{\beta=1}^{\nu}\gamma_\beta^a} e^{-\frac{\sum_{\beta=1}^{\nu} g_\beta(T,p)\gamma_\beta^a M_\beta}{RT}} \tag{3.15}$$

The right hand side is called the chemical constant and denoted by K_p^a. At first sight it might appear as if K_p^a were a function of p and T. However, if the form of $g_\beta(T,p)$ appropriate to an ideal gas is introduced, namely

$$g_\beta(T,p) = (z_\beta \frac{R}{M_\beta}T + \beta_\beta) - T(z_\beta \frac{R}{M_\beta}\ln T - \frac{R}{M_\beta}\ln \frac{p}{\frac{R}{M_\beta}T} + \alpha_\beta) + \frac{R}{M_\beta}T ,$$

we see that K_p^a has the form

$$K_p^a(T) = e^{-\sum_{\beta=1}^{\nu}\gamma_\beta^a(z_\beta + 1 - \ln\frac{R}{M_\beta})} e^{-\frac{\sum_{\beta=1}^{\nu}\gamma_\beta^a M_\beta(\beta_\beta - T\alpha_\beta)}{RT}} T^{\sum_{\beta=1}^{\nu}\gamma_\beta^a(z_\beta+1)} \tag{3.16}$$

and in particular that it is independent of p and depends only on T.

Thus we have

$$\prod_{\alpha=1}^{\nu} p^{\gamma_\alpha^a} = K_p^a(T) \tag{3.17}$$

in a mixture of ideal gases and this relation holds to a good approximation in reactions between vapors as well. For many important reactions the values of $K_p^a(T)$ have been experimentally determined and are represented in tables.

3.3.3. Mass Conservation for a Single Reaction

We shall exploit the law of mass action for homogeneous chemical reactions. In order to be able to do so we must first investigate the consequences of the mass balance for the constituents. In a homogeneous reaction we have

$$(\rho_\alpha V)^\cdot = \tau_\alpha V = \sum_{a=1}^{n} \gamma_\alpha^a M_\alpha \mu \Lambda^a V . \tag{3.18}$$

If $n = 1$, so that there is only one independent reaction, (3.18) can be rewritten for the particle numbers N_α in the form

$$\left(\frac{N_\alpha}{\gamma_\alpha}\right)^{\cdot} = \Lambda V \, , \tag{3.19}$$

whence we conclude that $\left(\frac{N_\alpha}{\gamma_\alpha}\right)^{\cdot}$ is independent of α. Upon integration we may thus write

$$\frac{N_\alpha}{\gamma_\alpha} - \frac{N_\alpha}{\gamma_\alpha}\bigg|_{t=0} = \frac{N_\beta}{\gamma_\beta} - \frac{N_\beta}{\gamma_\beta}\bigg|_{t=0} \, . \tag{3.20}$$

the equations (3.19) and (3.20) reflect the trivial observation that as a reaction proceeds, the atoms of all constituents are set free or absorbed at a rate determined by the stoichiometric coefficients.

3.3.4. Typical Example for the Application of the Law of Mass Action; Haber Bosch Synthesis

The formation of ammonia from nitrogen and hydrogen proceeds according to the reaction equation

$$N_2 + 3H_2 - 2NH_3 = 0 \, . \tag{3.21}$$

Let the pressure p and the temperature T be given and let us start with 1 Mol N_2 and 3 Mol H_2 so that the initial particle numbers are

$$\overset{o}{N}_{N_2} = L \, , \quad \overset{o}{N}_{H_2} = 3L \, , \quad \overset{o}{N}_{NH_3} = 0 \, . \tag{3.22}$$

This is not an equilibrium situation. We wish to calculate the fractions of the three constituents and the volume in chemical equilibrium.

For this purpose we rely upon

i.) the law of mass action which in the present case reads

$$\frac{p_{N_2} \, p_{H_2}^3}{p_{NH_3}^2} = K_p(T) \, , \tag{3.23}$$

ii.) the relations (3.20) which represent two equations in this case viz. by (3.22)

$$N_{N_2} - L = -\frac{N_{NH_3}}{2} \quad \text{or} \quad p_{N_2} - \frac{LkT}{V} = -\frac{p_{NH_3}}{2} ,$$

$$\frac{N_{H_2}}{3} - L = -\frac{N_{NH_3}}{2} \quad \text{or} \quad \frac{p_{H_2}}{3} - \frac{LkT}{V} = -\frac{p_{NH_3}}{2} , \tag{3.24}$$

iii.) the fact that the pressure of the mixture is the sum of the constituent pressures

$$p_{N_2} + p_{H_2} + p_{NH_3} = p \ . \tag{3.25}$$

By (3.23) through (3.25) we have four equations for the four unknowns p_{N_2}, p_{H_2}, p_{NH_3}, and V. A short calculation provides

$$K_p(T) = p^2 \frac{\frac{1}{3}(\frac{3}{4})^4(1 - \frac{p_{NH_3}}{p})^4}{(\frac{p_{NH_3}}{p})^2} , \quad \text{hence} \quad \frac{p_{NH_3}}{p} = 1 + \frac{1}{2}\sqrt{\frac{3K_p(T)}{(\frac{3}{4})^4 p^2}} - \sqrt{(1 + \frac{1}{2}\sqrt{\frac{3K_p(T)}{(\frac{3}{4})^4 p^2}})^2 - 1} \tag{3.26}$$

and an examination of the right hand side shows that the fraction of ammonia molecules

$$\frac{N_{NH_3}}{N} = \frac{p_{NH_3}}{p}$$

increasing with increasing pressure as shown in Figure 2, where $K_p(T)$ was chosen as $11,7 \cdot 10^6 \ (\text{bar})^2$, a value appropriate for room temperature. This observation is the basis for the synthezization of ammonia under high pressures.

The other two fractions N_{H_2}/N and N_{N_2}/N are easily expressed by N_{HN_3}/N by use of (3.24) and (3.25) and subsequently V can be calculated from (3.24).

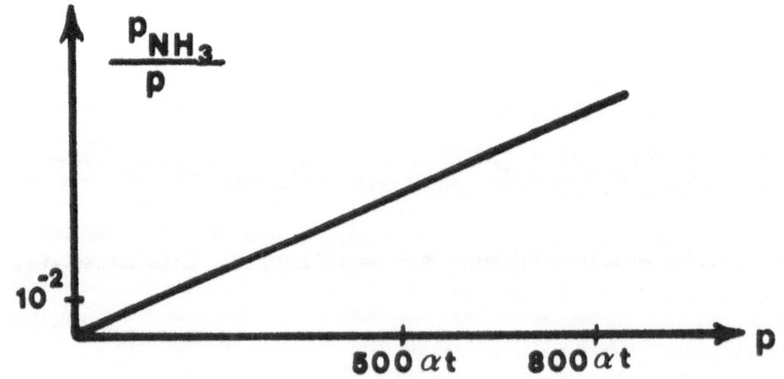

Figure 2

3.3.5. le Chatelier's Principle

The influence of changes in pressure and temperature on the shift in chemical equilibrium can be characterized by the following observations:

i.) The pressure influences the chemical equilibrium, when the volume changes in the reaction. Indeed, high pressure favours a decrease in volume.

ii.) An exothermal, i.e. heat emitting reaction proceeds less far at high temperature than at low temperature, while an endothermal, i.e. heat absorbing reaction proceeds further.

An example of the pressure influence upon a mixture of ideal gases has been discussed in the previous section where we considered the reaction (3.21) in which the volume of pure ammonia is only half of that of the mixture of hydrogen and nitrogen without ammonia.

In order to illustrate the influence of temperature on the chemical equilibrium we investigate the dependence of $K_p(T)$ in (3.16) on temperature by differentiation of $\ln K_p(T)$ with respect to T. We obtain

$$\frac{d \ln K_p^a(T)}{dT} = \frac{\sum_{\beta=1}^{\nu} \gamma_\beta^a M_\beta\left((z_\beta + 1)\frac{R}{M_\beta} T + \beta_\beta\right)}{RT^2}$$

$$\frac{d \ln K_p^a(T)}{dT} = \frac{\sum_{\beta=1}^{\nu} \gamma_\beta^a M_\beta h_\beta}{RT^2} \qquad (3.27)$$

where $h_\beta = \varepsilon_\beta + \frac{p_\beta}{\rho_\beta}$ is the specific enthalpy of the constituent β, so that $\sum \gamma_\beta^a M_\beta h_\beta$ is the change of enthalpy as reaction a occurs once. If $\sum \gamma_\beta^a M_\beta h_\beta$ is positive, the coefficients with positive stoichiometric coefficients have a greater enthalpy than those with negative coefficients and $K_p^a(T)$ grows with growing temperature. Therefore an increase in temperature will favour the constituents with a high enthalpy. Since heat absorbed at constant pressure increases the enthalpy, we conclude that a growth in temperature favours the endothermal reaction as stated in ii.) above.

The above observations about the influence of pressure and temperature on the chemical equilibrium have been generalized by le Chatelier who stated that any change in conditions results in a shift of equilibrium in the direction that will partially nullify the change. This generalization is known as le Chatelier's principle.

In the examples an increase in pressure is "partially nullified" because the reaction leads to a smaller volume and an increase of temperature is "partially nullified because the ensuing reaction absorbs heat.

4. DIFFUSION AND WAVE PROPAGATION IN SIMPLE MIXTURES

4.1. Transport Equations for Diffusion and Heat Conduction

4.1.1. Phenomenological Equations of Linear Irreversible Thermodynamics

Linear irreversible thermodynamics formulates constitutive relations for the diffusion fluxes $\rho_\alpha u_i^\alpha$ and the heat flux in a mixture. These fluxes come out as linear combinations of the so-called thermodynamic forces, namely the gradients of $\frac{1}{T}$ and of $\frac{\mu^\alpha - \mu^\nu}{T}$. Thus according to linear irreversible thermodynamics we have

$$\begin{aligned} \rho_\alpha u_i^\alpha &= \sum_{\beta=1}^{\nu-1} L_{\alpha\beta}\Big(-\frac{\partial \frac{\mu^\beta-\mu^\nu}{T}}{\partial x_i}\Big) + \tilde{L}_\alpha \frac{\partial \frac{1}{T}}{\partial x_i} , \\ q_i &= \sum_{\beta=1}^{\nu-1} L_\beta \Big(-\frac{\partial \frac{\mu^\beta-\mu^\nu}{T}}{\partial x_i}\Big) + \lambda \frac{\partial \frac{1}{T}}{\partial x_i} . \end{aligned} \tag{4.1}$$

The reader is referred to [9] or [10] for the derivation of (4.1) within linear irreversible thermodynamics. These equations are often called the laws of Fick and of Fourier for diffusion and heat conduction in a mixture respectively.

The coefficients are denoted as follows

$L_{\alpha\beta}$ - diffusion coefficients,

$\tilde{L}_\alpha$ - coefficients of thermal diffusion,

L_α - diffusion-thermo-coefficient,

and linear irreversible thermodynamics postulates symmetry relations for these coefficients, namely

$$L_{\alpha\beta} = L_{\beta\alpha} \quad \text{and} \quad \tilde{L}_\alpha = L_\alpha . \tag{4.2}$$

The symmetry relations represent an extrapolation of Onsager relations to the case of vectorial fluxes and forces.

We shall proceed to show how the equations come out of the present thermodynamic theory of mixtures. It will turn out that the present theory is superior to linear irreversible thermodynamics in two aspects:

i.) The present theory predicts a finite speed of propagation of disturbances in the density of the constituents, while linear irreversible thermodynamics predicts an infinite speed.

ii.) for a subclass of simple inviscid mixtures the present theory can derive the symmetry relations that were postulated by linear irreversible thermodynamics.

4.1.2. Fick's Law of Diffusion as a Mutilated Form of the Equations of Balance of Partial Momenta. Fourier's Law.

The equations of balance of partial momenta and the constitutive relations (2.37) are at the basis of the considerations of this section. We rewrite these equations ignoring body forces and chemical reactions; also, since we are now interested in a comparison with the formulae of linear irreversible thermodynamics, we leave out non-linear terms in relative velocities and velocity gradients

$$\frac{\partial \rho_\alpha v_i^\alpha}{\partial t} + \frac{\partial}{\partial x_i} (\rho_\alpha v_i^\alpha v_j^\alpha - t_{ij}^\alpha) = m_i^\alpha , \tag{4.3}$$

$$q_i = q_T T_{,i} + \sum_{\beta=1}^{\nu-1} q_v^\beta v_i^\beta ,$$

$$m_i^\alpha = M_T^\alpha T_{,i} + \sum_{\beta=1}^{\nu-1} M_v^{\alpha\beta} v_i^\beta + \sum_{\beta=1}^{\nu} \left(\frac{\partial p^\alpha}{\partial \rho_\beta} - \rho_\alpha \frac{\partial \mu_\alpha}{\partial \rho_\beta} \right) \rho_{\beta,i} . \tag{4.4}$$

By use of the balance equations of mass $(1.16)_1$ and ignoring the non-linear term $u_j^\alpha \frac{\partial v_i^\alpha}{\partial x_j}$ we may write (4.3) in the form

$$\rho_\alpha \dot{v}_i^\alpha - \frac{\partial t_{ij}^\alpha}{\partial x_j} = m_i^\alpha ,$$

or with $(4.4)_2$ and $t_{ij}^\alpha = -p^\alpha \delta_{ij}$ after dividing by ρ_α

$$\dot{v}_i^\alpha + \frac{1}{\rho_\alpha}\frac{\partial p^\alpha}{\partial x_i} = \frac{M_T^\alpha}{\rho_\alpha} T_{,i} + \sum_{\beta=1}^{\nu-1} \frac{M_v^{\alpha\beta}}{\rho_\alpha} V_i^\beta + \sum_{\beta=1}^{\nu}\left(\frac{1}{\rho_\alpha}\frac{\partial p^\alpha}{\partial \rho_\beta} - \frac{\partial \mu^\alpha}{\partial \rho_\beta}\right)\rho_{\beta,i} \,.$$

These are ν equations. We subtract the νth one from all others and obtain after a little rearrangement

$$\sum_{\beta=1}^{\nu-1}\left(\frac{M_v^{\alpha\beta}}{\rho_\alpha} - \frac{M_v^{\nu\beta}}{\rho_\nu}\right)V_i^\beta = T\,\frac{\partial \frac{\mu^\alpha-\mu^\nu}{T}}{\partial x_i} + T^2 D_\alpha \frac{\partial \frac{1}{T}}{\partial x_i} + \frac{dV_i^\alpha}{dt} \qquad (\alpha = 1,2,\ldots,\nu-1) \tag{4.5}$$

where

$$D_\alpha \equiv \left(\frac{M_T^\alpha}{\rho_\alpha} - \frac{M_T^\nu}{\rho_\nu}\right) + T\,\frac{\partial \frac{\mu^\alpha-\mu^\nu}{T}}{\partial T} - \frac{\partial\left(\frac{p^\alpha}{\rho_\alpha} - \frac{p^\nu}{\rho_\nu}\right)}{\partial T} \tag{4.6}$$

has been defined for abbreviation. We recall the relations (2.9) between V_i^α and u_i^α and write

$$V_i^\gamma = \sum_{\beta=1}^{\nu-1}\left(\frac{1}{\rho_\beta}\delta_{\gamma\beta} + \frac{1}{\rho_\nu}\right)\rho_\beta u_i^\beta \quad \text{and} \quad \rho_\beta u_i^\beta = \sum_{\beta=1}^{\nu-1}\left(\rho_\beta \delta_{\beta\gamma} - \frac{\rho_\beta \rho_\gamma}{\rho}\right)V_i^\gamma \tag{4.7}$$

$$V_i^\gamma = \sum_{\beta=1}^{\nu-1} F_{\gamma\beta}\rho_\beta u_i^\beta \quad \text{and} \quad \rho_\beta u_i^\beta = \sum_{\beta=1}^{\nu-1} F^{-1}_{\beta\gamma} V_i^\gamma \,.$$

Also, in the absence of chemical reactions $(1.46)_2$ shows that $M_v^{\nu\beta} = -\sum_{\alpha=1}^{\nu-1} M_v^{\alpha\beta}$ holds. Inserting all this in (4.5) and solving for the diffusion fluxes $\rho_\alpha u_i^\alpha$ we obtain

$$\rho_\alpha u_i^\alpha = \sum_{\beta=1}^{\nu-1}\underbrace{\left(-T[F^{-1}M_v^{-1}F^{-1}]_{\alpha\beta}\right)}_{L_{\alpha\beta}}\left(-\frac{\partial \frac{\mu^\beta-\mu^\nu}{T}}{\partial x_i}\right) + \underbrace{\left(T\sum_{\gamma=1}^{\nu-1}[F^{-1}M_v^{-1}F^{-1}]_{\alpha\gamma}D_\gamma\right)}_{\tilde{L}_\alpha}\frac{\partial \frac{1}{T}}{\partial x_i} + \sum_{\beta,\gamma=1}^{\nu-1}[F^{-1}M_v^{-1}F^{-1}]_{\alpha\beta}\frac{dF_{\beta\gamma}\rho_\gamma u_i^\gamma}{dt} \tag{4.8}$$

Apart from the acceleration on the right hand side this equation is of the form of Fick's law $(4.1)_1$, if the coefficients in the two equations are identified as indicated in (4.8). Note that the acceleration cannot be dropped from (4.8) on account of linearity. The absence of the acceleration represents the gravest shortcoming of the mixture theory in linear irreversible thermodynamics. We shall come back to the significance of this term, but for the present argument we drop it in order to continue the comparison with linear irreversible thermodynamics.

If (4.8), without acceleration term, is used to replace v_i^β in equation $(4.4)_1$, we obtain

$$q_i = \sum_{\beta=1}^{\nu-1} \underbrace{\left(-T \sum_{\gamma=1}^{\nu-1} q_v^\gamma [M_v^{-1} F^{-1}]_{\gamma\beta}\right)}_{L} \left(- \frac{\partial \frac{\mu^\beta - \mu^\nu}{T}}{\partial x_i}\right) + \underbrace{\left(T^2 \sum_{\alpha,\gamma=1}^{\nu-1} q_v^\gamma [M_v^{-1} F^{-1}]_{\gamma\alpha} D_\alpha - q_T T^2\right)}_{\lambda} \frac{\partial \frac{1}{T}}{\partial x_i} . \quad (4.9)$$

We compare this with the equation $(4.1)_2$ of linear irreversible thermodynamics and conclude that the coefficients L_β and λ must be identified as indicated in (4.9).

Thus with (4.8), (4.9) we have rederived the constitutive equations of linear irreversible thermodynamics. In the present theory we have also considerable knowledge about the coefficients $L_{\alpha\beta}$, $\tilde{L}_\alpha$ and L_β. We shall now proceed to exploit that knowledge in order to prove the symmetry relations (4.2) for a special case.

4.1.3. Symmetry Relations for Diffusion Coefficients

Inspection of (4.8) shows that $L_{\alpha\beta}$ is symmetric if $M_v^{\alpha\beta}$ is symmetric. While it is not possible to show the symmetry of $M_v^{\alpha\beta}$ in general, we shall now make reasonable assumptions upon $M_v^{\alpha\beta}$ which imply the symmetry of this matrix. In the proof of the symmetry of $M_v^{\alpha\beta}$ we follow [11].

First of all we introduce $M_v^{\alpha\nu} = -\sum_{\beta=1}^{\nu-1} M_v^{\alpha\beta}$ and rewrite the constitutive relation $(4.4.)_2$ in the form

$$m_i^\alpha = M_T^\alpha T_{,i} - \sum_{\beta=1}^{\nu} M_v^{\alpha\beta}(v_i^\alpha - v_i^\beta) + \sum_{\beta=1}^{\nu} \left(\frac{\partial p^\alpha}{\partial \rho_\beta} - \rho_\alpha \frac{\partial \mu^\alpha}{\partial \rho_\beta}\right)\rho_{\beta,i} .$$

The advantage of this form is that the coefficients $M_v^{\alpha\beta}$ (including $M_v^{\alpha\nu}$) now have a very suggestive meaning, because they determine how the relative motion of the constituents α and β affects the interaction force on constituent α. It is therefore reasonable to assume that for a large subclass of mixtures the following special properties hold

i.) $M_v^{\alpha\beta}$ is independent of ρ_γ for $\gamma \neq \alpha,\beta$,
ii.) $M_v^{\alpha\beta}$ tends to zero for $\rho_\beta \to 0$.

One says that the mixture exhibits binary drags only, if i.) is true. And ii.) states that there is no drag on constituent α from constituent β, if the latter is absent.

Because of conservation of momentum we must have

$$\sum_{\alpha,\beta=1}^{\nu} M_{\nu}^{\alpha\beta}(v_i^{\alpha} - v_i^{\beta}) = 0 \,, \qquad \text{or} \qquad \sum_{\alpha,\beta=1}^{\nu} M_{\nu}^{[\alpha\beta]} v_i^{\alpha} = 0 \,,$$

where the square brackets indicate the antisymmetric part of $M_{\nu}^{\alpha\beta}$. Since this relation must be valid for all v_i^{α} we have

$$\sum_{\beta=1}^{\nu} M_{\nu}^{[\alpha\beta]} = 0 \,.$$

This means that the sum over all elements in every row, and every column, must vanish. In particular, in a binary mixture

$$M_{\nu}^{[\alpha\beta]} = \begin{pmatrix} 0 & \alpha \\ -\alpha & 0 \end{pmatrix}$$

and since all rows must add up to zero we get $\alpha = 0$. In a mixture of three constituents we have

$$M_{\nu}^{[\alpha\beta]} = \begin{pmatrix} 0 & \alpha & -\alpha \\ -\alpha & 0 & \alpha \\ \alpha & -\alpha & 0 \end{pmatrix} .$$

Now, if the above conditions i.) and ii.) hold for the matrix $M_{\nu}^{\alpha\beta}$, they must also hold for its symmetric and antisymmetric parts. Thus $M_{\nu}^{[1,2]}$ is independent of ρ_3, $M_{\nu}^{[3,1]}$ is independent of ρ_2 and $M_{\nu}^{[2,3]}$ is independent of ρ_1. But all these components are equal to α. Hence α is a constant and by condition ii.) that constant is zero.

Mixture of more than three constituents can be reduced to the case of three constituents by condition ii.). Thus $M_{\nu}^{[\alpha\beta]} = 0$ and we have proved the symmetry of $M_{\nu}^{\alpha\beta}$ whence follows the symmetry of $L_{\alpha\beta}$ in (4.8) and (4.1).

4.1.4. Equality of Coefficients of Thermal Diffusion and Diffusion-Thermo-Coefficients

In order to obtain $(4.2)_2$ we have to prove

$$T^2 \sum_{\gamma=1}^{\nu-1} [F^{-1} M_{\nu}^{-1} F^{-1}]_{\alpha\gamma} D_{\gamma} = - T \sum_{\gamma=1}^{\nu-1} q_{\nu}^{\gamma} [M_{\nu}^{-1} F^{-1}]_{\gamma\beta} \,. \qquad (4.10)$$

Since we have just proved, under the conditions i.) and ii.) of Section 4.1.3), that

$M^{\alpha\beta}_{v}$ is symmetric, we may write (4.10) with (4.6) as

$$\sum_{\delta=1}^{\nu-1} [F^{-1}M_{v}^{-1}]_{\alpha\delta}\{q_{v}^{\delta} + T \sum_{\gamma=1}^{\nu-1} F_{\delta\gamma}^{-1}\left(\left(\frac{M_T^{\gamma}}{\rho_{\gamma}} - \frac{M_T^{\nu}}{\rho_{\nu}}\right) + T\,\frac{\partial\frac{\mu^{\gamma}-\mu^{\nu}}{T}}{\partial T} - \frac{\partial\left(\frac{p^{\gamma}}{\rho_{\gamma}} - \frac{p^{\nu}}{\rho_{\nu}}\right)}{\partial T}\right)\} = 0 \,. \tag{4.11}$$

Obviously this cannot be proved without further assumptions. We shall therefore restrict the attention to simple mixtures now and it seems reasonable to suppose that for a large subclass of simple mixtures it is true that

iii.) $$M_T^{\alpha} = 0 \,, \tag{4.12}$$

so that the temperature gradient does not affect the interaction force

iv.) q_{v}^{α} is given as

$$q_{v}^{\beta} = \sum_{\delta=1}^{\nu} F_{\beta\delta}^{-1}\,(\varepsilon_{\delta} + \frac{p_{\delta}}{\rho_{\delta}}) \qquad \text{or with} \qquad \sum_{\alpha=1}^{\nu} F_{\alpha\beta}^{-1} = 0 \tag{4.13}$$

$$= \sum_{\delta=1}^{\nu-1} F_{\beta\delta}^{-1}(\varepsilon_{\delta} - \varepsilon_{\nu} + \frac{p_{\delta}}{\rho_{\delta}} - \frac{p_{\nu}}{\rho_{\nu}}) \quad .$$

This assumption is motivated by a comparison of $(4.4)_1$ with the definition $(1.26)_2$ of q_i and it is tantamount to assuming that the partial flux of internal energy q_i^{α} does not depend on the relative velocities.

Together the two conditions iii.) and iv.) severely restrict the thermomechanical interaction. If they are valid, (4.11) reads

$$\sum_{\delta=1}^{\nu-1} [F^{-1}M_{v}^{-1}F^{-1}]_{\alpha\delta}\{(\varepsilon_{\delta} + \frac{p^{\delta}}{\rho_{\delta}} + T^2\,\frac{\partial\frac{\mu_{\delta}}{T}}{\partial T} - T\,\frac{\partial\frac{p_{\delta}}{\rho_{\delta}}}{\partial T}) - (\varepsilon_{\nu} + \frac{p_{\nu}}{\rho_{\nu}} + T^2\frac{\partial\frac{\mu_{\nu}}{T}}{\partial T} - T\,\frac{\partial\frac{p_{\nu}}{\rho_{\nu}}}{\partial T})\} = 0. \tag{4.14}$$

We have seen in Section 3.1.2 that in a simple mixture p_{α} and μ_{α} depend only on ρ_{α} and T and that $\rho\eta$ and $\rho\varepsilon$ are sums of terms $\rho_{\alpha}\eta_{\alpha}$ and $\rho_{\alpha}\varepsilon_{\alpha}$ respectively that depend only on ρ_{α} and T. Thus the constituents of a simple mixture behave like single fluids and there must be a Gibbs equation of the form (2.15) for each one of them:

$$d\eta_{\alpha} = \frac{1}{T}\,(d\varepsilon_{\alpha} - \frac{p_{\alpha}}{\rho_{\alpha}^2}\,d\rho_{\alpha}) \,. \tag{4.15}$$

This implies the integrability conditions

$$\rho_\alpha^2 \frac{\partial \varepsilon_\alpha}{\partial \rho_\alpha} = p_\alpha - T \frac{\partial p_\alpha}{\partial T}$$

and if that is introduced into (4.14), we get

$$\sum_{\delta=1}^{\nu-1} [F^{-1} M_v^{-1} F^{-1}]_{\alpha\delta} \left\{ \left(\frac{\partial \rho_\delta \varepsilon_\delta}{\partial \rho_\delta} + T^2 \frac{\partial \frac{\mu_\delta}{T}}{\partial T} \right) - \left(\frac{\partial \rho_\nu \varepsilon_\nu}{\partial \rho_\nu} + T^2 \frac{\partial \frac{\mu_\nu}{T}}{\partial T} \right) \right\} = 0 \,. \tag{4.16}$$

Now it takes only a glance at the integrability condition (2.35) together with the equation (3.4) for simple mixtures to see that (4.16) is indeed satisfied.

Thus we have proved the equality of $\tilde{L}_\alpha$ and L_α in (4.8), (4.9) or (4.1). It is to be stressed that in the proof no recourse was taken to statistical arguments which are usually considered as the basis of Onsager relations of the type (4.2).

4.2. Wave Propagation in a Binary Mixture of Non-Reacting Constituents

4.2.1. Linearized Equations of Balance for a Binary Non-Reacting Mixture

The following analysis on wave propagation is due to Müller and Villaggio [12] who also formulated conditions of stability.

For simplicity we shall now consider only simple mixtures of the type for which the conditions (4.12) and (4.13) hold that were previously used to prove the symmetry relations (4.2). Thus the constitutive relations have the forms

$$t_{ij}^\alpha = -p_\alpha(\rho_\alpha, T)\delta_{ij} \qquad \text{with} \qquad \frac{\partial p_\alpha}{\partial \rho_\alpha} = \rho_\alpha \frac{\partial \mu_\alpha}{\partial \rho_\alpha} \quad \text{and} \quad \mu_I^\alpha = \varepsilon_\alpha - T\eta_\alpha + \frac{p_\alpha}{\rho_\alpha} \tag{4.17$_1$}$$

$$m_i^\alpha - \tau_\alpha v_i^\alpha = \sum_{\beta=1}^{\nu-1} M_v^{\alpha\beta} V_i^\beta \tag{4.17$_2$}$$

$$q_i = \sum_{\beta=1}^{\nu-1} q_v^\beta V_i^\beta + q_T T_{,j} \qquad \text{with} \qquad q_v^\beta = \sum_{\delta=1}^{\nu-1} F_{\beta\delta}^{-1}\left(\varepsilon_\delta - \varepsilon_\nu + \frac{p^\delta}{\rho_\delta} - \frac{p^\nu}{\rho_\nu}\right) \tag{4.17$_3$}$$

$$\varepsilon_I = \sum_{\alpha=1}^{\nu} \frac{\rho_\alpha}{\rho} \varepsilon_\alpha(\rho_\alpha, T) \qquad \text{with} \qquad \rho_\alpha^2 \frac{\partial \varepsilon_\alpha}{\partial \rho_\alpha} = p_\alpha - T \frac{\partial p_\alpha}{\partial T} \tag{4.17$_4$}$$

and there is a residual inequality of the forms

$$\begin{bmatrix} -\frac{q_T}{T^2} & 0 \\ 0 & -\frac{1}{T} M_v^{\gamma\beta} \end{bmatrix} \begin{bmatrix} T_{,j} \\ V_j^{\gamma} \end{bmatrix} , \begin{bmatrix} T_{,j} \\ V_i^{\beta} \end{bmatrix} - \frac{1}{T} \sum_{\alpha=1}^{\nu-1} (\mu_\alpha^I - \mu_\nu^I + \frac{1}{2} V_\alpha^2)\tau_\alpha \geq 0 \,. \tag{4.18}$$

Note that - in contrast to the general case of relation (2.38) - there are no off-diagonal terms in the matrix (4.18). This is due to the assumptions (4.12), (4.13) and to the fact that we are now considering a simple mixture.

For the present context it is useful to consider

$$\rho,\ v_i,\ T,\ c \equiv \frac{\rho_1}{\rho},\quad J_i = \rho_1 (v_i^1 - v_i) \tag{4.19}$$

as the basic variables, instead of ρ_α, v_i^α and T. c is called the concentration of constituent 1 and J_i is its diffusion flux. It is thus also appropriate to replace the balance laws (1.30) by the following equations which have the advantage that the new variables occur explicitly in many places:

$$\begin{aligned}
&\frac{\partial \rho}{\partial t} + \frac{\partial \rho v_j}{\partial x_i} = 0 \,, \\
&\frac{\partial \rho v_i}{\partial t} + \frac{\partial}{\partial x_j} (\rho v_i v_j - t_{ij}) = 0 \,, \\
&\frac{\partial \rho \varepsilon}{\partial t} + \frac{\partial}{\partial x_j} (\rho \varepsilon v_j + q_j) = t_{ij} \frac{\partial v_i}{\partial x_j} \,, \\
&\rho \frac{\partial c}{\partial t} + \rho v_j \frac{\partial c}{\partial x_j} + \frac{\partial J_i}{\partial x_i} = 0 \\
&\frac{\partial J_i}{\partial t} + \frac{\partial}{\partial x_j} (J_i v_j^1 - (t_{ij}^1 - c t_{ij})) = t_{ij} \frac{\partial c}{\partial x_j} - J_j \frac{\partial v_i}{\partial x_j} + m_i^1 \,,
\end{aligned} \tag{4.20}$$

where chemical reactions are excluded. The first three equations are the balance equations of mass, momentum and (internal) energy of the mixture, the fourth equation is a balance of concentration and the last one represents the balance of relative momenta of the two constituents.

The constitutive equations are still given by (4.17), but the new choice of variables makes slight changes necessary. In particular, the Gibbs equation $(2.33)_2$ now reads

$$d\eta = \frac{1}{T} (d\varepsilon_I - \frac{p}{\rho^2} d\rho - (\mu_I^1 - \mu_I^2) dc) \tag{4.21}$$

and this implies the following integrability conditions

$$\rho^2 \frac{\partial \varepsilon_I}{\partial \rho} = -T^2 \frac{\partial \frac{p}{T}}{\partial T}, \qquad \frac{\partial \varepsilon_I}{\partial c} = -T^2 \frac{\partial \frac{\mu_I^1 - \mu_2^2}{T}}{\partial T}, \qquad \frac{\partial p}{\partial c} = \rho^2 \frac{\partial(\mu_I^1 - \mu_I^2)}{\partial \rho} \; . \tag{4.22}$$

We linearize the equations of balance (4.20) about a reference state in which both velocities vanish and densities and temperature are uniform. Thus we neglect non-linear terms in the gradients of densities, velocities and temperature as well as products of velocities among themselves and with those gradients.

We denote quantities that refer to the uniform reference state by a superposed tilde and derivatives such as $\frac{\partial \varepsilon}{\partial \rho}$ and $\frac{\partial p_\alpha}{\partial T}$ by ε_ρ and p_T^α respectively. Moreover we abbreviate $\mu_1^I - \mu_2^I$ by μ.

Insertion of the constitutive relations into the equations of balance (4.20) and use of the integrability conditions (4.22) leads to a set of linear field equations for ρ, v_i, T, c and J_i, viz.

$$\begin{aligned}
&\frac{\partial \rho}{\partial t} + \tilde{\rho} \frac{\partial v_i}{\partial x_i} = 0 \, , \\
&\tilde{\rho} \frac{\partial c}{\partial t} + \frac{\partial J_i}{\partial x_i} = 0 \, , \\
&\tilde{\rho} \frac{\partial v_i}{\partial t} + \tilde{p}_\rho \frac{\partial \rho}{\partial x_i} + \tilde{p}_c \frac{\partial c}{\partial x_i} + \tilde{p}_T \frac{\partial T}{\partial x_i} = 0 \, , \\
&\frac{\partial J_i}{\partial t} - \tilde{M}''_v \frac{\tilde{\rho}}{\tilde{\rho}_1 \tilde{\rho}_2} J_i - \frac{\tilde{\rho}_1 \tilde{\rho}_2}{\tilde{\rho}} \left(\tilde{\mu}_\rho \frac{\partial \rho}{\partial x_i} + \mu_c \frac{\partial c}{\partial x_i} + \left(\frac{\tilde{p}_T^1}{\tilde{\rho}_1} - \frac{\tilde{p}_T^2}{\tilde{\rho}_2}\right) \frac{\partial T}{\partial x_i}\right) = 0 \, , \\
&\frac{\partial T}{\partial t} + \frac{\tilde{\varepsilon}_c}{\tilde{\varepsilon}_T} \frac{\partial c}{\partial t} - \frac{\tilde{T}}{\tilde{\rho}^2} \frac{\tilde{p}_T}{\tilde{\varepsilon}_T} \frac{\partial \rho}{\partial t} + \frac{\tilde{T}}{\tilde{\rho} \tilde{\varepsilon}_T} \left(\frac{\tilde{p}_T^1}{\tilde{\rho}_1} - \frac{\tilde{p}_T^2}{\tilde{\rho}_2} + \frac{\tilde{\mu}}{\tilde{T}} - \tilde{\mu}_T\right) \frac{\partial J_i}{\partial x_i} + \tilde{q}_T \frac{\partial^2 T}{\partial x_i \partial x_i} = 0 \, .
\end{aligned} \tag{4.23}$$

From $(4.23)_3$ we conclude that curl $\underset{\sim}{v}$ is constant in time and $(4.23)_4$ implies that the time derivative of curl $\underset{\sim}{J}$ is proportional to curl $\underset{\sim}{J}$ itself. Thus, if we assume that curl $\underset{\sim}{v}$ and curl $\underset{\sim}{J}$ are zero initially, they will continue to be zero.

4.2.2. First and Second Sound

We investigate the system of equations (4.23) for adiabatic sound propagation, where the term with q_T in $(4.23)_5$ may be neglected. It is then an easy matter within the linear theory to eliminate J_i, v_i and T from the field equations (4.23) and thus we obtain after some calculation

$$\frac{\partial^2 \rho}{\partial t^2} - [p_\rho + \frac{T(p_T)^2}{\rho^2 \varepsilon_T}] \Delta\rho - [p_c + \frac{T}{\varepsilon_T} p_T (\frac{\overset{1}{p}_T}{\rho_1} - \frac{\overset{2}{p}_T}{\rho_2})] \Delta c = 0 ,$$

$$\frac{\partial^2 c}{\partial t^2} - M''_v \frac{\rho}{\rho_1 \rho_2} \frac{\partial c}{\partial t} - \frac{\rho_1 \rho_2}{\rho^2} [\mu_c + \frac{T}{\varepsilon_T} (\frac{\overset{1}{p}_T}{\rho_1} - \frac{\overset{2}{p}_T}{\rho_2})^2] \Delta c - \frac{\rho_1 \rho_2}{\rho^4} [p_c + \frac{T}{\varepsilon_T} p_T (\frac{\overset{1}{p}_T}{\rho_1} - \frac{\overset{2}{p}_T}{\rho_2})] \Delta\rho = 0 , \tag{4.24}$$

where for convenience in notation I have dropped the tildes on the coefficients.

Before we reduce these equations further, we shall discuss a special case which is quite instructive: Consider the case in which the coefficient

$$W \equiv p_c + \frac{T}{\varepsilon_T} p_T (\frac{\overset{1}{p}_T}{\rho_1} - \frac{\overset{2}{p}_T}{\rho_2}) \tag{4.25}$$

vanishes. In this case the two equations (4.24) are uncoupled. The first one is the well-known wave equation for adiabatic sound propagation with the velocity

$$V_\rho = \sqrt{p_\rho + \frac{T(p_T)^2}{\rho^2 \varepsilon_T}} . \tag{4.26}$$

Note that the radicand can be written as $(\partial p/\partial \rho)_\eta$. The second equation is a hyperbolic equation, the so-called telegraph equation, which describes the propagation of damped waves with the velocity

$$V_c = \sqrt{\frac{\rho_1 \rho_2}{\rho^2} (\mu_c + \frac{T}{\varepsilon_T} (\frac{\overset{1}{p}_T}{\rho_1} - \frac{\overset{2}{p}_T}{\rho_2}))} . \tag{4.27}$$

We may call this the velocity of "second sound", whereby (4.26) becomes the velocity of "first sound".

A hyperbolic equation for diffusion was first derived from a thermodynamic theory of mixtures in [13]. That paper is based upon a slightly less general entropy principle than the present one, however, it leads to much the same results.

We continue to discuss the case $W = 0$. The equation $(4.24)_2$ is usually written without the second time derivative of c. This corresponds to the dropping of the acceleration in the balance of relative momenta $(4.20)_5$ of the constituents. Thus one obtains

$$M''_v \frac{\partial c}{\partial t} + [\mu_c + \frac{T}{\varepsilon_T} (\frac{p^1_T}{\rho_1} - \frac{p^2_T}{\rho_2})^2] \Delta c = 0 \qquad (4.28)$$

and this is the ordinary diffusion equation which is parabolic. The speed of propagation according to this equation is infinite. This fact is known as the paradox of diffusion theory in linear irreversible thermodynamics, where the equation (4.28) occurs as a consequence of Fick's law of diffusion. In the present theory no such paradox occurs, because the inertia of the relative motion is taken into account.

The paradox of diffusion theory has counterparts in the theory of heat conduction and in the theory of shear waves. These paradoxa have long been a thorn in the flesh of thermodynamics and their solution requires a major readjustment of some basic ideas. I shall come back to these points in the last Section of this series.

We note that, while it is conceivable that the coefficient W in (4.24), (4.25) may vanish, for a given mixture, it does not in general vanish for a mixture of ideal gases. Indeed, with $p_\alpha = \rho_\alpha \frac{R}{M_\alpha} T$ and $\varepsilon_\alpha = z_\alpha \frac{R}{M_\alpha} T + \beta_\alpha$ we obtain

$$W = \rho RT \left(\frac{1}{M_1} - \frac{1}{M_2}\right) \left(1 + \frac{\frac{\rho_1}{M_1} + \frac{\rho_2}{M_2}}{z_1 \frac{\rho_1}{M_1} + z_2 \frac{\rho_2}{M_2}}\right) \qquad (4.29)$$

and this is zero, if and only if $M_1 = M_2$ holds. If indeed $M_1 = M_2$ holds in a mixture of ideal gases, we have by (4.26) and (4.27)

$$V_\rho = \sqrt{\frac{z+1}{z} \frac{R}{M} T} \qquad \text{and} \qquad V_c = \sqrt{\frac{R}{M} T}\,. \qquad (4.30)$$

If W does not vanish, we must reduce the equations (4.24) further. We eliminate ρ between the two equations and obtain

$$\frac{\partial^4 c}{\partial t^4} - M''_v \frac{\rho}{\rho_1\rho_2} \frac{\partial^3 c}{\partial t^3} - (V_\rho^2 + V_c^2) \frac{\partial^2 \Delta c}{\partial t^2} + M''_v \frac{\rho}{\rho_1\rho_2} V_\rho^2 \frac{\partial \Delta c}{\partial t} + (V_\rho^2 V_c^2 - \frac{\rho_1\rho_2}{\rho^4} W^2) \Delta\Delta c = 0. \qquad (4.31)$$

This is again a wave equation describing the propagation of disturbances with the velocities

$$V_{\pm} = \sqrt{2}\,\sqrt{\frac{V_\rho^2 V_c^2 - \frac{\rho_1 \rho_2}{\rho^4} W^2}{V_\rho^2 + V_c^2}}\;\frac{1}{\sqrt{1 \pm \sqrt{1 - 4\,\frac{V_\rho^2 V_c^2 - \frac{\rho_1\rho_2}{\rho^4} W^2}{(V_\rho^2 + V_c^2)^2}}}} \tag{4.32}$$

As W tends to zero, V_+ tends to V_c and V_- tends to V . Thus we may say that V_+ is "primarily" the speed of propagation of disturbances in concentration, i.e. the speed of second sound, while V_- is primarily the speed of propagation of disturbances in density, i.e. the speed of first sound.

4.2.3. Plane Harmonic Waves

A scalar field of the form

$$s = \hat{s}\, e^{k_i \underset{\sim}{n} \cdot \underset{\sim}{x}} \cos\,(\omega t - k_r \underset{\sim}{n} \cdot \underset{\sim}{x} + \varphi_s) \tag{4.33}$$

is called a scalar harmonic wave with the amplitude $\hat{s}$, the frequency ω, the attenuation k_i, the wave number k_r and the phase $\omega t - k_r \underset{\sim}{n} \cdot \underset{\sim}{x} + s$. The wave is called plane, since the phase is constant on a plane. This plane is perpendicular to the unit vector $\underset{\sim}{n}$.

Similarly a vector field of the form

$$v_k = \hat{v}_k\, e^{k_i \underset{\sim}{n} \cdot \underset{\sim}{x}} \cos\,(\omega t - k_r \underset{\sim}{n} \cdot \underset{\sim}{x} + \varphi_{v_k}) \tag{4.34}$$

is called a vectorial harmonic wave within amplitude $\hat{v}_k$. If $\underset{\sim}{v}$ and $\underset{\sim}{n}$ are parallel, the wave is called longitudinal and if $\underset{\sim}{v}$ and $\underset{\sim}{n}$ are perpendicular, it is called transversal.

The plane of constant phase propagates in a perpendicular direction with the speed

$$v_{ph} = \frac{\omega}{k_r}$$

which is called the phase speed of the wave.

It is useful to introduce the complex amplitudes

$$\bar{s} = \hat{s}\, e^{i\varphi_s} \qquad \text{and} \qquad \bar{v}_k = \hat{v}_k\, e^{i\varphi_{v_k}}$$

as well as the complex wave number

$$k = k_r + i\,k_i$$

because thus (4.33) and (4.34) reduce to the simple forms

$$s = \mathrm{Re}\,\{\bar{s}\, e^{i(\omega t - k\,\underset{\sim}{n}\cdot\underset{\sim}{x})}\} \quad \text{and} \quad v_k = \mathrm{Re}\,\{\bar{v}_k\, e^{i(\omega t - k\,\underset{\sim}{n}\cdot\underset{\sim}{x})}\}\,. \tag{4.35}$$

Indeed, often we write

$$s = \bar{s}\, e^{i(\omega t - k\,\underset{\sim}{n}\cdot\underset{\sim}{x})} \qquad \text{and} \qquad v_k = \bar{v}_k\, e^{i(\omega t - k\,\underset{\sim}{n}\cdot\underset{\sim}{x})} \tag{4.36}$$

just keeping it in mind that it is only the real part of these expressions which represents the waves.

4.2.4. Plane Harmonic Waves in a Binary Mixture of Non-Reacting Constituents

We consider plane harmonic waves of the form

$$\begin{aligned}
\rho &= \tilde{\rho} + \bar{\rho}\, e^{i(\omega t - k\,\underset{\sim}{n}\cdot\underset{\sim}{x})}\,, \\
c &= \tilde{c} + \bar{c}\, e^{i(\omega t - k\,\underset{\sim}{n}\cdot\underset{\sim}{x})}\,, \\
v_i &= \bar{v}_i\, e^{i(\omega t - k\,\underset{\sim}{n}\cdot\underset{\sim}{x})}\,, \\
J_i &= \bar{J}_i\, e^{i(\omega t - k\,\underset{\sim}{n}\cdot\underset{\sim}{x})}\,, \\
T &= \tilde{T} + \bar{T}\, e^{i(\omega t - k\,\underset{\sim}{n}\cdot\underset{\sim}{x})}
\end{aligned} \tag{4.37}$$

and first of all we inquire whether this is a solution of the equations (4.23). Insertion of (4.37) into (4.23) leads to a homogeneous system of equations for $\bar{\rho}$ through $\bar{T}$, viz.

$$
\begin{aligned}
&u\bar\rho \qquad - \tilde\rho n_i \bar v_i &= 0\,,\\
&+\tilde\rho u \bar c \qquad - n_i \bar J_i &= 0\,,\\
&-\tilde p_\rho n_i \bar\rho \quad - \tilde p_c n_i \bar c + \tilde\rho u \bar v_i \qquad - \tilde p_T n_i \bar T &= 0\,,\\
&-\frac{\tilde\rho_1\tilde\rho_2}{\tilde\rho}\tilde\mu_\rho n_i\bar\rho - \frac{\tilde\rho_1\tilde\rho_2}{\tilde\rho}\tilde\mu_c n_i \bar c \quad + u(1+iM''_v \frac{\tilde\rho}{\tilde\rho_1\tilde\rho_2}\frac{1}{\omega})\bar J_i - \frac{\tilde\rho_1\tilde\rho_2}{\tilde\rho}(\frac{\tilde p^1_T}{\tilde\rho_1}-\frac{\tilde p^2_T}{\tilde\rho_2}) n_i \bar T &= 0\,,\\
&-\frac{\tilde T}{\tilde\rho^2}\frac{\tilde p_T}{\tilde\varepsilon_T} u\bar\rho \qquad - \frac{\tilde T}{\tilde\rho\tilde\varepsilon_T}(\frac{\tilde p^1_T}{\tilde\rho_1}-\frac{\tilde p^2_T}{\tilde\rho_2}) n_i \bar J_i \qquad + u\bar T &= 0\,,
\end{aligned}
\tag{4.38}
$$

where the term with q_T has been neglected, because we are interested in adiabatic sound propagation. u in (4.38) is defined as ω/k .

It is obvious from (4.38) that $\bar v_i$ and $\bar J_i$ point in the direction of n_i so that v_i and J_i are longitudinal waves. We choose $n_i = (1,0,0)$ and write $\bar J_i = (\bar J,0,0)$ and $\bar v_i = (\bar v,0,0)$. Elimination of $\bar\rho$, $\bar c$ and $\bar T$ from (4.38) leads to the equations

$$
\begin{aligned}
&-\rho(p_\rho + \frac{T(p_T)^2}{\rho^2\varepsilon_T} - u^2)\bar v - \frac{1}{\rho}(p_c + T\frac{p_T}{\varepsilon_T}(\frac{p^1_T}{\rho_1}-\frac{p^2_T}{\rho_2}))\bar J = 0,\\
&-\frac{\rho_1\rho_2}{\rho^2}(\rho^2\mu_\rho - \frac{\rho^2}{p_T}(\frac{p^1_T}{\rho_1}-\frac{p^2_T}{\rho_2})(p_\rho - u^2))\bar v - (\frac{\rho_1\rho_2}{\rho^2}(\mu_c - \frac{p_c}{p_T}(\frac{p^1_T}{\rho_1}-\frac{p^2_T}{\rho_2})) - u^2(1+iM''_v\frac{\rho}{\rho_1\rho_2}\frac{1}{\omega}))\bar J = 0,
\end{aligned}
\tag{4.39}
$$

where the tildes have been dropped for easier notation. For simplicity in the argument we also assume that the first and second sound are uncoupled which means that the coefficient of interaction W, defined in (4.25) is equal to zero. In that case the equations (4.39) may be rewritten in the forms

$$
\begin{aligned}
&(p_\rho + \frac{T(p_T)^2}{\rho^2\varepsilon_T} - u^2)\bar v = 0\,,\\
&\frac{\rho_1\rho_2}{p_T}(\frac{p^1_T}{\rho_1}-\frac{p^2_T}{\rho_2})(p_\rho + \frac{T(p_T)^2}{\rho^2\varepsilon_T} - u^2)\bar v - (\frac{\rho_1\rho_2}{\rho^2}(\mu_c - \frac{p_c}{p_T}(\frac{p^1_T}{\rho_1}-\frac{p^2_T}{\rho_2})) - u^2(1+iM''_v\frac{\rho}{\rho_1\rho_2}\frac{1}{\omega}))\bar J = 0.
\end{aligned}
\tag{4.40}
$$

For the homogenous set of two algebraic equations to have a non-trivial solution the determinant must vanish which means either

$$\frac{1}{u^1} = \frac{k^1}{\omega} = \frac{1}{V_\rho} \quad , \qquad \text{or}$$

$$\frac{1}{u^2} = \frac{k^2}{\omega} = \frac{1}{V_c} \sqrt{1 + iM''_v \frac{\rho}{\rho_1\rho_2}\frac{1}{\omega}} \,. \tag{4.41}$$

We conclude that the frist sound is undamped, while the second sound has an attenuation, because the imaginary part of k_2 is non-zero. Figure 3 presents the phase speeds ω/k_1^r, ω/k_2^r and the attenuation k_2^i as functions of $Q = -M''_v \frac{\rho}{\rho_1\rho_2}\frac{1}{\omega}$. From (4.40) it follows that $\bar{J} = 0$ in the propagation of first sound and that $\bar{v} = 0$ in the propagation of second sound. Thus in the first sound the mixture moves as a whole, i.e. there is no relative motion of the constituents, while in the second sound one constituent moves with respect to the other one in such a manner that the mixture is at rest. Moreover, going back to the equations (4.38), we conclude that $\bar{c} = 0$ in first sound, while $\bar{\rho} = 0$ in second sound. From $(4.38)_{1,2,3}$ we obtain

$$T = \frac{1}{p_T}\left((u^2 - p_\rho)\bar{\rho} - p_c\bar{c}\right) \,, \tag{4.42}$$

whence it follows that a temperature wave accompanies both the first and the second sound.

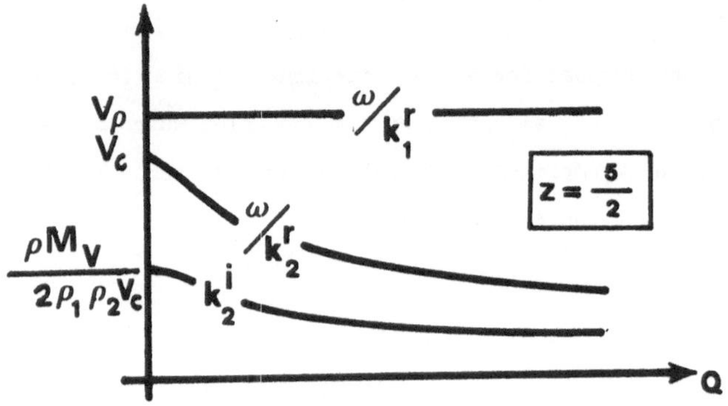

Figure 3

5. LIQUID HELIUM AS A SPECIAL BINARY MIXTURE

5.1 Landau's Theory as a Theory of a Binary Mixture

5.1.1. Assumptions leading to Landau's Theory of Liquid Helium

Tisza [14] and Landau [15] have formulated theories of liquid helium II, i.e. helium below a temperature of 2.19 K, which may be fitted into the thermodynamic theory of a binary mixture provided that the following three assumptions are valid.

i.) (Mixture Character)

The first assumption states that the single liquid to which He 4 condenses behaves like a mixture. This strange assumption is motivated by some of the observed phenomena in liquid helium and it receives some support by arguments in quantum statistics. According to these arguments it may be more proper to speak of two phases of helium rather than of two constituents of a mixture. Indeed, quantum mechanics and quantum statistics provide for the possibility of having a phase composed of atoms in their energetic ground state and another phase of atoms in energetically elevated states. These form the supercomponent and the normal component of Landau's mixture theory respectively. We shall denote these by the indices 1 or s for the supercomponent and 2 or n for the normal component. As is appropriate for a mixture of two phases we must assume that one phase can be converted into the other as the temperature and pressure might require. In other words, we must allow for "chemical" reactions in the theory.

It turns out that we may consider liquid helium as a special case of simple mixtures. Although some greater generality is possible, we shall assume here that helium represents a simple mixture of the type that was considered in Sections 4.1 and 4.2; thus $M_T = 0$, q_V^{α} is given by (4.13), and the residual entropy inequality is (4.18).

ii.) (No Dissipation)

The supercomponent is endowed with very remarkable properties. In particular it is assumed to be non-viscous and also its dynamical interaction with the normal component is assumed to be zero.

While the normal component has viscosity and heat conductivity, it is possible to ignore these features in the study of wave propagation just like this is done in the study of the leading terms of sound propagation in an ordinary mixture.

In fact, in the treatment of wave propagation in helium we assume that there is no dissipation at all, not even dissipation due to the conversion of one constituent into the other one. This means that the residual inequality (4.18) is satisfied with the equality sign. Inspection of that inequality shows that this assumption of vanishing dissipation in a binary mixture can be satisfied by

$$q_T = 0\ , \quad M_V^{ss} = 0\ , \quad \text{and} \quad \mu_s^I - \mu_n^I + \frac{1}{2} V_s^2 = 0\ . \tag{5.1}$$

Equation $(5.1)_3$ requires a comment. It would be possible to suppress the chemical dissipation by disallowing chemical reactions, i.e. by setting $\tau_\alpha = 0$ in (4.18). In fact we have done just that in the previous section. But in helium this assumption would contradict the observations which rather clearly indicate that there is a free and easy transition from the supercomponent to the normal component and vice versa. Therefore we must ensure the lack of chemical dissipation by setting $\mu_s^I - \mu_n^I + \frac{1}{2} V_s^2$ equal to zero.

Equation $(5.1)_3$ has rather severe implications, because the left hand side is a function of ρ_s, ρ_n, T and $\underset{\sim}{V}^s$ and its vanishing provides an algebraic dependence between these fields. This constraint enables us to ignore one of the field equations and we shall choose to ignore the mass balance for the superfluid. Another way of putting this is to say that - when all the fields have been determined from the remaining equations and from $\mu_s^I - \mu_n^I + \frac{1}{2} V_s^2 = 0$ - the mass balance of the superfluid may be used to calculate the "chemical" production τ_s of the superfluid.

iii.) (Vanishing Entropy)

We assume that

$$\eta_s = 0 \tag{5.2}$$

so that the supercomponent has no entropy. This assumption can be motivated by quantum statistics which suggests that the superfluid may be the phase of helium that prevails at absolute zero and there the entropy vanishes according to Nernst's and Planck's formulation of the third law of thermodynamics.

5.1.2. Landau's Equations

The basic equations of Landau's theory of liquid helium can be understood as the equations of balance of the masses and momenta of the constituents and the balance of internal energy of the mixture, just like in the theory of any other type of mixture presented heretofore. The only exception is that we ignore the balance of mass of the supercomponent, because we have the constraint $(5.1)_3$ to replace it. Also, instead of the remaining balance of mass of the normal component we choose to write the balance of total mass and instead of the two balance equations of partial momenta we choose to consider the balance of total momentaum and the balance of momentum of the superfluid. This is merely for later convenience. Thus the relevant balance equations read

$$\frac{\partial\rho}{\partial t}+\frac{\partial\rho v_i}{\partial x_i}=0\,, \qquad \text{balance of mass of mixture}$$

$$\frac{\partial\rho v_i}{\partial t}+\frac{\partial}{\partial x_i}(\rho v_i v_j - t_{ij})=0\,, \qquad \text{balance of momentum of mixture}$$

$$\frac{\partial\rho_s v_i^s}{\partial t}+\frac{\partial}{\partial x_j}(\rho_s v_i^s v_j^s - t_{ij}^s)=m_i^s\,, \qquad \text{balance of momentum of superfluid} \tag{5.3}$$

$$\frac{\partial\rho\varepsilon}{\partial t}+\frac{\partial}{\partial x_i}(\rho\varepsilon v_j + q_j)=t_{ij}\frac{\partial v_i}{\partial x_j}\,, \qquad \text{balance of internal energy}$$

With one field equation replaced by the condition $\mu_s^I - \mu_n^I + \frac{1}{2}v_s^2 = 0$ this is a permissible choice, i.e. we have the correct number of equations for the determination of the fields ρ_s, ρ_n, v_i^s, v_i^n, T.

But of course we also need constitutive equations and for those we pick the equations characterizing a simple mixture of the type treated in Section 4 for which the interaction force is independent of the gradient of temperature and for which the partial heat fluxes do not depend on the relative velocities.

In summary we rewrite the constitutive equations for such a mixture as

$$\eta = \sum_{\alpha=1}^{2} \frac{\rho_\alpha}{\rho} \eta_\alpha(\rho_\alpha, T)$$

$$\varepsilon = \sum_{\alpha=1}^{2} \frac{\rho_\alpha}{\rho} \varepsilon_\alpha(\rho_\alpha, T) + \frac{\rho_n \rho_s}{2\rho^2} V_s^2 \qquad \text{with} \quad \eta_\alpha = \frac{1}{T}(d\varepsilon_\alpha - \frac{p}{\rho_\alpha^2} d\rho_\alpha) \quad \text{and} \quad \rho_\alpha^2 \frac{\partial \varepsilon_\alpha}{\partial \rho_\alpha} = p_\alpha - T \frac{\partial p_\alpha}{\partial T}$$

$$t_{ij}^\alpha = -p^\alpha(\rho_\alpha, T)\delta_{ij} \qquad \text{with} \quad \frac{\partial p^\alpha}{\partial \rho_\alpha} = \rho_\alpha \frac{\partial \mu_\alpha^I}{\partial \rho_\alpha} \quad \text{and} \quad \mu_\alpha^I = \varepsilon_\alpha - T\eta_\alpha + \frac{p_\alpha}{\rho_\alpha} \tag{5.4}$$

$$m_i^s - \tau_s v_i^s = M_v^{ss} V_i^s + M_T^s T_{,i} \qquad \text{with} \quad M_v^{ss} = 0\ , \quad M_T^s = 0$$

$$q_i = q_v^s V_i^s + q_T T_{,i} + \frac{\rho_n \rho_s}{2\rho^2}(\rho_n - \rho_s) V_s^2 V_i^s \quad \text{with} \quad q_T = 0, \quad q_v^s = \frac{\rho_n \rho_s}{\rho}\left(\varepsilon_s + \frac{p_s}{\rho_s} - (\varepsilon_n + \frac{p_n}{\rho_n})\right) .$$

M_T^s is taken to be zero as explained before, and M_v^{ss} vanishes because of the assumption that there be no dissipation in helium. Accordingly there is no interaction force. Also q_T is zero, because of lack of dissipation so that the flux of internal energy is proportional to the relative velocity V_i^s .

Insertion of the constitutive relations (5.4) into the equations of balance (5.3) leads to the equations

$$\boxed{\frac{d\rho}{dt} + \rho \frac{\partial v_i}{\partial x_i} = 0\ ,}$$

$$\boxed{\rho \frac{dv_i}{dt} + \frac{\partial(p\delta_{ij} + \frac{\rho_s \rho_n}{\rho} V_i^s V_j^s)}{\partial x_j} = 0} \tag{5.5}$$

$$\rho_s \frac{\partial v_i^s}{\partial t} - \rho_s (\underset{\sim}{v}^s \times \text{curl}\ \underset{\sim}{v}^s)_i + \rho_s \frac{\partial \frac{1}{2} v_s^2}{\partial x_i} + \frac{\partial p^s}{\partial x_i} = 0\ ,$$

$$\rho \frac{d\varepsilon}{dt} + \frac{\partial}{\partial x_i}\left(q_v^s V_i^s + \frac{\rho_s \rho_n}{2\rho^2}(\rho_n - \rho_s) V_s^2 V_i^s\right) = -(p\delta_{ij} + \frac{\rho_s \rho_n}{\rho} V_i^s V_j^s) \frac{\partial v_i}{\partial x_j} .$$

These equations can be made more specific by use of the assumptions on the special character of the mixture that represents helium.

First of all, we use $(5.4)_3$ to write

$$\frac{\partial p^s}{\partial x_i} = \rho_s \frac{\partial \mu_s^I}{\partial x_i} - \left(\rho_s \frac{\partial \mu_s^I}{\partial T} - \frac{\partial p_s}{\partial T}\right) \frac{\partial T}{\partial x_i} ,$$

$$\frac{\partial p^s}{\partial x_i} = \rho_s \frac{\partial \mu_s^I}{\partial x_i} - \left(\rho_s \left(\frac{\partial \varepsilon_s}{\partial T} - T \frac{\partial \eta_s}{\partial T}\right) - \rho_s \eta_s\right) \frac{\partial T}{\partial x_i} .$$

The last two terms vanish, because of the Gibbs equation $(5.4)_2$ and because $\eta_s = 0$ holds in helium (see (5.2)). Also, since $\mu_s^I = \mu_n^I - \frac{1}{2} v_s^2$ we have by $(5.4)_3$ $\mu_s^I = \varepsilon_I - T\eta + \frac{p}{\rho} - \frac{\rho_n}{2\rho} v_s^2$ and hence follows

$$\frac{\partial p^s}{\partial x_i} = \rho_s \frac{\partial}{\partial x_i} \left(\varepsilon_I - T\eta + \frac{p}{\rho} - \frac{\rho_n}{2\rho} v_s^2\right) . \tag{5.6}$$

Elimination of $\partial p^s/\partial x_i$ between $(5.5)_3$ and (5.6) leads to the equation

$$\frac{\partial v_i^s}{\partial t} - (\underset{\sim}{v}^s \times \text{curl}\, \underset{\sim}{v}^s)_i + \frac{\partial \frac{1}{2} v_s^2}{\partial x_i} + \frac{\partial}{\partial x_i} \left(\varepsilon_I - T\eta + \frac{p}{\rho} - \frac{\rho_n}{2\rho} v_s^2\right) = 0$$

whence follows by taking the curl

$$\frac{\partial\, \text{curl}_i\, \underset{\sim}{v}^s}{\partial t} = \text{curl}_i\, (\underset{\sim}{v}^s \times \text{curl}\, \underset{\sim}{v}^s) .$$

We conclude from this that

$$\text{curl}\, \underset{\sim}{v}_s = 0 \tag{5.7}$$

must hold for all times, if it held at one time throughout the fluid. Indeed, (5.7) has motivated Landau to state that the supercomponent cannot participate in a rigid rotation. We shall come back to this point in Section 5.2.

The Gibbs equation (4.21) may be written in the form

$$d\left(\varepsilon_I - T\eta + \frac{p}{\rho}\right) = -\eta dT + \frac{1}{\rho} dp + (\mu_I^s - \mu_I^n) dc_s$$

or, because of $\mu_I^s - \mu_I^n = -\frac{1}{2} v_s^2$:

$$d\left(\varepsilon_I - T\eta + \frac{p}{\rho} - \frac{\rho_n}{2\rho} v_s^2\right) = -\eta dT + \frac{1}{\rho} dp - \frac{\rho_n}{\rho} d\left(\frac{1}{2} v_s^2\right) .$$

Thus finally the gradient of p^s assumes the form

$$\frac{\partial p^s}{\partial x_i} = -\rho_s \eta \frac{\partial T}{\partial x_i} + \frac{\rho_s}{\rho} \frac{\partial p}{\partial x_i} - \frac{\rho_s \rho_n}{2\rho} \frac{\partial V_s^2}{\partial x_i} \qquad \text{or by } (5.5)_2$$

$$\frac{\partial p^s}{\partial x_i} = -\rho_s \eta \frac{\partial T}{\partial x_i} - \rho_s \frac{dv_i}{dt} - \frac{\rho_s}{\rho} \frac{\partial \frac{\rho_s \rho_n}{\rho} V_i^s V_j^s}{\partial x_j} - \frac{\rho_s \rho_n}{2\rho} \frac{\partial V_s^2}{\partial x_i} \tag{5.8}$$

Next we consider q_v^s, which according to the assumptions about the character of the mixture is given by $(5.4)_5$. We have

$$q_v^s = \frac{\rho_s \rho_n}{\rho} \left(\varepsilon_s + \frac{p_s}{\rho_s} - \left(\varepsilon_n + \frac{p_n}{\rho_n}\right)\right) \qquad \text{with} \quad \mu_\alpha^I = \varepsilon_\alpha - T\eta_\alpha + \frac{p_\alpha}{\rho_\alpha} \quad \text{and} \quad \mu_s^I = \mu_n^I - \frac{1}{2} V_s^2$$

$$q_v^s = \frac{\rho_s \rho_n}{\rho} \left(T(\eta_s - \eta_n) - \frac{1}{2} V_s^2\right) \qquad \text{with} \quad \eta_s = 0 \quad \text{and} \quad \eta = \frac{\rho_n}{\rho} \eta_n \tag{5.9}$$

$$q_v^s = -T\rho_s \eta - \frac{\rho_s \rho_n}{2\rho} V_s^2 \ .$$

Finally we rewrite $\rho \frac{d\varepsilon}{dt} + p \frac{\partial v_i}{\partial x_i}$:

$$\rho \frac{d\varepsilon}{dt} + p \frac{\partial v_i}{\partial x_i} = \rho \frac{\partial \varepsilon_I}{\partial T} \frac{dT}{dt} + \rho \left(\frac{\partial \varepsilon_I}{\partial \rho} - \frac{p}{\rho^2}\right) \frac{d\rho}{dt} + \rho \frac{\partial \varepsilon_I}{\partial c_s} \frac{dc_s}{dt} + \rho \frac{d \frac{\rho_s \rho_n}{2\rho^2} V_s^2}{dt} \qquad \text{with (4.21), (4.22)}$$

$$= \rho T \frac{\partial \eta}{\partial T} \frac{dT}{dt} - \frac{1}{\rho} T \frac{\partial p}{\partial T} \frac{d\rho}{dt} + \rho \frac{\partial \varepsilon_I}{\partial c_s} \frac{dc_s}{dt} + \rho \frac{d \frac{\rho_s \rho_n}{2\rho^2} V_s^2}{dt}$$

$$= \rho T \frac{d\eta}{dt} - \left(\rho T \frac{\partial \eta}{\partial \rho} + \frac{T}{\rho} \frac{\partial p}{\partial T}\right) \frac{d\rho}{dt} - \rho \left(T \frac{\partial \eta}{\partial c_s} - \frac{\partial \varepsilon_I}{\partial c_s}\right) \frac{dc_s}{dt} + \rho \frac{d \frac{\rho_s \rho_n}{2\rho^2} V_s^2}{dt}$$

and since $\frac{\partial \eta}{\partial \rho} = -\frac{1}{\rho^2} \frac{\partial p}{\partial T}$ holds and $T \frac{\partial \eta}{\partial c_s} - \frac{\partial \varepsilon_I}{\partial c_s} = -(\mu_I^s - \mu_I^n) = \frac{1}{2} V_s^2$ we have

$$\rho \frac{d\varepsilon}{dt} + p \frac{\partial v_i}{\partial x_i} = \rho T \frac{d\eta}{dt} - \rho_s V_s^2 \frac{dc_s}{dt} + \frac{\rho_s \rho_n}{2\rho} \frac{dV_s^2}{dt} \ .$$

Again by $(5.1)_3$ c_s is a function of ρ, T and V_s^2, or of p, T and V_s^2 so that we may write

$$\frac{dc_s}{dt} = \frac{\partial c_s}{\partial T} \frac{dT}{dt} + \frac{\partial c_s}{\partial p} \frac{dp}{dt} + \frac{\partial c_s}{\partial V_s^2} \frac{dV_s^2}{dt} \ .$$

Thus finally $\rho \frac{d\varepsilon}{dt} + p \frac{\partial v_i}{\partial x_i}$ assumes the form

$$\rho \frac{d\varepsilon}{dt} + p \frac{\partial v_i}{\partial x_i} = \rho T \frac{d\eta}{dt} - \rho_s V_s^2 \frac{\partial c_s}{\partial T} \frac{dT}{dt} - \rho_s V_s^2 \frac{\partial c_s}{\partial p} \frac{dp}{dt} + \frac{\rho_s \rho_n}{2\rho} \frac{dV_s^2}{dt} , \tag{5.10}$$

where the term $\partial c_s / \partial V_s^2$ has been neglected as of fourth order in V_i^s.

We combine (5.8) through (5.10) with $(5.5)_{3,4}$ and obtain

$$\boxed{\rho_s \left(\frac{\partial V_i^s}{\partial t} - \frac{dv_i}{dt}\right) - \rho_s \eta \frac{\partial T}{\partial x_i} + \rho_s \frac{\partial \frac{1}{2} V_s^2}{\partial x_i} - \frac{\rho_s}{\rho} \frac{\partial \frac{\rho_s \rho_n}{\rho} V_i^s V_j^s}{\partial x_i} - \frac{\rho_s \rho_n}{2\rho} \frac{\partial V_s^2}{\partial x_i} = 0}$$

$$\boxed{\begin{aligned} &\rho T \frac{d\eta}{dt} - \frac{\partial}{\partial x_i} (T \rho_s \eta V_i^s) - \rho_s V_s^2 \left(\frac{\partial c_s}{\partial T} \frac{dT}{dt} + \frac{\partial c_s}{\partial p} \frac{dp}{dt}\right) - \\ &\qquad - \frac{\partial}{\partial x_i} \left(\frac{\rho_s \rho_n}{\rho} V_s^2 \frac{\rho_s}{\rho} V_i^s\right) + \frac{\rho_s \rho_n}{\rho} V_i^s V_j^s \frac{\partial v_i}{\partial x_j} + \frac{\rho_s \rho_n}{2\rho} \frac{dV_s^2}{dt} = 0 \end{aligned}} \tag{5.11}$$

From the framed equations in (5.5) and (5.11) we can easily derive Landau's linearized equations for liquid helium. They come out when all fields are supposed to be close to a reference state of constant densities and temperature and of vanishing velocity. In that case we neglect products of small quantities and obtain

$$\begin{aligned} &\frac{\partial \rho}{\partial t} + \tilde{\rho} \frac{\partial v_i}{\partial x_i} = 0 , \\ &\tilde{\rho} \frac{\partial v_i}{\partial t} + \frac{\partial p}{\partial x_i} = 0 , \\ &\frac{\tilde{\rho}_s \tilde{\rho}_n}{\tilde{\rho}} \frac{\partial V_i^s}{\partial t} - \tilde{\rho}_s \tilde{\eta} \frac{\partial T}{\partial x_i} = 0 , \\ &\frac{\partial \eta}{\partial t} - \frac{\tilde{\rho}_s}{\tilde{\rho}} \tilde{\eta} \frac{\partial V_i^s}{\partial x_i} = 0 . \end{aligned} \tag{5.12}$$

The tilde refers to the reference state.

The equations (5.12) are Landau's linear equations for liquid helium. In Landau's presentation it is not easy to see, how the equations fit into a general theory of simple mixtures, but here they have been derived from such a theory. The preceding treatment and the later section on rotating helium form part of a dissertation by Dreyer [16] that is now in progress.

5.1.3. Landau's Wave Equations

It is an easy matter to eliminate v_i between the two equations $(5.12)_{1,2}$ and v_i^s between $(5.12)_{3,4}$. The following equations result from this elimination: (For simplicity in notation the tildes are dropped in this section)

$$\frac{\partial^2 \rho}{\partial t^2} - \Delta p = 0 \qquad \text{and} \qquad \frac{\partial^2 \eta}{\partial t^2} - \eta^2 \frac{\rho_s}{\rho_n} \Delta T = 0 \, . \tag{5.13}$$

Normally of course, we should consider p and η as functions of ρ, c_s and T, but since $\mu_s^I - \mu_n^I + \frac{1}{2} v_s^2 = 0$ holds, which reduces to $\mu_s = \mu_n$ in the linear theory, we may drop c_s (say) from this list of variables and retain only ρ, T. Alternatively we may consider ρ and η as functions of p and T which, by tradition, is the preferred choice of variables. Thus we write in the linear case

$$\frac{\partial^2 \rho}{\partial t^2} = \frac{\partial \rho}{\partial T}\frac{\partial^2 T}{\partial t^2} + \frac{\partial \rho}{\partial p}\frac{\partial^2 p}{\partial t^2} \qquad \text{and} \qquad \frac{\partial^2 \eta}{\partial t^2} = \frac{\partial \eta}{\partial T}\frac{\partial^2 T}{\partial t^2} + \frac{\partial \eta}{\partial p}\frac{\partial^2 p}{\partial t^2}$$

and obtain by insertion into (5.13)

$$\begin{aligned} \frac{\partial^2 p}{\partial t^2} - \left(\frac{\partial p}{\partial \rho}\right)_T \Delta p + \left(\frac{\partial p}{\partial \rho}\right)_T \left(\frac{\partial \rho}{\partial T}\right)_p \frac{\partial^2 T}{\partial t^2} &= 0 \, , \\ \left(\frac{\partial \eta}{\partial p}\right)_T \frac{\partial^2 p}{\partial t^2} + \left(\frac{\partial \eta}{\partial T}\right)_p \frac{\partial^2 T}{\partial t^2} - \eta^2 \frac{\rho_s}{\rho_n} \Delta T &= 0 \, . \end{aligned} \tag{5.14}$$

Note that $(\partial\eta/\partial p)_T = \frac{1}{\rho^2}(\partial\rho/\partial T)_p$ holds which is an integrability conditions for $\varepsilon - T\eta + p/\rho$. Therefore the two equations (5.14) decouple, if the thermal expansion $(\partial\rho/\partial T)_p$ vanishes. In that case we have

$$\frac{\partial^2 p}{\partial t^2} - \left(\frac{\partial p}{\partial \rho}\right)_T \Delta p = 0 \qquad \text{and} \qquad \frac{\partial^2 T}{\partial t^2} - \left(\frac{\rho_s}{\rho_n}\frac{\eta^2}{(\partial\eta/\partial T)_p}\right) \Delta T = 0 \, . \tag{5.15}$$

These are both ordinary wave equations which describe a propagation with the speeds

$$V_1 = \sqrt{\left(\frac{\partial p}{\partial \rho}\right)_T} \quad \text{and} \quad V_2 = \sqrt{\frac{\rho_s}{\rho_n} \frac{\eta^2}{(\partial \eta/\partial T)_p}} \quad . \tag{5.16}$$

We call these the speeds of first and second sound in helium.

If the thermal expansion cannot be ignored, the two equations (5.14) combine to give a wave equation of the fourth order. Indeed, by elimination of T (say) we get after some standard rearrangement

$$\frac{\partial^4 p}{\partial t^4} - \left(\left(\frac{\partial p}{\partial \rho}\right)_\eta + \frac{T\eta^2}{(\partial \varepsilon/\partial T)_\rho} \frac{\rho_s}{\rho_n}\right) \frac{\partial^2 \Delta p}{\partial t^2} + \frac{T\eta^2}{(\partial \varepsilon/\partial T)_\rho} \frac{\rho_s}{\rho_n} \left(\frac{\partial p}{\partial \rho}\right)_T \Delta\Delta p = 0 \ , \tag{5.17}$$

where $\left(\frac{\partial \varepsilon}{\partial T}\right)_\rho$ is the specific heat at constant density. This is a wave equation which predicts the propagation of disturbances with the velocities

$$V_\pm = \sqrt{\frac{1}{2}\left[\left(\left(\frac{\partial p}{\partial \rho}\right)_\eta + \frac{T\eta^2}{\left(\frac{\partial \varepsilon}{\partial T}\right)_\rho} \frac{\rho_s}{\rho_n}\right] \pm \sqrt{\frac{1}{4}\left[\left(\frac{\partial p}{\partial \rho}\right)_\eta + \frac{T\eta^2}{\left(\frac{\partial \varepsilon}{\partial T}\right)_\rho} \frac{\rho_s}{\rho_n}\right]^2 - \frac{T\eta^2}{\left(\frac{\partial \varepsilon}{\partial T}\right)_\rho} \frac{\rho_s}{\rho_n} \left(\frac{\partial p}{\partial \rho}\right)_T}} \ , \tag{5.18}$$

where $\left(\frac{\partial p}{\partial \rho}\right)_\eta$ apart from a factor, is the reciprocal of the adiabatic compressibility. In the case where $\left(\frac{\partial \rho}{\partial T}\right)_p = 0$ holds, we have $\left(\frac{\partial \eta}{\partial T}\right)_p = \left(\frac{\partial \eta}{\partial T}\right)_\rho$ and therefore the adiabatic compressibility is equal to the isothermal compressibility. In that case $V_\pm$ reduce to the two velocities $V_{1,2}$ of equations (5.16).

5.1.4. Amplitude of First and Second Sound

Just like we have considered plane harmonic waves in Section 4 we shall now consider such waves in helium. As the reference state we consider a state of constant p and T and of vanishing v^1 and v^2, such that the assumed solution has the form

$$\begin{aligned} p &= \tilde{p} + \bar{p}\, e^{i(\omega t - k\, \underset{\sim}{n} \cdot \underset{\sim}{x})} \ , \\ v_i^1 &= \bar{v}_i^1\, e^{i(\omega t - k\, \underset{\sim}{n} \cdot \underset{\sim}{x})} \ , \\ v_i^2 &= \bar{v}_i^2\, e^{i(\omega t - k\, \underset{\sim}{n} \cdot \underset{\sim}{x})} \ , \\ T &= \tilde{T} + \bar{T}\, e^{i(\omega t - k\, \underset{\sim}{n} \cdot \underset{\sim}{x})} \ , \end{aligned} \tag{5.19}$$

where the barred quantities are small amplitudes whose products we neglect.

Insertion of (5.19) into Landau's equations (5.12) leads to a homogenous system of algebraic equations, viz.

$$
\begin{aligned}
&\left(\frac{\partial\tilde\rho}{\partial T}\right)_p u\,\bar T + \left(\frac{\partial\tilde\rho}{\partial p}\right)_T u\,\bar p - \tilde\rho_1 n_i \bar v^1_i - \tilde\rho_2 n_i \bar v^2_i = 0\,,\\
&- n_i\bar p + u\,\tilde\rho_1 \bar v^1_i + u\,\tilde\rho_2 \bar v^2_i = 0\,,\\
&- \tilde\eta\;{}_i\bar T - u\,\frac{\tilde\rho_2}{\tilde\rho}\bar v^1_i + u\,\frac{\tilde\rho_2}{\tilde\rho}\bar v^2_i = 0\,,\\
&\left(\frac{\partial\tilde\eta}{\partial T}\right)_p u\,\bar T + \frac{1}{\tilde\rho^2}\left(\frac{\partial\tilde\rho}{\partial T}\right)_p u\,\bar p + \tilde\eta\,\frac{\tilde\rho_1}{\tilde\rho} n_i \bar v^1_i - \tilde\eta\,\frac{\tilde\rho_1}{\tilde\rho} n_i \bar v^2_i = 0\,,
\end{aligned}
\tag{5.20}
$$

where $u = \omega/k$.

It follows from $(5.20)_{2,3}$ that $\bar v^1_i$ and $\bar v^2_i$ have the direction of n_i, so that both velocities are transversal waves. In that case we may take $n_i = (1,0,0)$ without loss of generality, so that four equations remain. Among these we eliminate $\bar T$ and $\bar p$ to obtain

$$
\begin{aligned}
&\left[\left(\frac{\eta}{\left(\frac{\partial\eta}{\partial T}\right)_p}\frac{\rho_1}{\rho^2}\left(\frac{\partial\rho}{\partial T}\right)_p - \frac{\rho_2}{\rho}\right)u^2 + \frac{\eta^2}{\left(\frac{\partial\eta}{\partial T}\right)_p}\frac{\rho_1}{\rho}\right]\bar v_1 + \left[\left(\frac{\eta}{\left(\frac{\partial\eta}{\partial T}\right)_p}\frac{\rho_2}{\rho^2}\left(\frac{\partial\rho}{\partial T}\right)_p + \frac{\rho_2}{\rho}\right)u^2 - \frac{\eta^2}{\left(\frac{\partial\eta}{\partial T}\right)_p}\frac{\rho_1}{\rho}\right]\bar v_2 = 0\,,\\
&\left[\left(\rho_1\left(\frac{\partial\rho}{\partial p}\right)_T - \frac{1}{\eta}\frac{\rho_2}{\rho}\left(\frac{\partial\rho}{\partial T}\right)_p\right)u^2 - \rho_1\right]\bar v_1 + \left[\left(\rho_2\left(\frac{\partial\rho}{\partial p}\right)_T + \frac{1}{\eta}\frac{\rho_2}{\rho}\left(\frac{\partial\rho}{\partial T}\right)_p\right)u^2 - \rho_2\right]\bar v_2 = 0\,.
\end{aligned}
\tag{5.21}
$$

As in the previous section we shall now restrict the attention to the case where the first and second sound are uncoupled; the condition for this is the vanishing of the thermal expansion $(\partial\rho/\partial T)_p = 0$ which is well satisfied in helium II. In that case (5.21) reduces to

$$
\begin{aligned}
&\left(\frac{\eta^2}{\left(\frac{\partial\eta}{\partial T}\right)_p}\rho_1 - \rho_2 u^2\right)(\bar v_1 - \bar v_2) = 0\,, \quad \text{and}\\
&\left(\left(\frac{\partial\rho}{\partial p}\right)_T u^2 - 1\right)(\rho_1\bar v_1 + \rho_2\bar v_2) = 0\,.
\end{aligned}
\tag{5.22}
$$

These equations have non-trivial solutions only if

$$u^2 = V_1^2 = \left(\frac{\partial p}{\partial \rho}\right)_T \qquad \text{or} \qquad u^2 = V_2^2 = \frac{\rho_1}{\rho_2} \frac{\eta^2}{\left(\frac{\partial \eta}{\partial T}\right)_p} \quad , \tag{5.23}$$

where V_1 and V_2 are the velocities of first and second sound. From $(5.22)_1$ it follows that a wave propagating with the velocity V_1 has $\bar{v}_1 = \bar{v}_2$ while a wave propagating with V_2 has $\rho_1\bar{v}_1 + \rho_2\bar{v}_2 = 0$. Thus we see that in a first sound wave the two constituents move with the same speed, i.e. the mixture moves as a whole and there is no relative motion of the constituent. On the other hand in a second sound wave the mixture is at rest and the constituents are in relative motion.

Also from $(5.20)_1$ with $(\partial\rho/\partial T)_p = 0$ it follows that $\bar{p} = 0$ in a second sound wave, and from $(5.20)_3$ we conclude that $\bar{T}$ is zero in a first sound wave.

These considerations have given rise to the expression that the first sound is a pressure wave, while the second sound is a temperature wave. Of course, if there is some thermal expansion, both sounds are connected both with pressure waves and temperature waves, even though, for small values of $(\partial\rho/\partial T)_p$ one will have a much smaller amplitude than the other.

Both sounds are not absorbed which obviously reflects the absence of dissipation from Landau's theory.

5.2. Helium II in Rotation

5.2.1. Two Observations in Rotating Helium II

We recall the consequence (5.7) of the momentum balance of the superfluid, according to which this constituent cannot participate in a rigid rotation. This prediction of the Landau theory was proved quite wrong by an experiment by Osborne. Osborne put a cylindrical vessel with helium II on a turntable and observed that the surface assumed a parabolical shape of exactly the form that would result from the rigid rotation of both constituents. The conclusion was inevitable: The superfluid rotated rigidly just like the normal fluid.

Another observation concerns the propagation of second sound. Experiments show that in rotating helium the second sound is attenuated and that the coefficient of absorption increases linearly with the angular velocity Ω_i of the rotation. This contradicts Landau's theory according to which no absorption should occur for either sound due to the lack of dissipation.

The question arises of how to modify the Landau theory in order to incorporate the above observations.

5.2.2. Vortex Lines and Rigid Rotation

Feynman [17] following a suggestion by Onsager has explained the rigid rotation of the superfluid and yet maintained Landau's statement that the velocity of the superfluid is curl-free. The apparent contradiction in this sentence is resolved by distinguishing between the microscopic "real" velocity of the superfluid and its macroscopic "smoothed-out" velocity. The argument runs as follows.

The real microscopic velocity field $\underset{\sim}{c}_s$ is a complicated superposition of the velocity fields of potential vortices. Because, according to Feynman the superfluid in a rotating cylinder forms potential vortices of the strength $\frac{h}{m}$, where h is Planck's constant and m is the mass of a helium atom. For an angular velocity of about 1/min there are only a few such vortices but for angular velocities of about 1/sec there are already thousands of vortices per cm^2. We may then define an axial vector field w_i pointing in the direction of the vortices and with a magnitude equal to the number of vortices per unit area.

The vortices have a very small nucleus with a radius of atomic dimensions and this nucleus is supposed to be free of the superfluid. Thus the velocity field c_s of the superfluid is a superposition of velocity fields of potential vortices and we have

$$\text{curl } \underset{\sim}{c}_s = 0 \,. \tag{5.24}$$

Since each vortex has a strength $\frac{h}{m}$ and since there are $\iint_F w_i \, da_i$ vortices intersecting a surface F, the circulation of the boundary L of F is given by

$$\oint_L c_i^s \tau_i \, d\ell = \frac{h}{m} \iint_F w_i e_i \, ds \,. \tag{5.25}$$

The velocity field $\underset{\sim}{c}_s$ - being the superposition of individual vortex fields - varies considerably on the microscopic scale within a single volume element and, what is more, it is defined on a multiply connected region, because the nuclei of the vortices are free of superfluid. Therefore we cannot apply the Stokes theorem to the left hand side of (5.25).

However, we may define a smoothed-out velocity field $\underset{\sim}{v}_s$ of the superfluid by requiring

$$\oint_L v_i^s \tau_i \, d\ell = \oint_L c_i^s \tau_i \, d\ell \tag{5.26}$$

to hold for all curves whose length is large compared to the spacing of the vortices. $\underset{\sim}{v}^s$ is supposed to be smooth and defined in all points of the fluid mixture; it is the macroscopically observed velocity field of the superfluid. Elimination of $\oint c_i^s \tau_i \, d\ell$ between (5.25) and (5.26) gives

$$\oint_L v_i^s \tau_i \, d\ell = \frac{h}{m} \iint_F w_i e_i \, ds$$

and now we may use Stokes' theorem to obtain

$$\iint_F (\mathrm{curl}_i \underset{\sim}{v}^s - \frac{h}{m} w_i) \, e_i \, ds = 0 \, . \tag{5.27}$$

This relation must hold for all surfaces large compared to the vortex spacing and therefore implies the local equation

$$\mathrm{curl}_i \underset{\sim}{v}^s = \frac{h}{m} w_i \, . \tag{5.28}$$

We conclude that, while on the microscopic level the superfluid velocity $\underset{\sim}{c}_s$ is curl-free,the macroscopic velocity $\underset{\sim}{v}_s$ possesses rotation.

In particular, if we assume w_i to be constant, (5.28) has the solution

$$\underset{\sim}{v}^s = \frac{1}{2} \frac{h}{m} \underset{\sim}{w} \times \underset{\sim}{x} \tag{5.29}$$

which is the field of a rigid rotation. In this manner we can retain the Landau theory with a curl-free $\underset{\sim}{c}_s$ and yet understand Osborne's observation of rigidly rotating helium II.

$\underset{\sim}{w}$ by its definition is an axial vector so that in a non-inertal frame we have

$$w_i^* = (\mathrm{sign}\ 0^*) \ 0_{ij}^* w_j \, . \tag{5.30}$$

The transformation law of $\mathrm{curl}_i \underset{\sim}{v}^s$ reads

$$\operatorname{curl}_i \underset{\sim}{v}^{*s} = (\operatorname{sign} O^*) O^*_{ij} \operatorname{curl}_j \underset{\sim}{v}^{s} - \varepsilon_{ijk} W^*_{ij} , \tag{5.31}$$

where $\underset{\sim}{W}^* = \dot{\underset{\sim}{O}}^* \underset{\sim}{O}^{*T}$ is the matrix of angular velocity of the frame.

Thus we conclude that in a non-inertial frame the equation (5.28) assumes the form

$$\operatorname{curl}_i v^{*s} + \varepsilon_{ijk} W^*_{jk} = \frac{h}{m} w^*_i . \tag{5.32}$$

When the vortex lines were first conceived they were regarded with scepticism. Since then, however, these lines have been made visible and their existence can now be taken for granted. Even before the actual observation of vortex lines considerable support for the idea came from the fact that the explanation of the attenuation of the second sound requires vortex lines. Before we can discuss this point properly we need some preparation.

5.2.3. Balance of Vortices

Once the vortex lines are conceived it becomes necessary to ascribe a motion to them, because they may move differently from either the superfluid or the normal fluid. The only interesting component of the velocity of the lines is perpendicular to $\underset{\sim}{w}$ and we denote that perpendicular velocity by $\underset{\sim}{v}_L$, We exclude the possibility that vortex lines are produced or annihilated inside the helium II and thus obtain

$$\frac{d}{dt} \int_w w_i e_i \, ds = 0 \tag{5.33}$$

for all surfaces ω - and boundaries φ - that move $\underset{\sim}{v}_L$. Equation (5.33) expresses the expectation that the number of vortices intersecting ω does not change when the boundary φ moves with $\underset{\sim}{v}_L$.

The local form of (5.33) can be derived by use of Reynolds' transport theorem for surfaces. By (5.28) the divergence of $\underset{\sim}{w}$ vanishes and we obtain

$$\int_\omega \frac{\partial w_i}{\partial t} e_i \, ds + \int_\varphi (\underset{\sim}{w} \times \underset{\sim}{v}_L)_i \tau_i \, d\ell = 0$$

or by use of Stokes' theorem and in local form

$$\frac{\partial w_i}{\partial t} + \operatorname{curl}_i (\underset{\sim}{w} \times \underset{\sim}{v}_L) = 0 . \tag{5.34}$$

Since by (5.28) w_i is equal to $\frac{m}{h}\,\mathrm{curl}_i \underset{\sim}{v}_s$, the equation (5.34) can be integrated to give

$$\frac{\partial v_i^s}{\partial t} + \frac{h}{m}\,(\underset{\sim}{w} \times \underset{\sim}{v}_L)_i = \frac{\partial \bar{\chi}}{\partial x_i} \quad , \tag{5.35}$$

where $\bar{\chi}$ is an arbitrary function of integration depending on $\underset{\sim}{x}$ and t. We shall make use of this relation later.

In a non-inertial frame the equation (5.34) is still valid but (5.35) reads

$$\frac{\partial v_i^{*s}}{\partial t} + \frac{h}{m}\,(w^* \times v_L^*)_i = \dot{W}_{ik}^* x_k^* + \frac{\partial \bar{\chi}^*}{\partial x_i^*} \quad , \tag{5.36}$$

because w_i^* is now given by (5.32). It is thus obvious that $\bar{\chi}$ is not an objective scalar. Later it will be useful to deal with scalars and therefore we replace the arbitrary function $\bar{\chi}^*$ by another arbitrary function χ defined as

$$\chi \equiv \chi^* + 2\, W_{ij}^* \dot{b}_j^* x_i^* + \frac{1}{2} W_{ij}^{*2}\, x_i^* x_j^* - W_{ij}^{*2}\, x_i^* b_j^* + \dot{W}_{ij}^*\, x_i^* b_j^* - \ddot{b}_j^* x_i^* + \frac{1}{2} v_s^{*2} \quad , \tag{5.37}$$

which is an objective scalar. The easiest way to see this is by elimination of $\bar{\chi}$ between (5.36) and (5.37) which results in the equation

$$\Big[\frac{\partial v_i^{*s}}{\partial t} + v_j^{*s} \frac{\partial v_i^{*s}}{\partial x_j^*} - 2W_{ij}^* (v_j^{*s} - \dot{b}_j^*) + W_{ij}^{*2} (x_j^* - b_j^*) - \dot{W}_{ij}^* (x_j^* - b_j^*) - \ddot{b}_j^* \Big] + $$
$$+ \frac{h}{m}\,(w^* \times (v_L^* - v_s^*))_i = \frac{\partial \chi}{\partial x_i} \quad . \tag{5.38}$$

The square bracket is an objective vector and this is also true for the difference $\underset{\sim}{v}_L - \underset{\sim}{v}_s$. Thus indeed χ is an objective scalar.

5.2.4. Contribution of Vortices to the Interaction Force

We have seen in $(5.4.)_4$ that in the Landau theory the interaction force is absent. This is no longer the case when there are vortices. Indeed, Hall has assumed that the normal component is subject to a force $m_i^n - \tau_n v_i^n$ which is due to friction with the vortices. This force is supposed to be

i.) perpendicular to the vortex lines,

ii.) proportional to the vortex density $|\underset{\sim}{w}|$, iii.) opposite to the relative velocity $\underset{\sim}{v}_n - \underset{\sim}{v}_L$,

iv.) linear in $\underset{\sim}{v}_n - \underset{\sim}{v}_L$.

Thus we may write $m_i^n - \tau_n v_i^n$ in the form

$$m_i^n - \tau_n v_i^n = M_w \left(\underset{\sim}{w} \times \frac{\underset{\sim}{w}}{|\underset{\sim}{w}|} \times (\underset{\sim}{v}_n - \underset{\sim}{v}_L)\right)_i - \tau_s v_i^s \; . \tag{5.39}$$

To understand the last term we need to realize, that even without vortices the force $m_i^n - \tau_n v_i^n$ does not vanish; rather it is equal to $-\tau_s v_i^s$. This is due to (1.38) and the fact that $m_i^s - \tau_s v_i^s$ vanishes in that case. The interaction force $m_i^s - \tau_s v_i^s$ on the superfluid results from $(1.38)_3$ and we have

$$m_i^s - \tau_s v_i^s = - M_w \left(\underset{\sim}{w} \times \frac{\underset{\sim}{w}}{|\underset{\sim}{w}|} \times (\underset{\sim}{v}_n - \underset{\sim}{v}_L)\right)_i \; . \tag{5.40}$$

This interaction force forms the right hand side of the momentum balance of the superfluid, viz. $(5.5.)_3$ and in a non-inertial frame this equation of balance reads

$$\rho_s \left[\left(\frac{\partial v_i^{*s}}{\partial t} + v_j^{*s} \frac{\partial v_i^{*s}}{\partial x_j^*}\right) - 2 W_{ij}^* (v_j^{*s} - b_j^*) + W_{ij}^{*2} (x_j^* - b_j^*) - \dot{W}_{ij}^* (x_j^* - b_j^*) - \bar{b}_i \right] + \frac{\partial p^s}{\partial x_i^*} =$$

$$= - M_w \left(w^* \times \frac{\underset{\sim}{w}^*}{|\underset{\sim}{w}^*|} \times (\underset{\sim}{v}^{*n} - \underset{\sim}{v}^{*L})\right)_i \; . \tag{5.41}$$

In order to get a field equation from this momentum balance, it is necessary to formulate constitutive equations for $\underset{\sim}{v}^L$.

The constitutive equation for $\underset{\sim}{v}^L$ is restricted in generality by the fact that $\underset{\sim}{v}^L$ must also satisfy the equation (5.38) which, by (5.41), may be written in the alternative form

$$\left(\frac{M_w}{\rho_s} \varepsilon_{ijk} w_j^* + \frac{h}{m} |w| \delta_{ij} \right) \left(\frac{\underset{\sim}{w}^*}{|\underset{\sim}{w}|} \times (v_n^* - v_L^*)\right)_k = - \frac{h}{m} (w^* \times v^s)_i - \frac{\partial \chi}{\partial x_i^*} - \frac{1}{\rho_s} \frac{\partial p^s}{\partial x_i^*} \; ,$$

or, by (5.6), in the form

$$\left(\frac{M_w}{\rho_s} \varepsilon_{ijk} \frac{w_j^*}{|\underset{\sim}{w}^*|} + \frac{h}{m} \delta_{ij} \right) \left(\underset{\sim}{w}^* \times (\underset{\sim}{v}_n - \underset{\sim}{v}_L)\right)_k =$$

$$= - \frac{h}{m} (\underset{\sim}{w}^* \times \underset{\sim}{v}^s)_i + \frac{\partial}{\partial x_i^*} \left(\chi + (\varepsilon_I - T\eta + \frac{p}{\rho} - \frac{\rho_n}{2\rho} v_s^{*2})\right) . \tag{5.42}$$

Solving for $\underset{\sim}{w}^* \times (\underset{\sim n}{v}^* - \underset{\sim L}{v}^*)$ we obtain

$$(\underset{\sim}{w}^* \times (\underset{\sim n}{v}^* - \underset{\sim L}{v}^*))_i = - \frac{1}{1 + (\frac{m}{h}\frac{M_w}{\rho_s})^2} \left[\delta_{ij} + \frac{m}{h}\frac{M_w}{\rho_s}\varepsilon_{ijk}\frac{w_k^*}{|\underset{\sim}{w}^*|} + (\frac{m}{h}\frac{M_w}{\rho_s})^2 \frac{w_i^* w_j^*}{w^{*2}}\right] \cdot$$

$$\cdot \left((\underset{\sim}{w}^* \times \underset{\sim}{v}^{*s})_j + \frac{m}{h}\frac{\partial}{\partial x_j^*}(\chi + \varepsilon_I - T\eta + \frac{p}{\rho} - \frac{\rho_n}{2\rho} v_s^{*2})\right) . \tag{5.43}$$

Inspection of (5.43) shows that

$$\frac{\partial}{\partial x_j^*}(\chi + (\varepsilon_I - T\eta + \frac{p}{\rho} - \frac{\rho_n}{2\rho} v_s^{*2}))$$

is perpendicular to w_i^* for w_i^*. Therefore we must have

$$\chi + (\varepsilon_I - T\eta + \frac{p}{\rho} - \frac{\rho_n}{2\rho} v_s^{*2}) = 0 \tag{5.44}$$

to within a constant which we ignore. We define $\delta \equiv \frac{m}{h}\frac{M_w}{\rho_s}$ and conclude from (5.43) that

$$\left(\frac{\underset{\sim}{w}^*}{|w^*|} \times (\underset{\sim n}{v}^* - \underset{\sim L}{v}^*)\right)_i = - \frac{\delta}{1 + \delta^2}\left(\delta_{in} + \frac{1}{\delta}\varepsilon_{i\ell n}\frac{w_\ell^*}{|w^*|} - \frac{w_i^* w_n^*}{w^{*2}}\right) v_n^{*s} \tag{5.45}$$

holds, so that v_L^* - or its only relevant component $\underset{\sim}{w}^* \times \underset{\sim L}{v}^*$ - is now determined by the fields $\underset{\sim}{w}^*$, and $\underset{\sim}{v}^{*s}$. Insertion of (5.45) into (5.40) gives

$$m_i^s - \tau_s v_i^s = \frac{M_w |w^*|}{1 + \delta^2}\left(- \delta_{in} + \delta\varepsilon_{i\ell n}\frac{w_\ell^*}{|w^*|} + \frac{w_i^* w_n^*}{|w^*|^2}\right) v_n^{*s} \tag{5.46}$$

and this must be put on the right hand side of the momentum balance of the superfluid to represent the interaction of the helium with its vortex lines.

5.2.5. Second Sound in Rotating Helium

The basic equations for the consideration of small amplitude waves are still given by (5.12), but these equations must be supplemented by a few terms.

i.) The balance of total momentum and the balance of momentum of the superfluid contain inertial forces in a non-inertial frame,

ii.) the balance of momentum of the superfluid contains the interaction force (5.46).

We shall be interested in a time independent slow rotation without translation. The centrifugal force is neglected, because it is of second order in the angular velocity. Therefore the intertial forces are represented by the Coriolis force only and we obtain by use of (5.32)

$$\frac{\partial \rho}{\partial t} + \tilde{\rho}\,\frac{\partial v_i^*}{\partial x_i^*} = 0 ,$$

$$\tilde{\rho}\,\frac{\partial v_i^*}{\partial t} + \frac{\partial p}{\partial x_i^*} = \tilde{\rho}\; 2W_{ij}^* v_j^* , \tag{5.47}$$

$$\frac{\tilde{\rho}_n\tilde{\rho}_s}{\tilde{\rho}}\,\frac{\partial v_i^{*s}}{\partial t} - \tilde{\rho}_s\tilde{\eta}\,\frac{\partial T}{\partial x_i^*} = \frac{\tilde{\rho}_n\tilde{\rho}_s}{\tilde{\rho}}\, 2W_{ij}^* v_j^{*s} - 2\,\frac{m}{h}\,\frac{M_W W}{1+\delta^2}\left(\delta_{ij} - \partial\varepsilon_{i\ell j}\,\frac{W_\ell}{W} - \frac{W_i W_j}{W^2}\right) v_j^{*s} ,$$

$$\frac{\partial \eta}{\partial t} - \frac{\tilde{\rho}_s}{\tilde{\rho}}\,\tilde{\eta}\,\frac{\partial v_i^{*s}}{\partial x_i^*} = 0 .$$

where $W_i \equiv \frac{1}{2}\,\varepsilon_{ijk} W_{ik}^*$ is the (axial) vector of angular velocity and W is its magnitude.

For simplicity we investigate only the second sound, for which $v_i^* = 0$, $p = \text{const}$ holds. The mass balance $(5.47)_1$ is satisfied, since thermal expansion is negligable and there remain only the two equations $(5.47)_{3,4}$. We let the rotation of the frame be about the 3-axis with the angular velocity W and consider waves in the 1-direction of the form

$$v_i^{*s} = \bar{v}_i^s\, e^{i(\omega t - kx)} , \qquad T = \tilde{T} + \bar{T}\, e^{i(\omega t - kx)} . \tag{5.48}$$

Insertion of (5.48) into $(5.47)_{3,4}$ leads to a homogeneous system of linear equations. We obtain

$$\begin{bmatrix} \left(i\omega\rho_s + \frac{\rho}{\rho_n}\, 2\,\frac{m}{h}\,\frac{M_W W}{1+\delta^2}\right) & -2\left(\rho_s - \frac{\rho}{\rho_n}\,\frac{m}{h}\,\frac{M_W}{1+\delta^2}\right) W & ik\rho_s\eta \\ 2\left(\rho_s - \frac{\rho}{\rho_n}\,\frac{m}{h}\,\frac{M_W}{1+\delta^2}\right) W & \left(i\omega\rho_s + \frac{\rho}{\rho_n}\, 2\,\frac{m}{h}\,\frac{M_W W}{1+\delta^2}\right) & 0 \\ ik\,\frac{\rho_s}{\rho_n}\,\eta & 0 & i\omega\,\frac{\partial\eta}{\partial T} \end{bmatrix} \cdot \begin{bmatrix} \bar{v}_1^s \\ \bar{v}_2^s \\ \bar{T} \end{bmatrix} = 0 \tag{5.49}$$

The third component of $(5.47)_3$ has been dropped, because its only consequence is

$$\bar{v}^s_3 = 0 .$$

Non trivial solutions exist, if the determinant vanishes and this requirement leads to the dispersion relation

$$\frac{k}{\omega} = \sqrt{\frac{\rho_n}{\rho_s}\frac{\partial\eta/\partial T}{\eta^2}}\left(1 - i\frac{\rho}{\rho_s\rho_n}\frac{m}{h}\frac{M_w W}{1+\delta^2}\frac{1}{\omega}\right) , \tag{5.50}$$

where again terms of order W^2 have been neglected. Thus there is an attenuation

$$k^i = - \sqrt{\frac{\rho_n}{\rho_s}\frac{\partial\eta/\partial T}{\eta^2}}\frac{\rho}{\rho_s\rho_n}\frac{m}{h}\frac{M_w W}{1+\delta^2} . \tag{5.51}$$

The attenuation comes out as observed, namely proportional to the angular velocity of the rotation and independent of the frequency ω .

The phase speed is not affected by the rotation, it come out as before in $(5.23)_2$ as

$$V_2 = \frac{1}{\mathrm{Re}\left(\frac{k}{\omega}\right)} = \sqrt{\frac{\rho_n}{\rho_s}\frac{\eta^2}{\partial\eta/\partial T}} . \tag{5.52}$$

In reality the phase speed does vary somewhat with the rotation, but that effect is not described by this theory.

6. OUTLOOK INTO THE FUTURE OF THERMODYNAMICS

6.1. Introduction

We have seen in Section 5 that disturbances in concentration of a mixture propagate at a finite speed. Thus the systematic thermodynamic theory presented heretofore has resolved what has sometimes been called the paradox of diffusion by which the speed of diffusion is infinite. We have identified Fick's law as the cause for the paradox and everything has come out quite satisfactory as far as diffusion is concerned.

However, heat conduction knows a similar paradox, due to the form of Fourier's law and that law has been unchanged in the foregoing analysis. Indeed, it is not easy to use macroscopic arguments to suggest a reasonable generalized Fourier law that allows disturbances in temperature to propagate at a finite speed.

But the kinetic theory of gases exhibits such a generalization. This microscopic theory points out the direction for the further development of thermodynamics. However, it also raises problems that must be faced:

i.) the kinetic theory contradicts the principle of material frame indifference for stress and heat flux, and

ii.) it contradicts the Clausius-Duhem inequality.

I shall proceed to explain these points in more detail.

6.2. Problems in Irreversible Thermodynamics

6.2.1. Choice of Constitutive Class

Let us recall the basic objective of classical thermodynamics of fluids viz. the determination of the fields

$$\begin{aligned} &\rho(\underset{\sim}{x},t) \text{ - density} \\ &v_i(\underset{\sim}{x},t) \text{ - velocity} \\ &T(\underset{\sim}{x},t) \text{ - absolute temperature} \end{aligned} \tag{6.1}$$

from the equations of balance of

$$\begin{aligned} &\text{mass} && \dot{\rho} + \rho \frac{\partial v_i}{\partial x_i} = 0 , \\ &\text{momentum} && \rho\dot{v}_i - \frac{\partial t_{ij}}{\partial x_j} = \rho(f_i + i_i) , \\ &\text{(internal) energy} && \rho\dot{\varepsilon} + \frac{\partial q_i}{\partial x_i} - t_{ij}\frac{\partial v_i}{\partial x_j} = 0 . \end{aligned} \tag{6.2}$$

Here, for simplicity, we consider only single fluids and we ignore the possibility of radiational heating. However, body forces f_i are taken into account and so are the inertial forces i_i composed of Coriolis force, centrifugal force, Euler force and force of relative translation according to

$$i_i = 2\, W_{ik}(v_k - \dot{b}_k) - W^2_{ik}(x_k - b_k) + \dot{W}_{ik}(x_k - b_k) - \ddot{b}_i \,. \tag{6.3}$$

In order to obtain field equations from (6.2) for the thermodynamic fields (6.1) we need constitutive equations which relate the stress t_{ij}, the heat flux q_i, and the internal energy ε to those fields. In a fluid the constitutive equations will have the forms

$$\begin{aligned} t_{ij} &= \hat{t}_{ij}\left(\rho, v_i, T;\ \frac{\partial \rho}{\partial t}, \frac{\partial v_i}{\partial t}, \frac{\partial T}{\partial t};\ \frac{\partial \rho}{\partial x_j}, \frac{\partial v_i}{\partial x_j}, \frac{\partial T}{\partial x_j};\ \frac{\partial^2 \rho}{\partial t^2}, \dots;\ \frac{\partial^2 \rho}{\partial x_i \partial t}, \dots;\ \frac{\partial^2 \rho}{\partial x_i \partial x_j}, \dots\right), \\ q_i &= \hat{q}_i\,(\dots\dots\dots\dots\dots\dots\dots\dots\dots\dots)\,, \\ \varepsilon &= \hat{\varepsilon}\,(\dots\dots\dots\dots\dots\dots\dots\dots\dots\dots)\,. \end{aligned} \tag{6.4}$$

Depending on how many derivatives of ρ, v_i, T occur in these equations we have fluids with more or less memory and more or less non-local action. In particular, an Eulerian fluid has only ρ, v_i and T as variables, while a Navier-Stokes-Fourier fluid has ρ, v_i, T, $\frac{\partial v_i}{\partial x_j}$ and $\frac{\partial T}{\partial x_j}$.

Constitutive equations of the type (6.4) with any finite set of derivations as variables are considered general enough by most people for the description of fluid properties. Indeed, for practical purposes the generality indicated by (6.4) is more than sufficient.

6.2.2. First Problem: The Speed of Heat Conduction

However, constitutive equations of the type (6.4) have invariably led to the so-called paradox of heat conduction. Consider for instance a one-atomic ideal gas as a Navier-Stokes-Fourier fluid with the constitutive equations

$$\begin{aligned} t_{ij} &= -\rho R T \delta_{ij} + 2\mu(\rho,T)\left(\frac{\partial v_{(i}}{\partial x_{j)}} - \frac{1}{3}\frac{\partial v_\ell}{\partial x_\ell}\delta_{ij}\right) \\ q_i &= -k(\rho,T)\,\frac{\partial T}{\partial x_i} \\ \varepsilon &= \frac{3}{2}\hat{R}T + \alpha\,. \end{aligned} \tag{6.5}$$

$(6.5)_1$ is the Navier-Stokes stress and $(6.5)_2$ is the Fourier law for the heat flux.

$\hat{R}$ stands for R/M.

We investigate the propagation of longitudinal small harmonic waves in the x-direction:

$$\rho = \tilde{\rho} + \bar{\rho}\, e^{i(\omega t - kx)} , \qquad v = \bar{v}\, e^{i(\omega t - kx)} , \qquad T = \tilde{T} + \bar{T}\, e^{i(\omega t - kx)} .$$

In that case the field equations can be linearized by the deletion of products of barred quantities and we obtain

$$\begin{bmatrix} \omega & -k\tilde{\rho} & 0 \\ -k\hat{R}\tilde{T} & \omega\tilde{\rho} - ik^2\dfrac{4\tilde{\mu}}{3} & -k\tilde{\rho}\hat{R} \\ 0 & -ik\tilde{\rho}\hat{R}\tilde{T} & i\omega\dfrac{3}{2}\hat{R}\tilde{\rho} + k^2\tilde{\kappa} \end{bmatrix} \begin{bmatrix} \bar{\rho} \\ \bar{v} \\ \bar{T} \end{bmatrix} = 0 . \tag{6.6}$$

Non-trivial solutions exist only, if the determinant of the matrix in (6.6) vanishes and this leads to the dispersion relation

$$\left(i\frac{3}{5}\frac{\tilde{\kappa}}{\tilde{\rho}\frac{3}{2}\hat{R}} - \omega\frac{\frac{4}{3}\frac{\tilde{\mu}}{\tilde{\rho}}\frac{\tilde{\kappa}}{\tilde{\rho}\frac{3}{2}\hat{R}}}{v^2(0)}\right)\left(\frac{k}{\frac{\omega}{v(0)}}\right)^4 - \left(\frac{v^2(0)}{\omega} + i\left(\frac{\tilde{\kappa}}{\tilde{\rho}\frac{3}{2}\hat{R}} + \frac{4}{3}\frac{\tilde{\mu}}{\tilde{\rho}}\right)\right)\left(\frac{k}{\frac{\omega}{v(0)}}\right)^2 + \frac{v^2(0)}{\omega} = 0 , \tag{6.7}$$

where $v(0)$ is defined as $\sqrt{\frac{5}{3}\hat{R}T}$. We solve this algebraic equation for the phase speed $V = \omega/k$ as a function of the frequency ω. The two limiting values for small and large frequencies are

$$V_{\pm} \xrightarrow[\omega \to 0]{} \begin{cases} \sqrt{\frac{5}{3}\hat{R}T} \\ 0 \end{cases} \qquad \text{and} \qquad V_{\pm} \xrightarrow[\omega \to \infty]{} \begin{cases} \sqrt{\hat{R}T} \\ \infty \end{cases} . \tag{6.8}$$

Equation $(6.8)_1$ gives the well known adiabatic speed of sound, while $(6.8)_2$ implies that a disturbance of ρ, $\underset{\sim}{v}$ or T is propagated at an infinite speed in the gas. This prediction is considered to be unrealistic and has been termed the paradox of heat conduction.

Cattaneo [18] was the first to modify the Fourier law $(6.5)_2$ so as to obtain a finite speed of heat conduction. Cattaneo used arguments from the kinetic theory of gases. Later Müller [19],[20] has systematized these arguments and incorporated them into an extended linear theory of irreversible thermodynamics. Since then the problem has been taken up by many authors using different ideas. E.g. Grioli [21] has formulated a Fourier law of integral type with fading memory which indeed can also be motivated in part by the kinetic theory of gases.

In Section 6.4 I shall give a brief account of the argument used by the kinetic theory of gases for the solution of the paradox.

6.2.3. Second Problem: Material Frame Indifference

As a restriction upon the generality of the constitutive functions $\hat{t}_{ij}$, $\hat{q}_i$, $\hat{\varepsilon}$ in (6.4) we often use the principle of material frame indifference which states that the constitutive functions are unchanged by a change of frame of the form

$$x_i^* = O_{ij}\, x_j + b_i \quad . \tag{6.9}$$

We shall see in Section 6.5 that the kinetic theory of gases does not support this principle. It is true that the kinetic theory allows us to derive the Navier-Stokes-Fourier equations which are frame-indifferent. But the theory also recognizes the Navier-Stokes-Fourier equations as a first approximation only and all higher approximations depend on frame.

This fact was pointed out by Müller [22] and was independently confirmed by Edelen & McLennan [23] and later by Söderholm [24]. It turned out that the heat flux for instance is related to the temperature gradient by the equation (see (6.35) below)

$$q_i = - \kappa \frac{\partial T}{\partial x_i} + \frac{2\kappa}{\rho B} \left(\frac{\partial v_{[i}}{\partial x_{j]}} - W_{ij} \right) \frac{\partial T}{\partial x_j} \tag{6.10}$$

in a stationary process in a rigidly rotating fluid, if the frame rotates with the angular velocity W_{ij} with respect to an inertial frame.

This observation has met much opposition and criticism. Truesdell [25] has declared "ex catedra" that it is wrong. Wang [26] has blamed it on the approximation within the kinetic theory overlooking that the relevant terms come from the <u>exact</u> equation of transfer. Finally Speziale [27] has not grasped the meaning of material frame indifference, because he argues that the constitutive function for the heat flux q_i cannot be frame dependent, since q_i is an objective vector.

It will turn out that the terms by which the kinetic theory violates material frame indifference are the same ones that allow for a finite speed of heat conduction.

6.2.4. Third Problem: Entropy Inequality

The entropy inequality in its most commonly applied form is the Clausius-Duhem inequality

$$\rho\dot{\eta} + \frac{\partial}{\partial x_i}\left(\frac{q_i}{T}\right) \geq 0 \,. \tag{6.11}$$

The specific entropy η is given by a constitutive equation of the same character as the one for the specific internal energy, see (6.4). But the entropy flux in (6.11) is not given by a constitutive equation in its own right; rather it is set equal to the heat flux divided by the absolute temperature.

Here again the kinetic theory disagrees. In that theory there is no a priori relation between entropy flux Φ_i and heat flux. It is only close to equilibrium that Φ_i equals q_i/T, and in fact that equality is only as good as the Navier-Stokes equations and Fourier's law. In Section 6.6 we shall calculate Φ_i in a better approximation, (see (6.43) below.

6.3. Kinetic Theory of Gases

6.3.1. Distribution Function and Boltzmann Equation

The objective of the kinetic theory of one-atomic ideal gases is the determination of the distribution function $f(x_i, c_i, t)$ which represents the number density of atoms at the point x_i with velocity c_i.

The distribution function has to be calculated from the Boltzmann equation

$$\frac{\partial f}{\partial t} + c_i \frac{\partial f}{\partial x_i} + (f_i + i_{ci}) \frac{\partial f}{\partial c_i} = \iint (f'f_1' - ff_1)\, \sigma g \sin\Theta d\Theta d\varepsilon dc^1 \tag{6.12}$$

where $i_{ci} = 2Wi_k(c_k - \dot{b}_k) - W^2_{ik}(x_k - b_k) + \dot{W}_{ik}(x_k - b_k) - \bar{b}_i$ is the specific inertial force on an atom. The expression on the right hand side is the collision integral whose composition I need not explain here. For the derivation of the Boltzmann equation I refer the reader to any book on the kinetic theory of gases (e.g. see [28]).

6.3.2. Moments of the Distribution Function

The thermodynamic fields ρ and v_i and the constitutive quantities ε, t_{ij} and q_i of thermodynamics can all be expressed in terms of the distribution function. Indeed, it is rather obvious that the following relations hold

$$\begin{aligned}
&\text{density} && \rho = \int mf d\underset{\sim}{c} \,, \\
&\text{velocity} && \rho v_i = \int mc_i f d\underset{\sim}{c} \,, \quad \text{and with } C_i = c_i - v_i \\
&\text{internal energy} && \rho\varepsilon = \int \frac{m}{2} C^2 f d\underset{\sim}{c} \,, \\
&\text{stress} && t_{ij} = \int mC_i C_j f d\underset{\sim}{c} \,, \\
&\text{heat flux} && q_i = \int \frac{m}{2} C^2 C_i f d\underset{\sim}{c} \,.
\end{aligned} \tag{6.13}$$

These are all of the quantities which occur in the thermodynamic equations of balance but in the kinetic theory we might of course continue and define higher moments. Such higher moments include

$$p_{ijk} = \int mC_i C_j C_k f d\underset{\sim}{c} \quad \text{and} \quad p_{ijk\ell} = \int mC_i C_j C_k C_\ell f d\underset{\sim}{c} \,. \tag{6.14}$$

For the definition of temperature the kinetic theory assumes that $\varepsilon = \frac{3}{2}\hat{R}T$ holds, as befits an ideal gas according to $(6.5)_3$. Thus we have

$$T = \frac{2}{3\hat{R}} \frac{1}{\rho} \int \frac{m}{2} C^2 f d\underset{\sim}{c} \,. \tag{6.15}$$

The pressure in the kinetic theory is defined by $p = -\frac{1}{3} t_{ii}$. Therefore it follows by comparison of $(6.13)_4$ and (6.15) that we have

$$p = \rho\hat{R}T \,. \tag{6.16}$$

6.3.3. General Equation of Transfer

Since the macroscopic quantities ρ, v_i, T, ε, q_i and t_{ij} are now all expressed as integrals of the distribution function f, whose evolution is governed by the Boltzmann equation, we may wonder whether perhaps the Boltzmann equation also dictates the evolution of the macroscopic quantities. We shall see that this is indeed the case. To prepare the argument we consider an arbitrary function ψ of x_i, c_i and t and define its mean value by

$$\bar{\psi} = \frac{m}{\rho} \int \psi f d\underset{\sim}{c} \ . \tag{6.17}$$

The Boltzmann equation can be used to derive an <u>equation of transfer</u> for ψ, viz.

$$\frac{\partial \rho\bar{\psi}}{\partial t} + \frac{\partial}{\partial x_j}(\rho\bar{\psi}v_j + \rho\overline{C_j\psi}) - \rho\overline{(f_i + i_i)\frac{\partial\psi}{\partial c_i}} = $$
$$= \rho\,\overline{\left(\frac{\partial\psi}{\partial t} + c_i\frac{\partial\psi}{\partial x_i}\right)} + \frac{1}{4}m\iiint (\psi + \psi^1 - \psi' - \psi^{1\prime})(f'f_1' - ff_1)\,\sigma g \sin\Theta d\Theta d\varepsilon d\underset{\sim}{c}^1 d\underset{\sim}{c}. \tag{6.18}$$

The second term on the right hand side of (6.18) is due to the collisions of the atoms.

6.3.4. Equations of Balance of Mass, Momentum and Energy

If in (6.18) ψ is chosen to be 1 or C_i or $\frac{1}{2}C^2$, the collision term in (6.18) vanishes, because mass, momentum and energy of an atom are conserved in a collision. Thus one obtains by use of (6.13)

$$\begin{aligned} \dot{\rho} + \rho\frac{\partial v_i}{\partial x_i} &= 0 \ , \\ \rho\dot{v}_i - \frac{\partial t_{ij}}{\partial x_j} &= \rho(f_i + i_i) \ , \\ \rho\dot{\varepsilon} + \frac{\partial q_i}{\partial x_i} &= t_{ij}\frac{\partial v_i}{\partial x_j} \ , \end{aligned} \tag{6.19}$$

and we conclude that the equations of transfer lead to the thermodynamic balance equations for the above choices of ψ. Thus the kinetic theory furnishes the equations of balance and one might wonder whether the theory also serves us with constitutive relations for t_{ij} and q_i .

6.3.5. Equations of Transfer for Stress, Heat flux and Higher Moments

For choices of ψ other than 1, c_i and $\frac{1}{2}C^2$ the collision term in (6.18) does not vanish and in order to evaluate it we must know the interatomic forces. If we assume a Maxwellian interaction law, by which the interaction potential is inversely proportional to the fourth power of the distance, the collision term in (6.18) can be evaluated for

$$\psi = C_iC_j, \quad \frac{1}{2}C_iC^2, \quad C_iC_jC_k, \quad C_iC_jC^2$$

and one obtains the following equation of transfer for the stress deviator

$$p_{ij} = -t_{ij} - p\,\delta_{ij} \tag{6.20}$$

the heat flux q_i, and the moments p_{ijk} and p_{ijkk} :

$$-\frac{3}{2}\rho B p_{ij} = \dot{p}_{ij} + 2p\left(\frac{\partial v_{(i}}{\partial x_{j)}} - \frac{1}{3}\frac{\partial v_\ell}{\partial x_\ell}\delta_{ij}\right) + p_{ij}\frac{\partial v_\ell}{\partial x_\ell} - \frac{2}{3}p_{k\ell}\frac{\partial v_k}{\partial x_\ell}\delta_{ij} + $$
$$+ 2p_{n(i}\frac{\partial v_{j)}}{\partial x_n} + \frac{\partial}{\partial x_n}\left(p_{ijn} - \frac{2}{3}q_n\delta_{ij}\right) - 4p_{n(i}W_{j)n} \tag{6.21$_1$}$$

$$-\rho B q_i = \dot{q}_i + q_i\frac{\partial v_n}{\partial x_n} + \frac{\partial}{\partial x_n}\left(\frac{1}{2}p_{inkk}\right) - \frac{1}{\rho}(p_{i\ell} + p\delta_{i\ell})\frac{\partial(p_{\ell n} + p\delta_{\ell n})}{\partial x_n} - \frac{3}{2}\frac{p}{\rho}\frac{\partial(p_{in} + p\delta_{in})}{\partial x_n} +$$
$$+ p_{nik}\frac{\partial v_k}{\partial x_n} + q_n\frac{\partial v_i}{\partial x_n} - 2q_nW_{in} \tag{6.21$_2$}$$

$$-\frac{3}{4}\rho B\left(3p_{ijk} - 2q_{(i}\delta_{jk)}\right) = \dot{p}_{ijk} - p_{ijk}\frac{\partial v_n}{\partial x_n} + \frac{\partial p_{ijkn}}{\partial x_n} - 3t_{(ij}\frac{1}{\rho}\frac{\partial t_{k)n}}{\partial x_n} +$$
$$+ 3p_{n(ij}\frac{\partial v_{k)}}{\partial x_n} - 6p_{n(ij}W_{k)n} \tag{6.22$_1$}$$

$$-\frac{3}{2}\rho B\left(\frac{7}{6}p_{ijkk} - \frac{1}{6}p_{kk\ell\ell}\delta_{ij} + \frac{5}{2}\frac{p}{\rho}t_{ij} + \frac{2}{3}\frac{1}{\rho}t_{im}t_{jm} - \frac{3}{2}\frac{p^2}{\rho}\delta_{ij}\right) =$$
$$= \dot{p}_{ijkk} + p_{ijkk}\frac{\partial v_\ell}{\partial x_\ell} + \frac{\partial p_{ijkkn}}{\partial x_n} + 2\left(p_{ijk}\frac{1}{\rho}\frac{\partial t_{kn}}{\partial x_n} + 2q_{(i}\frac{1}{\rho}\frac{\partial t_{j)n}}{\partial x_n}\right) +$$
$$+ \left(2p_{nijk}\frac{\partial v_k}{\partial x_n} + 2p_{n\ell\ell(i}\frac{\partial v_{j)}}{\partial x_n}\right) - 4p_{n\ell\ell(i}W_{j)n}\;. \tag{6.22$_2$}$$

B is a constant that depends on the strength of the Maxwellian interaction. Conserning this constant we need only know here that ρB is of the order of magnitude of the mean collision frequency of the atoms.

At first sight the equations $(6.21)_{1,2}$ look as if they might serve as constitutive relations for P_{ij} (or t_{ij}) and q_i. To be sure, such relations would be of the rate type, because $\dot{P}_{ij}$ and $\dot{q}_i$ occur on the right hand sides of $(6.21)_{1,2}$; but the situation is even worse; in fact, in the equation for P_{ij} the third moment p_{ijk} occurs and in the equation for q_i we have p_{inkk} and p_{ijn}. And this continues: the equation $(6.22)_1$ for p_{ijk} depends on p_{ijkn} and the equation $(6.22)_2$ for p_{ijkk} depends on the fifth moment p_{ijkkn}. Thus it happens that at no stage we obtain a closed system of equations. This is a situation which is frequently encountered in statistical mechanics and the physicists say that we have a "closure problem".

6.4. The Closure, Extended Thermodynamics, and the Propagation of Plane Waves of Small Amplitudes in Extended Thermodynamics

6.4.1. Closure by Grad's 13-Moment-Method

The easiest way to "close" the system of equations (6.19), (6.21) is to calculate the moments p_{ijk} and p_{ijkk} from the Maxwellian equilibrium distribution

$$f_M = \frac{\rho}{m} \frac{1}{\sqrt{2\pi\hat{R}T}^{\,3}} e^{-\frac{C^2}{2\hat{R}T}} \tag{6.23}$$

and insert the results into $(6.21)_{1,2}$. This would be very rough, however. Another, more attractive closure procedure makes use of Grad's non-equilibrium distribution

$$f_G = f_M\left(1 + \frac{1}{2p} P_{ij}\left[\frac{C_iC_j}{\hat{R}T} - \delta_{ij}\right] - \frac{1}{p}\left[\frac{C_i}{\sqrt{\hat{R}T}}\left(1 - \frac{1}{5}\frac{C^2}{\hat{R}T}\right)\right]\right) \tag{6.24}$$

which is determined by the 13 first moments of the distribution function, viz. ρ, v_i, T, P_{ij} and q_i. The Grad distribution represents the beginning of an expansion of f in terms of Hermite polynominals and it is to be expected that f_G is a good approximation to the actual distribution as long as the state of the gas is close to equilibrium; for details of the derivation of (6.24) and a discussion of the 13 moment method the reader is referred to Grad's articles [29] and [30].

If p_{ijk} and p_{ijkk} are calculated by insertion of (6.24) into the definitions we obtain

$$p_{ijk} = \frac{2}{5}(q_i\delta_{jk} + q_j\delta_{ki} + q_k\delta_{ij})$$
$$p_{ijkk} = 5\,\rho\hat{R}T\delta_{ij} + \frac{7}{2}P_{pq}\hat{R}T(\delta_{ip}\delta_{jq} + \delta_{iq}\delta_{jp}) \ . \tag{6.25}$$

When these expressions are introduced into $(6.21)_{1,2}$ the equations (6.19) and (6.21) become a closed system of differential equations for ρ, v_i, T, P_{ij} and q_i in an ideal gas, viz.

$$\dot{\rho} + \rho\frac{\partial v_i}{\partial x_i} = 0 \ ,$$

$$\rho\dot{v}_i - \frac{\partial t_{ij}}{\partial x_j} = \rho(f_i + i_i) \ ,$$

$$\rho\dot{\varepsilon} + \frac{\partial q_i}{\partial x_i} = t_{ij}\frac{\partial v_i}{\partial x_j} \ , \tag{6.26}$$

$$-\frac{3}{2}\rho B\,P_{ij} = 2p\left(\frac{\partial v_{(i}}{\partial x_{j)}} - \frac{1}{3}\frac{\partial v_\ell}{\partial x_\ell}\delta_{ij}\right) + \left[\delta_{ir}\delta_{js}\frac{d}{dt} + 2\delta_{s(j}\frac{\partial v_{[i)}}{\partial x_{r]}} - 4\delta_{s(i}W_{j)r}\right]P_{rs} +$$
$$+ 2P_{r(i}\frac{\partial v_{(j)}}{\partial x_{r)}} - \frac{2}{3}P_{rs}\frac{\partial v_s}{\partial x_r}\delta_{ij} + P_{ij}\frac{\partial v_\ell}{\partial x_\ell} + \frac{4}{5}\left(\frac{\partial q_{(i}}{\partial x_{j)}} - \frac{1}{3}\frac{\partial q_\ell}{\partial x_\ell}\delta_{ij}\right) \ ,$$

$$-\rho B\,q_i = \frac{5}{2}\hat{R}p\frac{\partial T}{\partial x_i} + \left[\delta_{in}\frac{d}{dt} + \frac{\partial v_{[i}}{\partial x_{n]}} - 2W_{in}\right]q_n + \frac{7}{5}q_i\frac{\partial v_\ell}{\partial x_\ell} +$$
$$+ \frac{9}{5}q_n\frac{\partial v_{(i}}{\partial x_{n)}} + \frac{7}{2}\hat{R}P_{in}\frac{\partial T}{\partial x_n} + \hat{R}T\frac{\partial P_{in}}{\partial x_n} - \frac{1}{\rho}P_{i\ell}\frac{\partial P_{\ell n}}{\partial x_n} - \frac{1}{\rho}P_{i\ell}\frac{\partial p}{\partial x_\ell} \ .$$

The equations are the basic equations of extended thermodynamics for a one-atomic ideal gas. Grad [29] derived these equations within the kinetic theory and Müller [19], [20] generalized it to arbitrary fluids by use of the arguments of linear irreversible thermodynamics.

It is true that the closed system (6.26) is not particularly simple but it is explicit. We may ask for solutions and in particular we shall now investigate wave solutions.

6.4.2. Plane Waves of Small Amplitude in an Ideal Gas

We recall that the previously investigated constitutive equations for an ideal gas as a Navier-Stokes-Fourier fluid were found deficient in Section 6.2.2., because they led to the paradox of heat conductivity. We shall now see that no such paradox occurs when we use the generalized theory that is expressed by the equations (6.26).

In an inertial frame and in the absence of body forces the equations (6.26) have a constant and time-independent solution

$$\rho = \tilde{\rho}\,, \qquad v^i = 0, \qquad T = \tilde{T}, \qquad P_{ij} = 0, \qquad q_i = 0 \;.$$

Upon this solution we superpose a plane wave

$$\rho = \tilde{\rho} + \bar{\rho}\, e^{i(\omega t - kx)}, \qquad v^i = \begin{pmatrix} \bar{v}\, e^{i(\omega t - kx)} \\ 0 \\ 0 \end{pmatrix}, \qquad T = \tilde{T} + \bar{T}\, e^{i(\omega t - kx)}$$

$$P_{ij} = \begin{pmatrix} \bar{P}\, e^{i(\omega t - kx)} & & \\ & -\frac{1}{2}\bar{P}\, e^{i(\omega t - kx)} & \\ & & -\frac{1}{2}\bar{P}\, e^{i(\omega t - kx)} \end{pmatrix}, \qquad q_i = \begin{pmatrix} \bar{q}\, e^{i(\omega t - kx)} \\ 0 \\ 0 \end{pmatrix} \tag{6.27}$$

of frequency ω which propagates in the x direction. $\bar{\rho}$, $\bar{v}$, $\bar{T}$, $\bar{P}$ and $\bar{q}$ are so small that products of these quantities can be neglected. Insertion of (6.27) into the equations (6.26) obtains

$$\begin{bmatrix} \omega & -k & 0 & 0 & 0 \\ -k\hat{R}T & \omega & -k & -k & 0 \\ 0 & -k\hat{R}T & \omega\frac{3}{2} & 0 & -k \\ 0 & -k\frac{4}{3}\hat{R}T & 0 & \omega - i\frac{3}{2}\tilde{\rho}B & -k\frac{8}{15} \\ 0 & 0 & -k\frac{5}{2}\hat{R}T & -k\hat{R}T & \omega - i\tilde{\rho}B \end{bmatrix} \begin{bmatrix} \bar{\rho} \\ \bar{v} \\ \bar{T} \\ \bar{P} \\ \bar{q} \end{bmatrix} = 0 \;. \tag{6.28}$$

This system of equations has a non-trivial solution only, if the determinant vanishes and that requirement leads to the dispersion relation

$$\left(i\frac{3}{2} - \frac{9}{5}\frac{\omega}{\tilde{\rho}B}\right)\left(\frac{k}{\omega/v(0)}\right)^4 - \left(\frac{5}{2}\frac{\tilde{\rho}B}{\omega} + 8i - \frac{26}{5}\frac{\omega}{\tilde{\rho}B}\right)\left(\frac{k}{\omega/v(0)}\right)^2 + \left(\frac{5}{2}\frac{\tilde{\rho}B}{\omega} + i\frac{25}{6} - \frac{3}{5}\frac{\omega}{\tilde{\rho}B}\right) = 0, \tag{6.29}$$

where $v(0)$ is given by $\sqrt{\frac{5}{3}\hat{R}\tilde{T}}$. When we solve for $\left(\frac{k}{\omega/v(0)}\right)^2$, we obtain after some calculation

$$\frac{\omega}{k} \xrightarrow[\omega \to 0]{} v(0) \qquad \text{and} \qquad \frac{\omega}{k} \xrightarrow[\omega \to 0]{} \frac{\sqrt{13+\sqrt{94}}}{3}\, v(0)\ . \tag{6.30}$$

We conclude that for low frequencies the phase velocity is the same in the extended theory as it was in the case when the ideal gas was considered as a Navier-Stokes-Fourier fluid. But for very high frequencies the new theory predicts a finite phase velocity. This means that - according to the generalized theory - a disturbance propagates into the gas with the finite speed

$$v = 1{,}58\ v(0)\ . \tag{6.31}$$

In other words: the kinetic theory of gases has suggested a new theory of an ideal gas, and this theory does not lead to the paradox of heat conduction.

6.5. Thermodynamic Approximation of the Generalized Theory

6.5.1. The Extended Theory and Thermodynamics

In the extended theory of an ideal gas - as presented in Section 6.4. - we have the 13 fields ρ, v_i, T, P_{ij}, and q_i rather than the five fields ρ, v_i, T of thermodynamics. Therefore the generalized theory cannot be called a thermodynamic theory in the sense of Section 6.2., but we may find its thermodynamic approximation by an iterative scheme that I proceed to explain.

We focus the attention upon the equations of transfer $(6.26)_{3,4}$ and calculate all moments on their right hand sides by use of the Maxwellian distribution function (6.23). Thus one obtains first iterates for the moments P_{ij} and q_i, viz.

$$P_{ij}^{(1)} = -2\left(\frac{2}{3B}\hat{R}T\right) E_{ij} \quad \text{with} \quad E_{ij}\ \ \frac{\partial v_{(i}}{\partial x_{j)}} - \frac{1}{3}\frac{\partial v_\ell}{\partial x_\ell}\delta_{ij}\ ,$$

$$q_i^{(1)} = -\frac{15}{4}\hat{R}\left(\frac{2}{3B}\hat{R}T\right)\frac{\partial T}{\partial x_i}\ . \tag{6.32}$$

Comparison of (6.32) with the expressions $(6.5)_{1,2}$ shows that the first iterates for stress and heat flux have the same structure as stress and heat flux in a Navier

Stokes-Fourier fluid. To emphasize this similarity we define

$$\mu = \frac{2}{3}\hat{R}T \qquad \text{and} \qquad \kappa = \alpha\mu = \frac{15}{4}\hat{R}\left(\frac{2}{3}\hat{R}T\right) \tag{6.33}$$

and call μ and κ the viscosity and the heat conductivity of an ideal gas respectively.

The iteration proceeds further as follows: The first iterates (6.32) are introduced into the right hand sides of the equation of transfer (6.26) and this leads to the second iterates for the moments P_{ij} and q_i, viz.

$$P_{ij}^{(2)} = -2\mu E_{ij} + 2\frac{\mu}{p}\left[\delta_{ir}\delta_{js}\frac{d}{dt} + 2\delta_{s(i}\frac{\partial v_{[j)}}{\partial x_{r]}} - 4\delta_{s(i}W_{j)r}\right](\mu E_{rs}) + 4\frac{\mu^2}{p}E_{r(i}E_{j)r} +$$

$$+ \frac{10}{3}\frac{\mu^2}{p}EE_{ij} - \frac{4}{3}\frac{\mu^2}{p}E_{sr}E_{sr}\delta_{ij} + \alpha\frac{\mu}{p}\frac{4}{5}\left(\frac{\partial}{\partial x_{(i}}\left(\mu\frac{\partial T}{\partial x_{j)}}\right) - \frac{1}{3}\frac{\partial}{\partial x_\ell}\left(\mu\frac{\partial T}{\partial x_\ell}\right)\delta_{ij}\right), \qquad (6.34)_1$$

$$q_i^{(2)} = -\alpha\mu\frac{\partial T}{\partial x_i} + \frac{3}{2}\alpha\frac{\mu}{p}\left[\delta_{in}\frac{d}{dt} + \frac{\partial v_{[i}}{\partial x_{n]}} - 2W_{in}\right]\left(\mu\frac{\partial T}{\partial x_i}\right) + 3\alpha\frac{\mu^2}{p}E\frac{\partial T}{\partial x_i} + \frac{165}{8}\hat{R}\frac{\mu^2}{p}E_{in}\frac{\partial T}{\partial x_n} +$$

$$+ 3\frac{\mu}{p}\hat{R}T\frac{\partial \mu E_{in}}{\partial x_n} + 6\frac{1}{\rho}\frac{\mu^2}{p}E_{in}\frac{\partial \mu E_{n\ell}}{\partial x_\ell} + \frac{1}{\rho}3\frac{\mu^2}{p}E_{i\ell}\frac{\partial p}{\partial x_\ell}. \qquad (6.34)_2$$

E stands for $\frac{\partial v_\ell}{\partial x_\ell}$. A moment's reflection will show that the nth iterate differs from the $(n-1)^{st}$ iterate by terms that contain n^{th} powers of the mean collision time $\tau = \frac{1}{\rho B}$. This is a very small time and therefore it is reasonable to break off the iteration, and we shall consider the second iterates $P_{ij}^{(2)}$ and $q_i^{(2)}$ as good approximations to the exact values of the stress deviator and the heat flux.

The iteration scheme presented here is a simplified version of the Maxwellian iteration (see [31]) which makes use of the full set (6.21), (6.22) of equations of transfer rather than of the 13-moment approximations $(6.26)_{4,5}$. It is true that the Maxwellian iteration is the more systematic scheme, but the terms that I wish to focus on come out in the same form in both schemes and therefore I have chosen the simpler one in this presentation.

The important thing to notice in (6.34) is that these equations are explicit constitutive equations of the type $(6.4)_{1,2}$. In other words, the iteration has led to ordinary thermodynamics with the equations of balance $(6.26)_{1,2,3}$ and the constitutive equation (6.34). While in thermodynamics the constitutive equations are assumed and their properties are postulated, the kinetic theory derives the constitutive relations and is therefore capable of checking the validity of the postulates.

In this manner the postulate of material frame indifference is put to the test and we shall see that it does not pass.

6.5.2. On the Frame Dependence of Stress and Heat Flux

We recall that the principle of material frame indifference requires that the constitutive functions for stress and heat flux be independent of the frame of reference. This assumption is not born out by the kinetic theory of gases, because the expressions $(6.34)_{1,2}$ for P_{ij} and q_i depend on the frame through the terms on the right hand sides that contain W_{ij}, the matrix of angular velocity of the frame with respect to an inertial frame. In other words, $(6.34)_2$ shows that the dependence of the heat flux on the gradient of temperature is different in a non-inertial frame and in an intertial frame, because in one the term with W_{ij} is present and in the other it is not. The same is true for the dependence of the stress deviator on the velocity gradient as shown by (6.34). To make things clearer we proceed by considering a special case.

In $(6.34)_2$ the frame dependent part of the heat flux is just one term among many and its physical significance is obscured by the complexity of the constitutive relation. Therefore we shall consider the special circumstance, in which

i) the gas performs a rigid rotation about the same axis about which the frame rotates,

ii) the temperature of a fluid particle is constant.

In this case $E_{ij} = 0$ and $\dot{T} = 0$ and W_{ij} is proportional to $\frac{\partial v_{[i}}{\partial x_{j]}}$ and $(6.34)_2$ reduces to

$$q_i = -\kappa\,[\delta_{ij} + \frac{2}{\rho B}\,(W_{ij} - \frac{\partial v_{[i}}{\partial x_{j]}})]\,\frac{\partial T}{\partial x_j} \; . \tag{6.35}$$

We conclude that q_i is not parallel to the gradient of temperature unless the gas is at rest in an inertial frame, so that $W_{ij} = \frac{\partial v_{[i}}{\partial x_{j]}}$. In this special circumstance the balance of internal energy reduces to the statement $\frac{\partial q_i}{\partial x_i} = 0$ and insertion of

(6.35) leads to the following equation of stationary heat conduction

$$\frac{\partial^2 T^2}{\partial x_i \partial x_i} = \frac{2}{\rho B} \left(W_{ij} - \frac{\partial v_{[i}}{\partial x_{j]}} \right) \frac{\partial \ln \rho}{\partial x_i} \frac{\partial T^2}{\partial x_j} \quad . \tag{6.36}$$

In deriving this equation I have used that κ, by $(6.33)_2$ is proportional to T. The equation tells us that the field of temperature is different when the fluid is at rest in an inertial frame and when it is not. There is no such inertial influence, however, when the gradients of ρ and T are parallel, or when either one of these gradients is parallel to the axis of rotation.

Speziale [27] has written that the constitutive equation for the heat flux cannot violate the principle of material frame indifference, because q_i is an objective vector. This is nonsense as we can see from $(6.34)_2$. The right hand side of that equation is an objective vector and still it is frame dependent. It is true that

$$\delta_{in} \frac{d}{dt} \left(\mu \frac{\partial T}{\partial x_i}\right) , \qquad \frac{\partial v_{[i}}{\partial x_{j]}} \left(\mu \frac{\partial T}{\partial x_i}\right) , \qquad -2W_{in} \left(\mu \frac{\partial T}{\partial x_i}\right)$$

are not objective vectors, but their sum is! Obviously Speziale has confused the notions of objective tensors and of material objectivity. (The latter expression is often used for material frame indifference).

6.5.3. A Suggestive Interpretation of the Non-Objective Term in the Heat Flux

We consider a cylinder filled with an ideal gas at rest in which we create a radial gradient of temperature by heating the axis. At first we assume that the cylinder is at rest in an inertial frame and we focus the attention upon a tiny volume element of the dimensions of a mean free path of an atom (see Figure 4a). Figure 4b represents a blown-up view of that element. The atoms at the lower edge of the element have a bigger mean energy than those at the upper edge, because the temperature is bigger below than above. Thus on the average an atom that moves upward from below carries a bigger energy than an atom which moves downward from above. In the kinetic theory these different energies transported through the plane s - s correspond to a heat flux and we conclude that the heat flux in the present case points in the direction opposite to the gradient of temperature just as is predicted by equation (6.35) when the gas rests in an inertial frame.

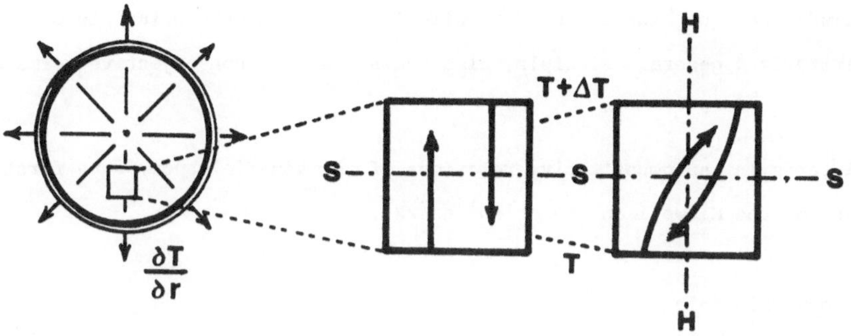

Figure 4a. 4b. 4c.

The situation is changed when the cylinder is taken to a non-inertial frame with an axis of rotation that coincides with the axis of the cylinder. Let the gas be at rest with respect to the non-inertial frame now. Figure 4c. represents a blown-up view of the small volume element that was considered before. Again, because of the temperature gradient, the atoms at the lower edge have a bigger mean energy than those at the upper edge. But now an atom that starts straight upward from below will be moving on a curved path under the influence of the Coriolis force and the same is true for an atom which starts downward from the upper edge, except that the Coriolis force has opposite directions in the two cases (see Figure 4c). Thus there is still a component of heat flux across the plane s - s, i.e. in the direction opposite to the temperature gradient, but there is now also a component of heat flux across the plane H - H, i.e. in a direction that is perpendicular to the temperature gradient and to the angular velocity of the frame. This situation is just what is described by equation (6.35).

The previous suggestive interpretation of the non-objective terms in the constitutive relation for the heat flux makes it obvious that this term will be small under normal circumstances, because it depends on the effect which the Coriolis force can achieve on the path of an atom during the time between collisions.

Another estimate on the size of the non-objective term is obtained directly from (6.35). Since $\frac{1}{\rho B}$ is of the order of magnitude of the mean time τ between the collisions and since the components of W_{ij} are equal to the reciprocal values of the period T of rotation of the frame, the non-objective term has a factor $\frac{\tau}{T}$. That factor is very small indeed for normally dense gases and for reasonable periods of rotation.

Thus we conclude that the kinetic theory, while it throws out the principle of material objectivity as a general principle, also shows that the non-objective terms are very small.

The inertial terms in the constitutive relations of the kinetic theory were first calculated by use of the Maxwellian iteration in [22].

6.6. The Entropy Principle

6.6.1. Boltzmann's H-Theorem

The counterpart of the entropy inequality of thermodynamics is Boltzmann's H-theorem in the kinetic theory of gases. This H-theorem results from the equation of transfer (6.18) by a special choice of ψ. We chose $\psi = -\hat{R}\ell nf$ and define

$$\eta \equiv \bar{\psi} = -\hat{R}\frac{m}{\rho}\int \ell nf\, f\, d\underset{\sim}{c} \qquad \text{and} \qquad \phi_i \equiv \overline{C_i\psi} = -\hat{R}m\int C_i \ell nf\, f\, d\underset{\sim}{c} \,. \tag{6.37}$$

Thus the equation of transfer assumes the form

$$\rho\dot{\eta} + \frac{\partial\phi_i}{\partial x_i} = -\frac{m}{4}\hat{R}\iint \ell n\frac{ff_1}{f'f'_1}(f'f'_1 - ff_1)\sigma g\, \sin\theta d\theta d\varepsilon d\underset{\sim}{c} d\underset{\sim}{c}^1 \,. \tag{6.38}$$

The right hand side of (6.38) is non-negative and we conclude that the kinetic theory of gases implies the inequality

$$\rho\dot{\eta} + \frac{\partial\phi_i}{\partial x_i} \geq 0 \qquad (6.39$$

which is called Boltzmann's H-theorem.

6.6.2. Entropy and η-Function

The obvious similarity of the inequality (6.39) with the Clausius-Duhem inequality (6.11) gives a first hint that η and ϕ_i, as defined by (6.37) should be viewed as the specific entropy and the entropy flux respectively within the kinetic theory of gases. And this idea gains conviction when we calculate the near-equilibrium values of η and ϕ_i from (6.37). Indeed, if we replace ℓnf by ℓnf_M in the integrals (6.37) where f_M is the Maxwellian distribution (6.23), we obtain

$$\eta = \hat{R}[\frac{3}{2} + \ln(\frac{m}{\rho}\sqrt{2\pi\hat{R}T}^3), \quad \text{hence} \quad d\eta = \frac{1}{T}(d\varepsilon - \frac{p}{\rho^2}d\rho), \tag{6.40$_1$}$$

$$\phi_i = \frac{q_i}{T} . \tag{6.40$_2$}$$

These are indeed the specific entropy in fluid and the usual assumption of thermodynamics concerning the entropy flux.

6.6.3. Entropy and Entropy Flux in the Generalized Theory of an Ideal Gas

Thus we have learned about the motives for the identification of the integrals (6.37) with the entropy and the entropy flux. These motives rest on the fact that the kinetic theory "close to equilibrium" gives the same results as thermodynamics of fluids. The results (6.40) are reliable only in the neighbourhood of equilibrium, because they were derived by replacing $\ln f$ in (6.37) by $\ln f_M$. We are bound to get better results for the entropy and the entropy flux, if we insert $\ln f_G$ into (6.37) where f_G is Grad's non-equilibrium distribution function (6.24). We set

$$\begin{aligned} \ln f_G &= \ln f_M + \ln\,(1 + \frac{1}{2p}p_{ij}[\frac{C_iC_j}{\hat{R}T} - \delta_{ij}] - \frac{1}{p}\frac{q_i}{\sqrt{\hat{R}T}}[\frac{C_i}{\sqrt{\hat{R}T}}(1 - \frac{1}{5}\frac{C^2}{\hat{R}T})]) , \\ &\approx \ln f_M + \frac{1}{2p}p_{ij}[\frac{C_iC_j}{\hat{R}T} - \delta_{ij}] - \frac{1}{p}\frac{q_i}{\sqrt{\hat{R}T}}[\frac{C_i}{\sqrt{\hat{R}T}}(1 - \frac{1}{5}\frac{C^2}{\hat{R}T})] , \end{aligned} \tag{6.41}$$

and obtain

$$\eta = \hat{R}[\frac{3}{2} + \ln(\frac{m}{\rho}\sqrt{2\pi\hat{R}T}^3)] - \frac{p_{ij}p_{ij}}{2p\rho T} - \frac{2}{5}\frac{q_iq_i}{p^2T} , \tag{6.42}$$

$$\phi_i = \frac{q_i}{T} - \frac{2}{5}\frac{p_{ij}q_j}{pT} . \tag{6.43}$$

The equation (6.42) shows that in the generalized theory the entropy is no longer simply a function of ρ and T, but it depends nonlinearly on the stress-deviator and the heat flux. Also the entropy flux is no longer just equal to the heat flux divided by the temperature; rather it too depends nonlinearly on p_{ij} and q_i. In the calculation of (6.42), (6.43) p_{ijk} and p_{ijkk} were chosen to be given by (6.25); this is proper in a calculation of η and ϕ_i with Grad's distribution function.

6.7. Outlook

It may be considered important, to generalize thermodynamics to make it more similar to the kinetic theory. Müller's papers [19], [20] were early attempts in that direction based upon linear irreversible thermodynamics, and recently Ruggeri [32] has taken up the problem within the context of rational thermodynamics with Lagrange multipliers.

The task is not easy because - as I have tried to show above - the kinetic theory invalidates the two most powerful restrictive principles of rational thermodynamics, viz. the Clausius-Duhem inequality and the principle of material frame indifference.

Of course, the entropy inequality remains valid in the form

$$\rho\dot{\eta} + \frac{\partial \phi_i}{\partial x_i} \geq 0 \ ,$$

but as we have seen in the treatment of mixtures, this inequality must be supplemented by a jump condition on the entropy flux. It is unclear at present under which condition the continuity of $\phi_i e_i$ is not the continuity of temperature as it used to be in the ordinary theory that was presented in the first sections of this treatise. We must ask: What replaces temperature as the continuous quantity on an ideal wall?

Note added in proof

In the time that has elapsed since the presentation of these lectures there has been some progress toward an answer for the above questions. In fact Liu & Müller [33] have formulated a new phenomenological theory of extended thermodynamics of gases, applicable to dilute and degenerate gases. In that theory it becomes obvious that temperature loses much of its usefulness in the treatment of non-equilibrium processes. The question of frame-dependence of stress and heat flux in seen in a new light in that paper. Also Greco & Müller [34] have shown that there may be an alternative to Landau's bi-fluid theory in the new version of extended thermodynamics. Most recently Müller & Ruggeri [35] have formulated the relativistic counterpart of extended thermodynamics and it turns out in their paper that the concept of an entropy inequality in non-equilibrium must be qualified.

References

[1] Müller, I. Thermodynamik. Grundlagen der Materialtheorie
Bertelsmann Universitätsverlag Düsseldorf (1973)

[2] Truesdell, C. Sulle Basi della Termomeccanica
Acc. Naz. dei Lincei 8 Vol. XXII p. 33 (1957)

[3] Truesdell, C. Rational Thermodynamics
McGraw-Hill Book Co., N.Y. (1969)

[4] Smith, G.F. On Isotropic Integrity Bases
Archive Rat. Mech. Anal. 18 p. 282 (1965)

[5] Wang, C.C. A New Representation Theorem for Isotropic Functions: Answer to Professor G.F. Smith's Criticism of my Papers on Representations for Isotropic Functions. Part I: Scalar Valued Isotropic Functions
Arch. Rational Mech. Anal. 36 p. 166 (1970).
Part II: Vector Valued Isotropic Skew-Symmetric Tensor - Valued Isotropic Functions, ibid. p. 198

[6] Müller, I. A New Approach to Thermodynamics of Simple Mixtures
Zeitschrift für Naturforschung 28a p. 1801 (1973)

[7] Liu, I-Shih Method of Lagrange Multipliers for Exploitation of the Entropy Principle
Archive Rat. Mech. Anal. 46 p. 131 (1972)

[8] Prigogine, I., Defay, R. Chemische Thermodynamik
Deutscher Verlag für Grundstoffindustrie (1962)

[9] Meixner, J., Reik, H.G. Die Thermodynamik der irreversiblen Prozesse in kontinuierlichen Medien mit inneren Umwandlungen.
Handbuch der Physik III/2 p. 417 Springer Berlin, Heidelberg, New York (1959)

[10] de Groot, S.R., Masur, J. Non-Equilibrium Thermodynamics
North Holland Publ. Co., Amsterdam (1962)

[11] Truesdell, C. Mechanical Basis of Diffusion
J. Chem. Phys. 37 p. 2336 (1962)

[12] Müller, I., Villaggio, P. Conditions of Stability and Wave Speeds for Fluid Mixtures
Meccanica 11 p. 191 (1976)

[13] Müller, I. A Thermodynamic Theory of Mixtures of Fluids
Archive Rat. Mech. Anal. 28 p. 1 (1968)

[14] Tisza, L. Nature, 141 p. 913 (1938)

[15] Landau, L.D. The Theory of Superfluidity of Helium II
J. Phys. (U.S.S.R.) 5 p. 72 (1941)

[16] Dreyer, W. Zur Thermodynamik von Helium II - Superfluides Helium mit und ohne Wirbellinien als binäre Mischung.
Dissertation TU Berlin (1983).

[17] Feynman, R.P. Atomic Theory of the Two Fluid Model of Liquid Helium
Phys. Rev. 94 p. 262 (1954).

[18] Cattaneo, C. Sulla Conduzione del Calore
Atti del Seminario Matematico e Fisico della Univ. di Modena
3 p. 3 (1948)

[19] Müller, I. Zur Ausbreitungsgeschwindigkeit von Störungen in kontinuierlichen Medien
Aachener Dissertation (1966)

[20] Müller, I. Zum Paradox der Wärmeleitungstheorie
Z. f. Physik 198 p. 329 (1967)

[21] Grioli, G. Sulla propagazione di onde termomeccaniche nei continui
Atti della Acc. Naz. dei Lincei 67 p. 332 (1979)

[22] Müller, I. On the Frame Dependence of Stress and Heat Flux
Arch. Rational Mech. Anal. 45 p. 241 (1972)

[23] Edelen, D.B.G., McLennan, T.A. Material Indifference: A Principle or a Convenience
Int. J. Engng. Sci. 11 p 813 (1973)

[24] Söderholm, L.H. The Principle of Material Frame Indifference and Material Equations in Gases
Int. J. Engng. Sci. 14 (1976)

[25] Truesdell, C. Correction of two Errors in the Kinetic Theory that have been used to cast unfounded doubt upon the Principle of Material Frame Indifference
Meccánica 11 p. 196 (1976)

[26] Wang, C.C. On the Concept of Frame Indifference in Continuous Mechanics and in the Kinetic Theory of Gases
Arch. Rational Mech. Anal. 58 381 (1975)

[27] Speziale, C.G. On the Frame Indifference and Iterative Procedures in the Kinetic Theory of Gases
Int. J. Engng. Sci. 19 p. 63 (1981)

[28] Chapman, S., Cowling, T.G. The Mathematical Theory of Non-Uniform Gases
Cambridge University Press (1961)

[29] Grad, H. On the Kinetic Theory of Rarefied Gases
Comm. Pure Appl. Math. 2 p. 331 (1949)

[30] Grad, H. Principles of the kinetic Theory of Gases
Handbuch der Physik Bd. XII p. 205 Springer (1958)

[31] Ikenberry, E. & Truesdell, C. On the Pressures and the Flux of Energy in a Gas according to Maxwell's Kinetic Theory I.
J. Rational Mech. Anal. 5 p. 1 (1956)

[32] Ruggeri, T. Symmetric Hyperbolic System of Conservative Equations for a Viscous Heat Conductor Fluid (in preparation)

[33] Liu, I-Shih, Müller, I., Extended Thermodynamics of Classical and Degenerate Ideal gases.
Arch. Rational Mech. Anal. 83 (1983).

[34] Greco, A., Müller, I., Extended Thermodynamics of Superfluidity.
Arch. Rational Mech. Anal. 85 (1984).

[35] Müller, I., Ruggeri, T., Extended Thermodynamics of Relativistic Gases.
(in preparation).

ON NONSTATIONARY THERMODYNAMICS, RELATIVISTIC ACCELERATION WAVES IN IDEAL FLUIDS

Aldo Bressan
University of Padova

N.1 Introduction(*)

In [2], and especially [4], some questions are treated, that concern nonstationary thermodynamics, i.e. the modern branch of thermodynamics that aims at avoiding the paradoxes of heat propagation: how the objectivity principle - briefly OP - is to be used in this thermodynamics, and whether Eckart's law of heat conduction has to be modified in order to take (all) relativistic corrections into account - see below. Furthermore a new relativistic law of heat conduction is proposed in the paper [2] where the main results of [4] are summarized.

In the present paper relativistic acceleration waves travelling in an ideal fluid $\mathbb{F}$ are studied, also under a strong heat flux, in connection with any of the afore-mentioned laws. The results are used also to deal with hyperbolicity properties of $\mathbb{F}$'s evolution systems.

* * *

In 1948 Cattaneo avoided the paradoxes of heat propagation in the case of a rigid conductor within classical physics, by replacing Fourier's law with a first order differential equation $\mathcal{E}_q$ in the heat flux $\underline{q}$, which substantially takes the inertia of heat into account; Cattaneo also supported his law on the basis of the kinetic theory - see [7](1).

Later several other authors avoided the same paradoxes in more and more general cases - see e.g. [5], [6], [10] to [27], and [29] -; among these proposals the so

---

(*) This paper has been performed in the sphere of activity of the research group n.3 of the Consiglio Nazionale delle Ricerche in the academic year 1980-81.

(1) Cattaneo's equation was considered in the past century by Maxwell, who discarded it immediately. It was substantially rediscovered by Vernotte - see [29] - in order to solve the paradox of heat conduction, admittedly without any kinetic justification.

called Müller-Israel approach is well known. Furthermore the support given to nonstationary thermodynamics (in its various versions) on the basis of the kinetic theory, has been improved considerably (Müller, Israel and Carrassi).

In 1968 Gurtin and Pipkin replaced the differential equation $\mathcal{E}_q$ of heat conduction with a constitutive equation $\mathcal{E}'_q$ for $\underline{q}$, of the visco-elastic type - see [13]. Both $\mathcal{E}_q$ and $\mathcal{E}'_q$ were extended by Lianis to very general cases, in the relativistic paper [20]. These generalisations comply with the OP totally, i.e. as constitutive equations generally do in contrast with e.g. [15] and [12].

In order to discuss the role of the OP in nonstationary thermodynamics, in [4] heat propagation is considered from a dynamic point of view, based on the mass-energy equivalence principle - briefly MEEP. This point of view is present also in Ekart's fundamental paper [9] on relativistic thermodynamics(1). Other authors, Kalinski and expecially Massa and Morro, used later the same point of view to reach e.g. a dynamic justification of Cattaneo's nonstationary law of heat conduction in classical physics - see [14] and [22].

In [4] the dynamic point of view is used in a partially different version, and especially with different aims. By means of it, first, a nonstationary relativistic equation of heat conduction is reached, which (involves only magnitudes usually used in the theory of continuous media, and) differs from Eckart's also in the stationary case. The former law appears preferable to the latter, as far as relativistic corrections are concerned, also when the criteria used to reach it are disregarded; in fact Eckart's relativistic thermodynamic theory has a rather contradictory feature - see [2] or [4](2).

A second aim reached in [4] is a support based on the MEEP, given to thesis T_1 below - also asserted in [3].

T_1: *In nonstationary thermodynamics, where the inertia of heat is taken into account, it is more natural to regard (any more or less general version of) the equation* $\mathcal{E}_q$ *for heat conduction as a (universal) fundamental law such as Cauchy's*

---

(1) In fact, in order to justify the contribution of TA_α to this law of heat conduction Eckart used in [9] the inertia of heat and precisely the dragging force acting on the heat flux.

(2) As is shown in [4], N6, in Eckart's relativistic thermodynamics Coriolis complementary force on the heat flux is present in the Cauchy dynamic equation of continuous media, while it does not appear in the equation for heat conduction.

dynamic equation for continuous media, rather than a genuine constitutive equation.

Consequently

T_2: *It is preferable to regard the constitutive equation* $\mathcal{E}'_q$, *substituted for* $\mathcal{E}_q$, *as a reasonable particular solution of the differential equation* $\mathcal{E}_q$ *rather than a genuine constitutive equation.*

Then, since the OP applies only to constitutive equations,

T_3: *The nonstationary equations* $\mathcal{E}_q$ *and* $\mathcal{E}'_q$ *of heat conduction in various published versions that do not comply with the* OP *totally - see e.g.* [15], [12], *and* [3] - *do not violate the* OP.

The afore-mentioned equations comply with the OP only partially in that some magnitudes involved by them, such as those related with the Fourier coefficient, are objective - see [28].

The acceptance of the law proposed in [2] and [4], whose version for fluids is (2.5) for $\zeta = 1$, implies the following thesis:

T_4: *The total compliance of equation* $\mathcal{E}_q$ *for heat conduction with the* OP *not only fails to be compulsory, but it is wrong as far as relativistic corrections are concerned; and also in classical physics the above compliance, though tolerable, renders the description of physical reality worse, especially when the heat flux* $\underline{q}$ *is large.*

This conclusion may have a pratical interest, in that, as is emphasised in [3], the behaviour of a nonstationary material can be treated in a simpler way when it complies with the OP partially rather than totally (expecially for $\underline{q}$ large). Thus an easier development of nonstationary thermodynamics is favoured also in the linear case; in fact the total compliance with the OP fails to influence a linearization of this thermodynamics only if it is performed around a state $S_{(e)}$ of thermodynamic equilibrium ($\underline{q} \equiv 0$).

The first problems to which the new thermodynamics was applied concerned $S_{(e)}$, but now the general case is being studied, e.g. in connection with acceleration waves and hyperbolicity questions - see e.g. [12] and [3].

Let us add that in [9] Eckart requires his law of heat conduction to comply with the Clausius Duhem inequality. However, as is now well known especially on the basis of papers of Müller, in nonstationary thermodynamics the second principle does not always hold exactly. Furthermore, in connection with this, in [4] a unified theory of piezo-elastic bodies is considered, where some of these bodies are <u>q-dependent</u> in the sense that their specific internal energy depends also on the heat flux $\underline{q}$, whereas others are not so and are called <u>q-independent</u>. Among other things in [4] one shows that the ordinary second principle holds for every process that some q-independent piezo-elastic material can undergo; and some of these processes appear incompatible with the Clausius Duhem inequality. The opposite situation holds for q-dependent piezo-elastic materials.

By the considerations above the best support given to Eckart's heat conduction law appears to be a well known argument based on pocket temperature and electromagnetic energy in thermodynamic equilibrium - see e.g. [1], p. 119. Since this argument concers equilibrium, it supports equally well the law proposed in [2] and [4].

* * *

In this paper thermo-elastic acceleration waves are considered within a relativistic nonstationary theory which depends on a parameter ζ and is agreement with [4] for $\zeta = 1$ and with [9] for $\zeta = 0$. Among them the normal waves, for $\underline{q}$ arbitrary and in particular for $\underline{q} = 0$, have received special attention.

N.2 Some kinematic and thermodynamic preliminaries

Let C be a thermo-elastic body, for which S^* is a reference state in a relativistic space time S_4^*, possibly different from the actual one S_4 - see [1], p.135. I consider an admissible co-ordinate system $(x) = (x_0,\dots,x_3)$ in S_4, an event point $\mathcal{E}$ on the world line W_{P^*} of the typical material point P^* of C, the space time metric ds^2, the 4-velocity u^α (at $\mathcal{E}$), intrinsic acceleration A_α, invariant mass density (i.e. reference gravitational rest mass per unit proper volume) k, actual gravitational mass $c^{-2}\rho$ (c = speed of light in vacuum), specific internal energy w, specific entropy η, absolute temperature T, (spatial) Römer heat flux vector q_α, (spatial) stress tensor $X_{\alpha\beta}$ $(= X_{\beta\alpha})$, pressure p, spatial projector $\overset{\perp}{g}_{\alpha\beta}$, the spatial part $T^{\cdots}_{\cdots\overset{\perp}{\alpha}}$ of any Tensor $T^{\cdots}_{\cdots}$ with respect to its index α, and absolute derivation (along W_{P^*}) $DT^{\cdots}_{\cdots}/Ds$. Thus we have in particular(1)

---

(1) Greek [Latin] indices run from 0 [1] to 3. Furthermore $T^{\cdots}_{\cdots/\alpha}$ is the covariant derivative of the arbitrary tensor $T^{\cdots}_{\cdots}$.

$$ds^2 = -g_{\alpha\beta}dx^\alpha dx^\beta \qquad (g_{oo} < 0) \quad ; \qquad \rho = k(c^2+w) \ , \tag{2.1}$$

$$\overset{\perp}{g}_{\alpha\beta} = g_{\alpha\beta} + u_\alpha u_\beta , \qquad T^{\cdots}_{\cdots\overset{\perp}{\alpha}} = T^{\cdots}_{\cdots\beta}\, \overset{\perp\beta}{g}_\alpha \qquad u^\alpha{}_{/\alpha} = u^\alpha{}_{/\overset{\perp}{\alpha}} \ , \tag{2.2}$$

and

$$\grave{k} + k\, u^\alpha{}_{/\alpha} = 0 \ , \qquad A_\alpha = \grave{u}_\alpha \ , \quad \text{where} \quad \grave{T}^{\cdots}_{\cdots} = DT^{\cdots}_{\cdots}/DS \ . \tag{2.3}$$

Eckart's conservation equations are

$$U_\alpha{}^\beta{}_{/\beta} = 0 \ , \quad \text{where} \quad U_{\alpha\beta} = \rho\, u_\alpha u_\beta + X_{\alpha\beta} + u_\alpha q_\beta + q_\alpha u_\beta \ . \tag{2.4}$$

A quantity of the same order of magnitude as c^{-r} is denoted by ⓡ. E.g. $u_{\alpha/\overset{\perp}{\beta}}$ and q_α are ①.

Equation (5.6) in [4], proposed there as a relativistic nonstationary law of heat conduction, reads for ideal fluids

$$\overset{\perp\beta}{g}_\alpha \grave{q}_\beta + \zeta\, [\overset{\perp}{g}_{\alpha\rho} (\frac{q^\rho q^\sigma}{\varepsilon})_{/\overset{\perp}{\sigma}} + u^\rho{}_{/\rho}\, q_\alpha + u_{\alpha/\rho} q^\rho] + \varepsilon_1 (\varkappa^{-1} q_\alpha + T/\overset{\perp}{\alpha}) + \psi_\alpha{}^\sigma A_\sigma = 0 \tag{2.5}$$

for $\zeta = 1$, where(1)

$$\psi_\alpha{}^\sigma = \psi_\alpha{}^\sigma (T,k,q_\beta,T/\overset{\perp}{\beta}) = \zeta\, \frac{q_\alpha q^\sigma}{\varepsilon} + (1-\zeta+\zeta v^2)\varepsilon_1 T\, \delta_\alpha{}^\sigma \tag{2.5'}$$

and where - see (5.3-4) and (5.7) in [4]

(i) $\varkappa$ is the Römer-Fourier coefficient,

(ii) ε_1 and ε, as well as the quantity $\varepsilon_{(1)}$ introduced by $(2.6)_2$ below, are functions of T, k, and possibly of $m = q_\alpha q^\alpha/2$, but their dependence on m can reasonably by regarded to be ② and

(iii) we have - see $(3.16)_2$, $(5.3)_1$, and (5.4) in [4] - the equalities

---

(1) The (Römer) Fourier tensor $\varkappa^{\rho\sigma}$ used in [4], becomes $\varkappa\, \overset{\perp}{g}^{\rho\sigma}$ for fluids, so that by [4, (5.7-8)] $K_{\alpha\beta} = \varepsilon_1 \varkappa^{-1} \overset{\perp}{g}_{\alpha\beta}$. Then, by [4, (4.5)], equation [4, (5.6)] becomes (2.5).

$$2V^2 = 1 + \left(1 + \frac{8m}{\varepsilon_1^2 T^2}\right)^{1/2} , \quad \varepsilon = \varepsilon_1 T V^2 = \varepsilon_{(1)} T = \varepsilon_1 T + \textcircled{2} \quad (m = q^\alpha q_\alpha / 2) \tag{2.6}$$

which, among other things, introduce $\varepsilon_{(1)}$ and espress V^2 in such a way that

$$\varepsilon_{(1)} T A_\alpha = \varepsilon_1 T A_\alpha + \textcircled{4} \quad \text{and} \quad V^2 = 1 + \textcircled{2} . \tag{2.6'}$$

For $\zeta = 0$ (2.5) *affords, so to say, the simplest non-stationary analogue of Eckart's constitutive equation for* q_α .

By (2.6) and (2.5') $\Psi_\alpha{}^\sigma = \bar{\Psi}_\alpha{}^\sigma + \textcircled{2}$, where

$$\bar{\Psi}_\alpha{}^\sigma = \bar{\Psi}_\alpha{}^\sigma(T,k,q_\beta) = \zeta \frac{q_\alpha q^\alpha}{\varepsilon_1} + \varepsilon_1 T \delta_\alpha^\sigma \quad (\text{hence } \Psi_\alpha{}^\sigma A_\sigma = \bar{\Psi}_\alpha{}^\sigma A_\sigma + \textcircled{4}). \tag{2.5"}$$

In case C is an ideal fluid $\mathbb{F}$, the following constitutive equations can be considered for it:

$$w = w(T,k), \qquad \eta = \eta(T,k), \qquad p = p(T,k), \qquad X_{\alpha\beta} = p \overset{\perp}{g}_{\alpha\beta}, \tag{2.7}$$

$$\varepsilon_1 = \varepsilon_1(T,k,m), \quad \varepsilon = \varepsilon(T,k,m), \quad \varkappa = \varkappa(T,k,m,\vartheta) \quad (\vartheta = T^{/\alpha} T_{/\alpha}^{\perp}/2). \tag{2.8}$$

where P^* can be regarded as an implicit variable (the only one), and where $(2.8)_2$ follows from $(2.8)_1$ and $(2.6)_{1-2}$.

In connection with q-independent constitutive equations for w and η such as $(2.7)_{1-2}$, according to [4] the 2nd principle for C can be postulated in its ordinary form, which is the first of the relations

$$0 \le Tk\grave{\eta} + q^\alpha{}_{/\alpha}^{\perp} + 2q^\alpha A_\alpha = 0 \tag{2.9}$$

- see (24.5) and (25.1) for $r = 0$ in [1] .

Suitable relations among the constitutive functions $(2.7)_{1-2}$ can be found, by which the 1st principle, i.e. the composition of $(2.4)_1$ with u^α, is equivalent to $(2.9)_2$.

<u>Remark</u>. In the stationary case within classical physics, a possibly non-linear version of the Fourier law holds:

$$cq_r = -c\varkappa T_{/r}, \quad \text{hence} \quad \varkappa(T,k,m,\vartheta)^2\vartheta = m \; (= q^r q_r/2) \; . \tag{2.10}$$

It is natural to assume the function $\varkappa(\ldots)$ in $(2.10)_2$ to be regular enough so that ϑ can be defined as a function of T, k, and m, even if $(2.10)_1$ is not linear in $\vartheta = T_{/r}\,T^{/r}/2$.

This shows that it *is only slightly restrictive to assume* $\varkappa_\vartheta \equiv 0$, *hence* $\varkappa_{T/s} \equiv 0$, *for a non-stationary extension of the classical non-linear (Fourier) law of heat conduction.*

N.3 Thermo-elastic acceleration waves in an ideal fluid $\mathbb{F}$. Equation for the propagation speeds.

The dynamic equation for C, i.e. the composition of (2.4) with $\overset{\perp}{g}_\rho{}^\alpha$, reads - see [1, (24.8)] and $(2.3)_{1,3}$

$$\rho(\overset{\perp}{g}_\alpha{}^\beta + X_\alpha{}^\beta)A_\beta + g_\alpha{}^\rho X_\rho{}^\sigma{}_{/\overset{\perp}{\sigma}} + g_\alpha{}^\beta \grave{q}_\beta + u^\rho{}_{/\rho}q_\alpha + u_{\alpha/\beta}q^\beta = 0 \; . \tag{3.1}$$

For $C = \mathbb{F}$ $(X_{\alpha\beta} = p\,\overset{\perp}{g}_{\alpha\beta})$ it simplifies to

$$(\rho + p)A_\alpha + p_{/\overset{\perp}{\alpha}} + \overset{\perp}{g}_\alpha{}^\beta \grave{q}_\beta + u^\rho{}_{/\rho}q_\alpha + u_{\alpha/\beta}\,q^\beta = 0 \; . \tag{3.2}$$

Let us now consider a thermo-mechanic acceleration wave $\sum$ travelling in $\mathbb{F}$, of spatial normal N_α and Römer propagation speed V, so that across it u^α, T, k, and q^α are continuous, while their first partial derivatives suffer discontinuities of the first kind. Let λ_α, λ_4, λ_5, and τ_α be the corresponding discontinuity parameters. Furthermore choose the frame (x) in such a way that locally, i.e. at the event point $\overset{\omega}{\mathcal{E}} \in \sum$, it is natural and proper, and its first [third] spatial axis is [spatially] normal to $\sum$ [to q^α]:

$$g_{\alpha r} = \delta_{\alpha r}, \quad g_{oo} = -1, \quad \partial g_{\alpha\beta}/\partial x^\gamma = 0; \qquad u^\alpha = \delta^\alpha{}_o, \; N_\alpha = \delta_\alpha{}^1, \; q_3 = 0 \; . \tag{3.3}$$

Denoting discontinuities across $\sum$ by $[\![\ldots]\!]$, from $(3.3)_{1-4}$ it follows that (at $\mathcal{E}$)

$$\begin{cases} [u_{r/s}] = \lambda_r N_s , & [u_{r/o}] = -V\lambda_r, \quad [T_{/r}] = \lambda_4 N_r , \\ [q_{r/s}] = \tau_r N_s, & [q_{r/o}] = -V\tau_r, \quad [k_{/r}] = \lambda_5 N_r . \end{cases} \tag{3.4}$$

By (2.5') and the constitutive relations (2.7-8), equalities (3.2), $(2.9)_2$, $(2.3)_1$, and (2.5), which are equivalent to 8 independent scalar differential equations in the 8 unknown scalar functions u^α, T, k, and q_r of x^α, constitute an evolution system for $\mathbb{F}$. By (3.4) and (2.7-8), the corresponding discontinuity equations give rise to the following 8 scalar equations in the 8 parameters λ_r, λ_4, λ_5, and τ_r :

$$-V(\rho + p)\lambda_r + p_T N_r \lambda_4 + p_k N_r \lambda_5 - V\tau_r + q_r\lambda_1 + q_1\lambda_r = 0 , \tag{3.5}$$

$$VTk(\eta_T\lambda_4 + \eta_k\lambda_5) + 2Vq^s\lambda_s - \tau_1 = 0 , \tag{3.6}$$

$$k\lambda_1 - V\lambda_5 = 0 , \tag{3.7}$$

$$\begin{aligned} -V\tau_r + \zeta[\frac{q_1}{\varepsilon}\tau_r + \frac{q_r}{\varepsilon}\tau_1 - \frac{q_r q_1}{\varepsilon^2}(\varepsilon_T\lambda_4 + \varepsilon_k\lambda_5 + \varepsilon_m q^\ell \tau_\ell) - \frac{V}{\varepsilon} q_r q^s \lambda_s + \\ + q_r\lambda_1 + q_1\lambda_r] + \varepsilon_1 N_r \lambda_4 - (1-\zeta+V^2\zeta)V\varepsilon_1 T\lambda_r - \frac{\varepsilon_1 q_r}{\varkappa^2}\varkappa_{T/1}\lambda_4 = 0 . \end{aligned} \tag{3.8}$$

Remark that only by the term in (3.8) containing $\varkappa_{T/s}$ $(= \varkappa_\vartheta T_{/s})$ *the above discontinuities must be supposed very small;* indeed (3.8) holds only in the first approximation, unless that term be replaced by $\varepsilon_1 q_r\{[\varkappa(T,k,m,\vartheta)^+]^{-1} - [\varkappa(T,k,m,\vartheta)^-]^{-1}\}$. Hence *in the case* $\varkappa_m \equiv 0$, *corresponding to a possibly non-linear Fourier law, equations* (3.5-8) *hold exactly for finite* λ_r, λ_4, λ_5, *and* τ_r .

Equations (3.5-8) form the system

$$\sum_{\ell=1}^{8} A_{h\ell}\lambda^\ell = 0 \quad (h = 1,\dots,8) \quad \text{with} \quad \lambda_\ell = \lambda^\ell \quad \text{and} \quad \lambda_{5+s} = \tau_s \quad (s = 1,2,3), \tag{3.9}$$

where by $(3.3)_5$ and (2.6), for r, s = 1,2,3 we have

$$\begin{cases} A_{rs} = - V(\rho + p)\delta_{rs} + q_r N_s + q_1\delta_{rs} , \\ A_{r4} = p_T N_r, \quad A_{r5} = p_k N_r, \quad A_{r,5+s} = -V\delta_{rs} , \end{cases} \tag{3.10}$$

$$A_{4s} = 2Vq_s, \quad A_{44} = VTk\eta_T, \quad A_{45} = VTk\eta_k, \quad A_{4,5+s} = -N_s, \tag{3.11}$$

$$A_{5s} = kN_s, \quad A_{54} = 0, \quad A_{55} = -V, \quad A_{5,5+s} = 0, \tag{3.12}$$

$$\begin{cases} A_{5+r,s} = \zeta[-\varepsilon^{-1}Vq_rq_s + q_rN_s + q_1\delta_{rs}] - (1+\zeta V^2 - \zeta)V\varepsilon_1 T\delta_{rs}, \\ A_{5+r,4} = -\dfrac{\zeta q_r q_1}{\varepsilon^2}\varepsilon_T + \varepsilon_1 N_r - \dfrac{\varepsilon_1 q_r}{\varkappa^2}\varkappa_{T/1}, \quad A_{5+r,5} = -\zeta\dfrac{q_r q_1}{\varepsilon^2}\varepsilon_k, \\ A_{5+r,5+s} = -V\delta_{rs} + \zeta(\varepsilon^{-1}q_1\delta_{rs} + \varepsilon^{-1}q_rN_s - q_1\varepsilon^{-2}\varepsilon_m q_r q_s). \end{cases} \tag{3.13}$$

The real solutions of the 8-th order polynomial equation in V

$$\det(A_{h\ell}) = \|A_{h\ell}\| = 0 \qquad (h,\ell = 1,\ldots,8) \tag{3.14}$$

are the possible propagation speeds for $\sum$.[1]

N.4 On the classical limit

Note that $c^{-3}\rho$, $u_{r/s}$, q_r, and $\varkappa$ are ①; hence, by (3.4), the same holds for τ_r, λ_r, and V, provided usual quantities (considered in classical physics) have ordinary sizes. Furthermore ε_o $(= \varepsilon_1 T)$ - see N.3, (3.8), (3.14), and (5.4) in [4] - has the order of the density (or pressure) of the energy responsible for the heat flux, which is reasonably comparable with the density $\bar{\rho}$ (or pressure $\bar{p}$) of electromagnetic energy in equilibrium at the same temperature. Thus ε_1, $\varepsilon_{(1)}$, and ε - see (2.6) - can be regarded to have the same order as $\bar{\rho}$. As a consequence the error $\varkappa^{-1}q_r + T_{/r}$ of the Fourier law (in describing the heat flux being considered) has an ordinary size if and only if so does the 3rd term in (2.5). The 2nd term in (2.5) and the 4th are ②. Now suppose that, initially, k, T, $T_{/r}$, q_r, $q_{r/s}$, and $u_{r/s}$ have ordinary (nonzero) sizes and that so does the error above (which conditions are mutually compatible). Then by the considerations above and by (2.5), initially $\grave{q}_r$

(1) The elimination of λ_5 for $V = 0$, by means of (3.7), causes us to multiply the 6 scalar equation (3.5) and (3.8) by V; then the corresponding analogue of (3.13) turns out to be a polynomial equation in V, of the 13-th order.

also has an ordinary size. On the other hand q_r appears to be ②. This is just the peculiarity of non-stationary thermodynamics. Thus $c^2\grave{q}_r$ is very large and such that q_r rapidly reaches values near those compatible with the (relativized) Fourier law, i.e. admissible stationary values of q_r (for which (2.5) holds with $\grave{q}_\alpha = 0$).

In classical nonstationary thermodynamics the fact above is very important in that it explains why the error of Fourier's law is practically zero. However the detailed way in which q_r reaches small values has no interest. Hence the 2nd term in (2.5) and the 4th, c^2 time smaller than the 2nd, have no interest (in classical physics) when $\varkappa^{-1}q_r+T_{/r}$ is large.

On the other hand, when $\varkappa^{-1}q_r+T_{/r}$ is (very) small, $c^2\grave{q}_r$, i.e. the first term in (2.5), has an ordinary size determined by all other terms in (2.5). Thus the 2nd term in (2.5) and the 4th are relevant only to determine the small fluctuations of q_r when $\varkappa^{-1}q_r+T_{/r}$ is very small. Hence they can be disregarded in classical physics.

According to a standard procedure, $\grave{q}_r$ is regarded to be ②, so that the classical limit of the non-stationary heat conduction law (2.5) (multiplied by c^2) is the classical Fourier law (2.10). In fact, for most purposes this law is satisfactory. Therefore, when the classical limit of (2.5) is required to be a non-stationary law, the limit procedure must be chosen suitably, taking into account the fact emphasized in [14], [22], and [4], that non-stationary thermodynamic effects are linked to the inertia of heat, via the mass-energy equivalence principle. - Thus they are essentially relativistic and hence strongly connected with the real value $\bar{c}$ of c (to be regarded as a constant when $c \to \infty$).

Let us distinguish classical quantities by an accent. Thus

$$q'_r = cq_r, \quad \dot{q}'_r = c^2\grave{q}_r, \quad A'_r = c^2A_r, \quad u'_{r/s} = cu_{r/s}, \quad \varepsilon'_1 = (\bar{c})^2\varepsilon_1, \tag{4.1}$$

and

$$V' = cV, \quad \lambda'_r = c\lambda_r, \quad \lambda'_4 = \lambda_4, \quad \lambda'_5 = \lambda_5, \quad \tau'_r = c\tau_r . \tag{4.2}$$

We can regard $(4.1)_5$ as a definition of ε'_1. Incidentally, if the classical limit of (2.5) is required to be the Fourier law, then instead of $(4.1)_5$, we ought to use the equality $\varepsilon'_1 = c^2\varepsilon_1$. Definition $(4.1)_5$ can be justified by remarking that, by it, only if ε'_1 tends to a finite limit when $c \to \infty$, the classical limit of (2.5) can have the desired non-stationary form

$$\dot{q}'_r + \varepsilon'_1\left(\frac{q'_r}{\varkappa'} + T_{/r}\right) = 0 \quad \textit{with} \quad \varepsilon'_1 = (\bar{c})^2\varepsilon_1 \ \textit{very large} \tag{4.3}$$

at least as an acceptable approximation - see [4, (8.1-2)].

Thus, in connection with the nonstationary classical limit, ε_1^{-1} ($= T\,\varepsilon^{-1}$) for $c = \infty$ has to be regarded as a finite non-vanishing quantity (of the same order as $(\bar{c})^2$).

That (4.3) really is a reasonable classical limit of (2.5) can be shown by noting that (2.5) affords an expression for dq'_r/dt $(= c^2 \overset{2\perp}{g}_{r\rho} \dot{q}^{\rho})$; this expression consists in ordinary size terms plus $(\bar{c})^2 \varepsilon_1 (\varkappa^{-1} q_r + T_{/r})$, of the same order as $(\bar{c})^2$. Hence the former are negligible with respect to the latter.

A parallel limit reasoning can be made on the discontinuity equation (3.8) corresponding to (2.5), in order to justify the nonstationary limit equation $(4.4)_4$ below. However it is quicker to justify $(4.4)_4$ by noting that it is the discontinuity equation corresponding to the nonstationary classical limit (4.3) of (2.5). Thus it is now clear that the nonstationary classical limits of equations (3.5-8) (each multiplied by a suitable power of c) are equivalent to

$$\begin{cases} -V'k\lambda'_r + (p_T\lambda_4 + p_k\lambda_5)N_r = 0, & V'kT(\eta_T\lambda_4 + \eta_k\lambda_5) - \tau'_1 = 0, \\ k\lambda'_1 - V'\lambda_5 = 0, & -V'\tau'_r + \varepsilon'_1\left(N_r - \dfrac{q'_r}{\varkappa'^2}\varkappa'_{T/1}\right)\lambda_4 = 0. \end{cases} \tag{4.4}$$

Let us assume the satisfactory condition $\varkappa_{T/s} \equiv 0$. Then a proper solution of system (4.4) in λ_1 to λ_5 and τ_1 to τ_5 exists if

$$V'^4[V'^4 - (V_\eta^2 + b)V'^2 + V_T^2 b] = 0 \qquad \left(b = \frac{\varepsilon'_1}{kT\eta_T},\quad V_\eta^2 = \frac{p_k\eta_T - p_T\eta_k}{\eta_T},\quad V_T^2 = p_k\right), \tag{4.5}$$

where, by $(4.5)_{3,4}$, V_η $[V_T]$ (> 0) is the adiabatic [isothermic] speed of acceleration waves in the purely mechanic case.[1]

---

(1) That V_η see $(4.5)_3$ - is the adiabatic propagation speed $(\partial p/\partial k)_{\eta = \text{const}}$ is shown in [3] - see (3.5) there. Furthermore the asserted property of (4.4) holds because by $(4.5)_2$ the elimination of λ'_1 and τ'_1 turns (4.4) into the equivalent system

$$\text{(a)}\quad p_T\lambda_4 + (p_k - V'^2)\lambda_5 = 0 = (V'^2 - b)\lambda_4 + V'^2\frac{\eta_k}{\eta_T}\lambda_5, \quad V'\lambda_5 = 0 = V'\tau_s \quad (s = 1,2)$$

of 6 equations in the 6 unknowns λ_4, λ_5, λ'_2, λ'_3, τ'_2, and τ'_3; and for it the determinant D of the coefficient matrix is clearly

$$\text{(b)}\quad D = V'^4\left[p_T\frac{\eta_k}{\eta_T}V'^2 - (p_k - V'^2)(V'^2 - b)\right] = V'^4\left[V'^4 + \frac{p_T\eta_k - p_k\eta_T}{\eta_T}V'^2 + bp_k\right].$$

Hence by $(4.5)_{3-4}$ the condition $D = 0$ is (4.5).
Let us add that the above result in classical physics is substantially included in [3]. In fact the classical limit $\mathbb{F}'$ of $\mathbb{F}$, for which (4.3) holds, is a fluid considered in [3], unobjective $(\zeta_o = 1)$, and thermically simple $(\mathring{\gamma} \equiv 1)$

The non-zero solutions of (4.5) V'_1, V'_2, $V'_3 = -V'_1$ (> 0), and $V'_4 = -V'_2$ (> 0) are given by

$$\left\{\begin{matrix} 2\,V_1'^2 \\ \\ 2\,V_2'^2 \end{matrix}\right. = V_\eta^2 + b \pm \sqrt{\Delta} \quad \text{with} \quad \Delta = b^2 + (2\,V_\eta^2 - 4V_T^2)b + V_\eta^4 \,, \tag{4.6}$$

so that, as was remarked in [3, (7.3)] - see the preceding footnote -

$$0 < V'_2 < V_\eta < V'_1 \qquad (V'_3 = -V'_1, \quad V'_4 = -V'_2) \,. \tag{4.7}$$

Let us add that for $V' = 0$ (and $\varkappa_{T/s} \equiv 0$) system (4.3) has ∞^4 solutions: $\lambda'_1 = \tau'_1 = \lambda'_4 = \lambda'_5 = 0$, while λ_2, λ_3, τ_2, and τ_3 are arbitrary. Hence the evolution system of $\mathbb{F}'$ (which is the non-stationary classical limit of (3.2), $(2.9)_2$, $(2.3)_1$, and (2.5)) is hyperbolic as soon as V'_1 to V'_4 are real, i.e. as soon as we have the first of the conditions

$$\Delta > 0 \,, \qquad V_n^2 + b + \Delta < 2c^2 \,. \tag{4.8}$$

Since ε'_1 is very large - see (4.3) - by $(4.5)_{2-4}$ the inequality $b^2 > 2V_\eta^2 - 4V_T^2$ is expected to hold; and it implies just $(4.8)_1$ by $(4.6)_2$. *Condition* $(4.8)_2$ *is necessary and sufficient for* $|V_r|$ *to be* $< c$ $(r = 1,\dots,4)$.

N.5 <u>On some special relativistic cases including the general 1-dimensional one. Hyperbolicity</u>.

Let us examine the relativistic cases when the discontinuities are normal or the heat flux vanishes.

If we have $0 \neq \underline{\lambda} \parallel \underline{N} \parallel \underline{\tau}$, *besides* (3.3), *then* (3.5-8) *become*

$$[2q_1 - (\rho + p)V]\lambda_1 + p_T\lambda_4 + p_k\lambda_5 - V\tau_1 = 0 \,, \qquad q_2\lambda_1 = 0 \,, \tag{5.1}$$

(and with $\overset{\circ}{\beta} = \varepsilon'_1$). For it [3, Theor. 9.1] holds and the non-zero values of V satisfy equation [3, (6.11)] with the terms in V^1 and V^3 cancelled. Then [3, (7.2-3)] hold; hence so do (4.6) and (4.7) below.

$$V(2q_1\lambda_1 + Tk\eta_T\lambda_4 + Tk\eta_k\lambda_5) - \tau_1 = 0 , \qquad k\lambda_1 - V\lambda_5 = 0 , \tag{5.2}$$

$$-V\tau_1 + \zeta [2\varepsilon^{-1}q_1\tau_1 - \varepsilon^{-2}(q_1)^2(\varepsilon_T \lambda_4 + \varepsilon_k\lambda_5 + \varepsilon_m q_1\tau_1) - \varepsilon^{-1}(q_1)^2 V\lambda_1 +$$

$$+ 2q_1\lambda_1] + \varepsilon_1\lambda_4 - V(1 + \zeta V^2 - \zeta)\varepsilon_1 T\lambda_1 - \varepsilon_1 q_1 \varkappa^{-2} \varkappa_{T/1} \lambda_4 = 0 \tag{5.3}$$

in that, by the consequence $q_2 = 0$ of (5.1), equation (3.8) holds identically for $r = 2,3$. Furthermore by (5.1) and $(3.3)_6$ *the assumptions above imply* $\underline{q} \parallel \underline{N}$.

Lastly let us note that, if $\underline{\lambda} \parallel \underline{N} \parallel \underline{\tau}$ besides (3.3), then (5.1-3) hold again, so that in the typical case also $0 \neq \underline{\lambda}$ holds again.

By $(2.6)_{1,5}$ *for* $q_r = 0$ *we have* $V = 1$, *since* (5.3) *simplifies to*

$$- V\tau_1 - V\varepsilon_1 T\lambda_1 + \varepsilon_1\lambda_4 = 0 \qquad (q_r = 0) . \tag{5.3'}$$

Under definitions (4.1-2), by eliminating λ_1 and τ_1 from (5.1-2) and (5.3'), one obtains for $q_r = 0$,[1]

$$[(\alpha + \beta)V'^2 - p_k]\lambda_5 + (\gamma V'^2 - p_T)\lambda_4 = 0 = \delta V'^2\lambda_5 + (\gamma V'^2 - \varepsilon_1)\lambda_4 , \tag{5.4}$$

where

$$\alpha = \frac{\rho + p}{c^2 k} , \qquad \beta = \frac{Tk\eta_k}{c^2} , \qquad \gamma = \frac{Tk\eta_r}{c^2} , \qquad \delta = \beta + \frac{\varepsilon_1 T}{c^2 k} . \tag{5.5}$$

System (5.4) in λ_5 and λ_4 has a proper solution iff

$$(\alpha + \beta - \delta)\gamma V'^4 - [(\alpha + \beta)\varepsilon_1 + \gamma p_k - p_T\delta] V'^2 + p_k\varepsilon_1 = 0 . \tag{5.6}$$

---

(1) For $q_r = 0$ (5.1-2) imply $[k^{-1}(\rho + p)V^2 - p_k]\lambda_5 - p_T\lambda_4 = -V\tau_1 = -V^2 Tk(\eta_T\lambda_4 + \eta_k\lambda_5)$, whence $[k^{-1}(\rho + p)V^2 - p_k + Tk\eta_k V^2]\lambda_5 + (V^2 Tk\eta_T - p_T)\lambda_4 = 0$. Thus by $(5.5)_{1-3}$, $(5.4)_1$ holds. Furthermore, for $q_r = 0$ (5.2) and (5.3') imply that

$$TkV^2(\eta_T\lambda_4 + \eta_k\lambda_5) = V\tau_1 = -V\varepsilon_1 T\lambda_1 + \varepsilon_1\lambda_4 = - V^2 k^{-1}\varepsilon_1 T\lambda_5 + \varepsilon_1\lambda_4 .$$

Thence, by $(5.5)_{2,3}$, $V'^2(\beta + k^{-1}\varepsilon_1 T)\lambda_5 + (\gamma V'^2 - \varepsilon_1)\lambda_4 = 0$.
Thus, by $(5.5)_4$ we have $(5.4)_2$.

Let us set - see $(2.1)_3$, $(4.5)_2$, and $(5.5)_{1-2}$

$$\bar{b} = \Big(1 + \frac{kw + p - T\,p_T + Tk^2\,\eta_k}{c^2 k}\Big)b = \Big(\alpha + \beta - \frac{T\,p_T}{c^2 k}\Big)\frac{\varepsilon'_1}{k\,T\eta_T} \quad . \tag{5.7}$$

Then by $(4.5)_{2,3}$ and $(4.1)_5$,

$$[(\alpha + \beta)\varepsilon_1 + \gamma\, p_k - p_T\delta]\gamma^{-1} = \frac{\alpha + \beta}{T\,k\,\eta_T}\,\varepsilon'_1 + \frac{\eta_T\,p_k - p_T\,\eta_k}{\eta_T} - \frac{p_T\,\varepsilon_1}{k^2\,\eta_T} = \bar{b} + v_\eta^2 \,. \tag{5.8}$$

Furthermore, by $(5.5)_4$ and $(2.1)_3$,

$$\alpha + \beta - \delta = \alpha - \frac{\varepsilon_1 T}{c^2 k} = 1 + \frac{kw + p - T\varepsilon_1}{c^2 k} \tag{5.9}$$

and (since $\varepsilon'_1 = c^2\varepsilon_1$) by $(4.5)_{2,4}$ and $(5.5)_3$, $p_k\varepsilon_1\gamma^{-1} = v_T^2 b$. Hence (5.6) (divided by γ) becomes the equation

$$\Big(1 + \frac{kw + p - T\varepsilon_1}{c^2 k}\Big)\,V'^4 - (\bar{b} + v_\eta^2) + v_T^2 b = 0 \ , \tag{5.6'}$$

so that its positive solutions $\bar{V}_1$ and $\bar{V}_2$ are given by

$$\left\{\begin{matrix} 2\bar{V}_1^{-2} \\ \\ 2\bar{V}_2^{-2}\end{matrix}\right. = \frac{\bar{b} + v_\eta^2 \pm \sqrt{\bar{\Delta}}}{1 + c^{-2}k^{-1}(kw + p - T\varepsilon_1)} \quad \text{with} \quad \bar{\Delta} = (\bar{b} + v_\eta^2)^2 - 4\Big(1 + \frac{kw + p - T\varepsilon_1}{c^2 k}\Big)v_T^2 b \ . \tag{5.10}$$

By (5.7), these relativistic speeds for normal waves in the case $q_r = 0$ can be easily compared with their classical analogues V'_1 and V'_2 in the general case - see (4.6).

* * *

Now let us consider the case $\underline{\lambda} \parallel \underline{N} \parallel \underline{\tau}$ $(q_r \neq 0)$ again, and let us set

$$\left\{\begin{array}{lll} c^2 B'_{11} = 2q'_1 - V'(kw + p), & c^2 B'_{16} = -V', & B'_{14} = B'_{15} = 0 \ , \\ c^2 B'_{41} = 2q'_1 V', & B'_{4i} = 0 = B'_{5i} = B'_{51} & (i = 4,5,6) \ , \end{array}\right. \tag{5.11}$$

$$\begin{cases} B'_{61} = -\zeta[\varepsilon^{-1}(q'_1)^2 V' + 2q'_1 - V'(V^2-1)\varepsilon_1 T] , \\ B'_{64} = -\zeta \dfrac{(q'_1)^2}{\varepsilon^2}\varepsilon_T , \quad B'_{65} = -\zeta \dfrac{(q'_1)^2}{\varepsilon^2}\varepsilon_k , \quad B'_{66} = -\zeta \dfrac{(q'_1)^2}{\varepsilon^2}\varepsilon_m q_1 . \end{cases} \tag{5.12}$$

Remark. By $(2.6')_2$, $V^2 - 1 = 0$. *Hence for* $\zeta = 0$ *we have* $B'_{h\ell} = ②$ $(h,\ell = 1,4,5,6)$.

Let $(A'_{h\ell})$ be the matrix of the coefficients for λ'_1 to λ'_5 and $\lambda'_{5+r} = \tau_r$ in the classical limit (4.4). Then *for* $q_2 = 0$ *(but possibly* $q_1 \neq 0$*) equations* (5.1-3) *regarded as equations in* λ'_1, λ'_4, λ'_5, *and* $\lambda'_6 = \tau'_1$ *(each multiplied by* c^2 *or* 1*) have a nonvanishing (common) solution iff*

$$\|A'_{h\ell} + B'_{h\ell}\|_{(h,\ell)\in\Gamma} = 0 \quad \text{with} \quad \Gamma = \{1,4,5,6\} , \tag{5.13}$$

in that $(A'_{h\ell} + B'_{h\ell})_{(h,\ell)\in\Gamma}$ is the matrix of their coefficients (the identity $(5.1)_2$ being obviously disregarded).

For $q_1 = 0$, the 4th order equation $(5.13)_2$ in V' has the solutions (5.10). By (5.11-12), if $q'_1 \neq 0$ but q_1 is not too large, the solutions of (5.13) are close to the classical solutions (4.6). Furthermore, by (4.7) these are simple and under condition (4.8) - which is expected to be abundantly satisfied in usual physical situations - they are less than c. Therefore *at least in standard physical situations, and also when* q_r *is not too large if* $\zeta = 1$, *the propagation speeds of relativistic normal acceleration waves travelling in* $\mathbb{F}$ *are real and* $< c$.

Let us now consider general 1- dimensional Cauchy problems for $\mathbb{F}$, in which the data are expressed by functions of x_1. Then the corresponding acceleration waves are normal and the italicized conclusion above holds for them. Hence *those reduced problems are strictly hyperbolic and (being symmetric) for them well known theorems of existence, uniqueness, and dependence (influence, and determinacy) hold - see* [8] *- provided standard physical conditions hold, and, if* $\zeta = 1$, q_r *be not too large.*

REFERENCES

[1] Bressan, A. (1978) Relativistic theories of materials, Springer Verlag, Berlin, 290 p.

[2] Bressan, A. (1980) a) On relativistic heat conduction in the stationary and non-stationary cases, the objectivity principle, and piezo-elasticity. *Lettere al Nuovo Cimento* 33, 108 (1982).

b) Addendum: On relativistic heat conduction in the nonstationary cases, the objectivity principle, and piezo-elasticity. *Lettere al Nuovo Cimento* 34, 63.

[3] Bressan, A. (1983) Non-stationary ideal fluids in classical physics under a possibly strong heat flux, printed as a memoir on *Atti Accad. Naz. Lincei* (VIII), 17, pp. 59-100.

[4] Bressan, A. (1983) On the relativistic non-stationary law of heat conduction and the objectivity principle,Piezo-elasticity, Fisica Mat., Suppl. *Boll. U.M.I.*, 2, 19-55.

[5] Carrassi,M. (1978) Linear heat equation from the kinetic theory, *Il Nuovo Cimento* 46 B, 363.

[6] Carrassi,M. and Morro, A. (1972) A modified Navier-Stokes equation and its consequence. *Il Nuovo Cimento* 9 B, 321.

[7] Cattaneo, C. (1948) Sulla conduzione del calore. *Atti del Seminario Mat. Fis. dell'Università di Modena*, 3, 3.

[8] Courant, R. and Hilbert, D. (1962) Methods of mathematical physics. Vol. 2, Interscience publishers, New York, 830 pp.

[9] Eckart, C. (1940) The thermodynamics of irreversible processes, III. *Phys. Rev.* 58, 267.

[10] Fox, A. (1969) Generalized thermoelasticity, *Int. J. Ingn. Sc.*, 7, 437.

[11] Green, A.E. and Lindsay, K.A. (1972) Thermoelasticity, *J. of Elasticity* 2.

[12] Grioli, G. (1979) Sulla propagazione di onde termomeccaniche nei continui, *Rend. Acc. Naz. Lincei* (VIII) 67, 332 (Nota I) and 67, 426 (Nota II).

[13] Gurtin, M.E., and Pipkin, A.C. (1968) A general theory of heat conduction with finite wave speed, *Arch. Rat. Mech. Anal.* 31, 139.

[14] Kalinski, S. (1965) *Bull. Acad. Polo. Sci. Sér. sci., techn.* 13, 211.

[15] Kranyš, M. (1966) Relativistic hydrodynamics with irreversible thermodynamics without the paradox of infinite velocity of heat conduction, *Il Nuovo Cimento* 62 B, N1, 51.

[16] Kranyš, M. (1977) Relativistic elasticity of dissipative media and its wave propagation modes, *J. Phys. A: Math. Gen.* 10, 1847.

[17] Kranyš, M. (1980) Relativistic electrodynamics of dissipative media, *Canadian J. of Phys.* 58, 666.

[18] Israel, W. (1976) Nonstationary irreversible thermodynamics, *Annals of Physics*, 100, 310.

[19] Israel, W. and Stewart, J.M. (1979) Transient relativistic thermodynamics and kinetic theory, *Annals of Physics*, 118, 341.

[20] Lianis, G. (1974) Relativistic thermodynamics of viscoelastic dielectrics, *Arch. Rat. Mech. Anal.*, 55, 300.

[21] Lord, H.W. and Schulman, Y. (1967) A generalized dynamical theory of thermoelasticity, *J. Mech. Phys. Solids*, 15, 299.

[22] Massa, E. and Morro, A. (1968) A dynamical approach to relativistic continuum mechanics, *Ann. Inst. H. Poincaré, Sect. A*, 29, 423.

[23] Maugin, G. (1974) Constitutive equations of heat conduction in general relativity, *J. Phys. A: Math. N. cl. Gen.*, 7, 465.

[24] Müller, I. (1967) Zum Paradoxon der Wärmeleitungstheorie, *Z. Physik*, 198, 329.

[25] Müller, I. (1969) Toward relativistic thermodynamics, *Arch. Rat. Mech. Anal.*, 34, 259.

[26] Pavon, D., Jou,D., and Casas-Vazquez, J. (1980) Heat conduction in extended relativistic thermodynamics, *J. Phys. A: Math. Gen.*, 13, L 77.

[27] Pavon, D., Jou,D., and Casas-Vazquez, J. (1982) On a covariant formulation of dissipative phenomena, *Ann. Inst. H. Poincaré*, 36, 79.

[28] Truesdell, C. (1977) A first course in rational continuous mechanics, Vol. I, Academic Press, New York.

[29] Vernotte, P. (1958) Les paradoxes de la théorie continue de l'équation de la chaleur, *C.R.Acad.Sci., Paris*, 246, 3154.

SOME THERMODYNAMICAL PROBLEMS IN MAGNETIC THEORY

Dario Graffi

University of Bologna

0) As is known, the application of modern thermodynamics to magnetic theory is due to B. Coleman and E. Dill [1],[2]; additional investigation in this field has been performed by M. Fabrizio [3],[4],[5] and the main results on thermodynamics in the electromagnetic theory are described in the Appendix I of [6].
However, already in the nineteenth century P. Duhem was able to obtain some properties of the magnetic bodies on the basis of thermodynamical considerations. His results can be found in livre IX of his treatise on electromagnetism which goes back to 1892. They are also described in a more modern language in the treatise of M. Jouguet [8]. In this talk (which however will not cover materials with memory) I will first briefly recall the Coleman Dill procedure in order to show how it makes possible the deduction of Duhem formulae. Then some minimum properties of the free energy will be recovered and discussed, properties which were already considered by Duhem (in his language the free energy is the internal potential); and finally I will describe a maximum property of free enthalpy.

1) Let us consider a domain D, of volume V, bounded by the regular surface σ; let x denote the generic point in D, and let the time be denoted by t. For every $t > 0$, let an electromagnetic field be given in D, i.e. for every x and every $t > 0$ at least one of electromagnetic vectors $\vec{E}(x,t)$, $\vec{H}(x,t)$, $\vec{D}(x,t)$, $\vec{B}(x,t)$, $\vec{u}(x,t)$ (i.e. the electric field, the magnetic field, the electric displacement vector, the magnetic induction vector, and the current density respectively) is different from zero. In what follows we shall often write $\vec{E}$ etc., for $\vec{E}(x,t)$ etc. wherever this does not generate misunderstanding; likewise we will omit the explicit indica-

tion of the (x,t) dependence for the additional quantities to be introduced later on. The electromagnetic vectors satisfy the Maxwell's equations:

$$\nabla \times \vec{H} = \dot{\vec{D}} + \vec{u} , \qquad \nabla \times \vec{E} = - \dot{\vec{B}} . \tag{1.1}$$

We can now proceed essentially following Coleman and Dill to describe the principles of thermodynamics in the form best suited for the electromagnetic field. Let be the bodies within D at rest, and the field sources outside D. Assume that the energy coming into D is equal to dt times the incoming flux of the Poynting vector $\vec{E} \times \vec{H}$ and of the heat flux vector $\vec{q}$. Since no mechanical work is performed in D, the energy coming into D will be equal to the increase of the internal energy U. Hence the first principle of thermodynamics is formulated (if $\vec{n}$ stands for the outward normal to σ) as follows:

$$-dt \int_{\sigma} (\vec{E} \times \vec{H} + \vec{q}) \cdot \vec{n} \, d\sigma = dU . \tag{1.2}$$

Now, if we denote by $\varepsilon(x,t)$ the (volume) density of internal energy at the point x, and at the time t, one has:

$$U = \int_V \varepsilon dV . \tag{1.3}$$

Substituting this in (1.2), dividing by dt and taking into account (1.1), by well known results we obtain:

$$\int_V (\vec{E} \cdot \dot{\vec{D}} + \vec{H} \cdot \dot{\vec{B}} + \vec{E} \cdot \vec{u} - \nabla \cdot \vec{q} - \dot{\varepsilon}) \, dV = 0 . \tag{1.4}$$

Since this relation holds for any domain contained in D as well, one has:

$$\vec{E} \cdot \dot{\vec{D}} + \vec{H} \cdot \dot{\vec{B}} + \vec{E} \cdot \vec{u} - \nabla \cdot \vec{q} - \dot{\varepsilon} = 0 . \tag{1.5}$$

This formula, valid for every $t > 0$ and every $x \in D$, gives, in our case, the local expression of the first principle of thermodynamics.

The second principle of thermodynamics can be expressed locally through the Clausius-Duhem inequality (recalled in Dafermos lectures) i.e.

$$\frac{\nabla \cdot \vec{q}}{\vartheta} - \frac{\vec{q} \cdot \nabla\vartheta}{\vartheta^2} + \dot{\eta} \geq 0 \tag{1.6}$$

where $\vartheta > 0$ is the absolute temperature and η the entropy density.
Let us introduce the free energy density:

$$\psi = \varepsilon - \vartheta\eta \,. \tag{1.7}$$

Then, subtracting (1.5) to (1.6) multiplied by $-\vartheta$, we get:

$$\dot{\psi} - \vec{E} \cdot \dot{\vec{D}} - \vec{H} \cdot \dot{\vec{B}} - \vec{E} \cdot \vec{u} + \frac{\vec{q} \cdot \nabla\vartheta}{\vartheta^2} + \eta\dot{\vartheta} \leq 0 \quad . \tag{1.8}$$

Since we have excluded from our considerations the materials with memory, $\psi(x,t)$ will depend only on the state of the medium at x. Specifically it will be assumed that such a state depends only on the $\vec{D}$, $\vec{B}$, ϑ (of course computed at x, t) since the other parameters needed for its determination (such as the material density) will be supposed time indipendent. Then, one has:

$$\dot{\psi} = \frac{\partial\psi}{\partial\vec{D}} \cdot \dot{\vec{D}} + \frac{\partial\psi}{\partial\vec{B}} \cdot \dot{\vec{B}} + \frac{\partial\psi}{\partial\vartheta}\dot{\vartheta} \quad . \tag{1.9}$$

Upon substitution in (1.8), the arbitrarity of $\dot{\vec{D}}$, $\dot{\vec{B}}$, $\dot{\vartheta}$ yields, by well known arguments (see [6]):

$$\vec{E} = \frac{\partial\psi}{\partial\vec{D}} \quad , \tag{1.10}$$

$$\vec{H} = \frac{\partial\psi}{\partial\vec{B}} \quad , \tag{1.11}$$

$$\eta = -\frac{\partial\psi}{\partial\vartheta} \,, \tag{1.12}$$

$$-\vec{E} \cdot \vec{u} + \frac{\vec{q} \cdot \nabla\vartheta}{\vartheta} \qquad 0 \quad . \tag{1.13}$$

The meaning of the notation is well known. In any case let us recall that if E_r, D_r, H_r, B_r are the components of $\vec{E}$, $\vec{D}$, $\vec{H}$, $\vec{B}$ respectively $(r = 1,2,3)$, one has (let $\psi = \psi(D_r, B_r, \vartheta)$):

$$E_r = \frac{\partial\psi}{\partial D_r} \quad , \tag{1.10'}$$

$$H_r = \frac{\partial\psi}{\partial B_r} \quad . \tag{1.11'}$$

2. Formulae (1.10) and (1.11) establish the relations between $\vec{E}, \vec{H}$ and $\vec{D}, \vec{B}$, i.e. they represent the constitutive equations of the electromagnetic field. In order to obtain a more explicity form for these relations let us try to obtain an expression of ψ having more concrete physical meaning. Let us begin by the case in which the medium in D is the vacuum. In this case the constitutive equations are:

$$\vec{D} = \varepsilon_o \vec{E} \quad , \tag{2.1}$$

$$\vec{B} = \mu_o \vec{H} \quad , \tag{2.2}$$

where ε_o is the vacuum dielectric constant and μ_o the vacuum magnetic permeability. Moreover, since it is natural to take $\vec{q} = 0$ in the vacuum, (1.6) implies $\dot{\eta} = 0$, and then η is a constant which can be chosen equal to zero. Then if we denote by ψ_v the free energy density in the vacuum, (1.7) yields $\psi_v = \varepsilon$. Hence (1.5) yields, taking into account (2.1) and (2.2)

$$\dot{\psi}_v = \varepsilon_o \vec{E} \cdot \dot{\vec{E}} + \mu_o \vec{H} \cdot \dot{\vec{H}} \quad , \tag{2.3}$$

whence, integrating both sides and supposing $\psi_v = 0$ for $\vec{E} = \vec{H} = 0$:

$$\psi_v = \frac{\varepsilon_o E^2}{2} + \frac{\mu_o H^2}{2} \quad . \tag{2.4}$$

That is, according to the natural intuition, the free energy density in the vacuum coincides with the Maxwell electromagnetic energy.

Assume now that D contains some material bodies. We will limit ourselves to the case of a permanent magnet and a body magnetizable by induction, referred as the magnet and the body in what follows.

Let the magnet occupy a domain D_p, and the body a domain D_i, $D_p \subset D$, $D_i \subset D$, $D_i \cap D_p = \emptyset$, and the complement of D with respect $D_i \cup D_p$ be filled up with vacuum. Let $\vec{J}_p(y_p)$ denote the magnetization vector at the point $y_p \in D_p$, $\vec{J}_p$ will be time independent. On the contrary let $\vec{J}_i(y_i,t)$ the magnetization intensity vector at the point y_i of the body $(y_i \in D_i)$, this last vector may be time dependent. Let us furthermore recall that the following relation holds:

$$\vec{B} = \mu_o \vec{H} + \vec{J} \tag{2.5}$$

where $\vec{J} = \vec{J}_p$ in the magnet, $\vec{J} = \vec{J}_i$ in the body, $\vec{J} = 0$ outside $D_p \cup D_i$ according to (2.2).

Now obviously the free energy in the bodies will be different from its value ψ_v in the vacuum. Since from now on the temperature will be assumed time independent and uniform, and, unless otherwise specified, the electrical polarization in the magnet and in the body will be equal to zero, it is natural to assume that the variation of ψ at some point of space depends only on the magnetization intensity at that point. Therefore we will write in the magnet and in the body respectively

$$\psi = \psi_v + \varphi_p(\vec{J}_p) \quad , \tag{2.6}$$

$$\psi = \psi_v + \varphi_i(\vec{J}_i) \quad , \tag{2.7}$$

where $\varphi_p(\vec{J}_p)$, $\varphi_i(\vec{J}_i)$ are scalar functions of $\vec{J}_p$, $\vec{J}_i$ respectively.

Let us consider (1.8) at some point of the magnet, and replace ψ by (2.6). Since $\vec{J}_p$, and hence $\varphi_p(\vec{J}_p)$, is time independent, as well as ϑ, and taking in account (2.1) and (2.6), (1.8) becomes (1.13) and therefore no constitutive relations.

If on the contrary we consider a point of body, we have to insert (2.7) in (1.8). By (2.1), (2.2) we get (ϑ is constant)

$$\frac{\partial\varphi(J_i)}{\partial \vec{J}_i} \cdot \dot{\vec{J}}_i - \vec{H} \cdot \dot{\vec{J}}_i - \vec{E} \cdot \vec{u} \le 0 \tag{2.8}$$

whence

$$\vec{H} = \frac{\partial\varphi_i(\vec{J}_i)}{\partial \vec{J}_i} \quad . \tag{2.9}$$

(2.9) is a constitutive equation because it establishes a relation between the magnetization intensity vector in the body and the magnetic field generating the magnetization. (2.9) is essentially due to Duhem. It can be also written in a way analogous to (1.10'), (1.11') introducing the components H_r, J_{ir} of $\vec{H}$ and $\vec{J}_i$

$$H_r = \frac{\partial \varphi(\vec{J}_i)}{\partial J_{ir}} \quad . \tag{2.10}$$

Let us discuss (2.9) and (2.10) in the particular case in which the body is an isotropic medium. Then φ depends only on the length $J_i = \sqrt{J_{i1}^2 + J_{i2}^2 + J_{i3}^2}$ of the vector J_i. Therefore one has:

$$H_r = \varphi_i'(J_i) \; \frac{\partial J_i}{\partial J_{ir}} = \frac{\varphi_i'(J_i)}{J_i} \; J_{ir} \quad . \tag{2.11}$$

Hence if we set

$$K(J_i) = \frac{J_i}{\varphi_i'(J_i)} \tag{2.12}$$

it follows:

$$\vec{H} = \frac{\vec{J}_i}{K(J_i)} \quad . \tag{2.13}$$

As an example in the case of weak magnetic bodies one has:

$$\varphi(J_i) = \frac{J_i^2}{2\chi} \tag{2.14}$$

where χ is the (constant) magnetic susciptivity of the body, positive in the paramagnetic case and negative in the diamagnetic one.

Substituting into (2.13), we get the well known relation:

$$\vec{H} = \frac{\vec{J}_i}{\chi} \quad . \tag{2.15}$$

However (2.13) with K dependent on J_i can also represent the ferromagnetic bodies provided the hysteresis can be neglected. In this case the function $K(J_i)$ is certainly positive. Furthermore, since the curve representing the relation between J_i and H can be considered very close to the first magnetization curve, J_i will be an increasing function of H, tending asymptotically to the saturation value of J_i for $H \to \infty$. Hence, in this case, (2.13) is invertible (as well as (2.15) obviously) and Moreover $\frac{\partial H}{\partial J_i} > 0$. More generally if we now assume that (2.9) is always invertible, inserting in (2.5) we have in the body:

$$\vec{B} = \mu_o \vec{H} + \vec{J}_i(\vec{H}) = \vec{B}(\vec{H}) \quad . \tag{2.16}$$

Remark that (2.16) holds also in the magnet, provided $J_i(\vec{H})$ is replaced by $\vec{J}_p$. In other words $\vec{B} = \vec{B}(\vec{H})$ for any medium provided there is no hysteresis. According to experimental evidence we will assume that $\vec{B}$ is an increasing vector function of $\vec{H}$ i.e. for any $\vec{H}' \neq \vec{H}$

$$(\vec{B}(\vec{H}') - \vec{B}(\vec{H})) \cdot (\vec{H}' - \vec{H}) > 0 \; . \tag{2.17}$$

Remark that (2.17) is a consequence of (2.16) because $\vec{J}_i(\vec{H})$ is increasing, and also a consequence of (2.15) because also for diamagnetic body $\mu_o + \chi$ is always positive as shown by the experimental evidence. Before concluding, remark that if in D there are also bodies with an electrical polarization $\vec{P}$ different from zero, a function of $\vec{P}$ has to be added in (2.6), (2.7). By what described above, the investigation in this case does not give rise to any further conceptual difficulty.

3) In what follows we will take D to be the whole space and, except in the last section, let in D be contained only the magnet and the body. Let the magnetization distribution be known (i.e. the function $\vec{J}_p(y_p)$) in the magnet. The body is, on the contrary, magnetized by induction; as soon as the equilibrium condition is reached, there will be in the body a time indipendent distribution of the magnetization intensity vector. This last function is, a priori, unknown. In this section and in the next one, we will recall the method used to determine this distribution. It is however convenient first to recall some notions of static magnetization that we prove useful in what follows.

Let S_i and S_p respectively be the volumes of the domains D_i and D_p occupied by the body and the magnet. It is well known (see e.g. [9] section 385, notice that the notation and the unit there are different from ours) that the fields $\vec{H}_i$ and $\vec{H}_p$ generated in x by the body and the magnet are given by (here the operator ∇ is always taken with respect to x):

$$\vec{H}_i = - \nabla V_i \tag{3.1}$$

$$\vec{H}_p = -\nabla V_p \tag{3.1'}$$

$$V_i = -\frac{1}{4\pi\mu_o} \int_{S_i} \vec{J}_i(y_i) \cdot \nabla \frac{1}{|x-y_i|} dS_i \tag{3.2}$$

$$V_p = -\frac{1}{4\pi\mu_o} \int_{S_p} \vec{J}_p(y_p) \cdot \nabla \frac{1}{|x-y_p|} dS_p \tag{3.2'}$$

Therefore the total field at any point x of space is given by:

$$\vec{H} = \vec{H}_i + \vec{H}_p = -\nabla V , \qquad V = V_i + V_p . \tag{3.3}$$

Consider now (2.5); by (3.1) and (3.2) (see [9] section 402, the same result can be obtained from the last formula of [9] section 385, on account of well known properties of the Newtonian potential) one has that the vector $\vec{B}$ is solenoidal, i.e.

$$\nabla \cdot \vec{B} = 0 \tag{3.4}$$

$$[B_n] = 0 . \tag{3.5}$$

(3.4) holds at all points of space, excluding at most the surfaces to be mentioned in a moment. (3.5) holds on the surface of the body and on the surface of the magnet, (3.5) means that on these surfaces the normal component of $\vec{B}$ is continuous.
It is worth remarking that if $\vec{J}_i(y_i)$ is increased by $\Delta J(y_i)$, by (3.1) and (3.2), V, $\vec{H}_i$ and hence $\vec{H}$ are increased by ΔV_i, $\Delta\vec{H}_i$ and $\Delta\vec{H}$ respectively. ΔV_i and $\Delta\vec{H}_i$ can be determined by (3.1), (3.2) replacing $\vec{J}_i(y_i)$ with $\Delta\vec{J}(y_i)$. Also $\vec{B}$ is in turn increased by

$$\Delta\vec{B} = \mu_o \, \Delta\vec{H} + \Delta\vec{J} \tag{3.6}$$

and $\Delta\vec{B}$ is solenoidal as well i.e. satisfies (3.4), (3.5).
A final remark is that, by well known properties of the Newtonian potential V_i, V_p, $\vec{H}_i$, $\vec{H}_p$ (and consequently ΔV, $\vec{H}$, $\vec{B}$, $\Delta\vec{H}$, $\Delta\vec{B}$) are convergent at infinity that is, if x_o stands for a reference point, $V_i(x)$ and $V_p(x)$ and the remaining scalar quantities considered above are, at least, $O((x-x_o)^{-2})$ as $x \to \infty$, and the vector quantities $O((x-x_o)^{-3})$. Therefore the following theorem can be applied to all quantities considered; if $g(x)$ and $\vec{u}(x)$ are a scalar and a vector quantities respectively and $\vec{u}(x)$ is solenoidal and S_∞ is the whole space, then:

$$\int_{S_\infty} \nabla g \cdot \vec{u}\, d\, S_\infty = 0 \ . \tag{3.7}$$

4) Let us now turn to write the determining equations for V_i, where of course, $\vec{J}_p$ and hence V_p and $\vec{H}_p$ are to be considered as assigned.
Once V_i is known, by (3.1), $\vec{H}_i$ is determined and hence, by (2.9), $\vec{J}_i$, i.e. the magnetization of the body. To this end remark that, by (2.16), (3.3) and (3.1) one has:

$$\vec{B} = \vec{B}(\vec{H}) = \vec{B}(\vec{H}_p - \nabla V_i) \ . \tag{4.1}$$

Inserting in (3.4) we have the differential equation (non linear in the ferromagnetic case) satisfied by V_i. The function $\vec{B}$ given by (4.1) must furthermore satisfy (3.5) and be convergent at infinity. These conditions uniquely determine $V_i(x)$, since any other function $V_i'(x)$ fulfilling them coincides with $V_i(x)$.
For, if we set:

$$\vec{H}' = \vec{H}_p - \nabla V'_i \tag{4.2}$$

$$\vec{B}' = \vec{B}(\vec{H}') \tag{4.3}$$

since $\vec{B}'$ is solenoidal as well as $\vec{B}$ and V_i, V'_i have the same properties at infinity, by (3.7), (4.2), (3.1), (3.3) one has:

$$0 = -\int_V \nabla(V_i' - V_i) \cdot (\vec{B}'(\vec{H}') - \vec{B}(\vec{H}))dS_\infty = \int_{S_\infty} (\vec{B}(\vec{H}') - \vec{B}(\vec{H})) \cdot (\vec{H}' - \vec{H})dS_\infty \tag{4.4}$$

However (4.4) is compatible with (2.17) only if $\vec{H} = \vec{H}'$ i.e. (taking into account V_i and V_i' both vanish at the infinity) $V_i = V_i'$ according to the stated theorem.
Remark that the same argument would prove the uniqueness of V_i and $\vec{J}_i$ provided the inducting field is generated by a stationary electrical current (see also [10], §II - 6).

5) The uniqueness theorem of the former section for V_i, $\vec{J}_i$, $\vec{H}$ (that we call equilibrium solution) requires the validity of the fundamental laws of magnetostatic (i.e. (3.4), (3.5)) associated with (3.1), (3.2), (3.3) and the constitutive equations (2.16) which are essentially a consequence of (2.9). If however (2.16) is not taken into account, there are infinitely many solutions $\vec{J}_i(y_i) + \Delta\vec{J}(y_i)$, $V_i(x) + \Delta V_i(x)$, $\vec{H}(x) + \Delta\vec{H}(x)$ of the other equations (hence the relations (3.1), (3.2), (3.3) among $\Delta\vec{H}$, ΔV, $\Delta\vec{J}$ remain valid as remarked in section 3) that we call admissible solutions. Corresponding to each admissible solution there is by (2.6), (2.7) (in which we may suppose $\vec{E} = 0$) a value for the density of free energy and therefore for the total free energy:

$$F(\vec{J}_i(y_i) + \Delta\vec{J}(y_i)) = \int_{S_\infty} \frac{\mu_o}{2} (\vec{H} + \Delta\vec{H})^2 d\,S_\infty + \int_{S_i} \varphi_i[\vec{J}_i(y_i) + \Delta\vec{J}(y_i)]d\,S_i + \\ + \int_{S_p} \varphi_p(\vec{J}_p(y_p))d\,S_p \quad . \tag{5.1}$$

That is the total free energy is a functional of $\vec{J}_i(y_i) + \Delta\vec{J}(y_i)$ since such is $\vec{H}(x) + \Delta\vec{H}(x)$ for every x.

Let us prove the theorem: the total free energy is stationary in correspondence to equilibrium solution, i.e. if $\Delta\vec{J}(y_i)$ is infinitesimal of order h, the total free energy is increased of order h^2. For, one has:

$$F(\vec{J}_i(y_i) + \Delta\vec{J}(y_i)) - F(\vec{J}_i(y_i)) = \int_{S_\infty} \mu_o\vec{H} \cdot \Delta\vec{H}\,d\,S_\infty + \\ + \int_{S_i} [\varphi_i(\vec{J}_i(y_i) + \Delta\vec{J}(y_i)) - \varphi_i(\vec{J}_i(y_i))]d\,S_i + \frac{1}{2}\int_{S_\infty} (\Delta\vec{H})^2 d\,S_\infty \tag{5.2}$$

Now for y_i fixed a truncated Taylor expansion yields:

$$\varphi_i(\vec{J}_i + \Delta\vec{J}_i) - \varphi_i(\vec{J}_i) = \frac{\partial\varphi_i}{\partial J_i} \cdot \Delta\vec{J}_i + \alpha\Delta\vec{J}_i \cdot \Delta\vec{J}_i + O(\Delta\vec{J}_i^3) \tag{5.3}$$

where α is a rank two tensor of components

$$\alpha_{rs} = \frac{\partial^2\varphi(\vec{J}_i)}{\partial J_{it}\partial J_{is}} \quad . \tag{5.4}$$

Hence, taking into account (2.9), (3.6), (3.7) and the fact that outside the body $\Delta\vec{J} = 0$,

$$\int_{S_\infty} \left(\mu_o \vec{H}\cdot\Delta\vec{H} + \frac{\partial\varphi(\vec{J}_i)}{\partial\vec{J}_i}\cdot\Delta\vec{J}\right) dS_\infty = \int_{S} \vec{H}\cdot\Delta\vec{B}\, dS_\infty = 0 \tag{5.5}$$

The inserting (5.3) in (5.2) and taking into account (5.5) we have:

$$F(\vec{J}_i(y_i) + \Delta\vec{J}_i(y_i)) - F(\vec{J}_i(y_i)) = \frac{1}{2}\int_{S_\infty} (\Delta\vec{H})^2 dS_\infty + \int_{S_i} (\alpha\Delta\vec{J}\cdot\Delta\vec{J} + O(\Delta\vec{J}^3))\, dS_i \ . \tag{5.6}$$

Now, if $\Delta\vec{J}$ is infinitesimal of order h, the same is true for $\Delta\vec{H}$ (as mentioned section (3), $\Delta\vec{H}$ is obtained by (3.1) and (3.2) replacing $\vec{J}_i$ by $\Delta\vec{J}$). Hence by (5.6) the increasing of total free energy is infinitesimal as h^2, in agreement with the stated theorem.

But there is more. If the tensor α is positive (i.e. $\alpha\Delta J \cdot \Delta J > 0$ for $\Delta J \neq 0$) for ΔJ small enough we have:

$$F(J(y_i) + \Delta J(y_i)) > F(J_i(y_i)) \tag{5.7}$$

i.e. the total free energy is a minimum on the solution of equilibrium.

Let us look under what condition α is positive. When the bodies are weakly magnetic, (2.14) immediatly implies:

$$\alpha_{rs} = \frac{1}{\chi}\delta_{rs} \tag{5.8}$$

and α is positive for paramagnetic bodies.

In the case of isotropic ferromagnetic bodies, differentiating (2.11) one has:

$$\alpha_{rs} = \frac{\partial H_r}{\partial J_s} = \frac{\varphi_i'(J_i)}{J_i}\delta_{rs} - \frac{K'(J_i)}{J_i K^2(J_i)} J_{i/r} J_{i/s} \ . \tag{5.9}$$

Hence

$$\alpha\Delta J\cdot\Delta J = \frac{(\Delta J)^2}{K(J_i)} - \frac{K'(J_i)}{J_i K^2(J_i)}(\Delta\vec{J}_i\cdot\vec{J}_i)^2 \tag{5.10}$$

Now if $K'(J_i) < 0$, $\alpha\Delta J\cdot\Delta J$ is positive; if $K'(J_i) > 0$ by (2.13) and what remarked in section 2 one has:

$$\alpha\Delta J \cdot \Delta J \geq (\Delta J)^2 \left(\frac{1}{K(J_i)} - \frac{K'(J_i)}{J_i K^2(J_i)} J_i^2\right) = (\Delta J)^2 \frac{d}{dJ_i}\left(\frac{J_i}{K(J_i)}\right) =$$

$$= (\Delta J)^2 \frac{dH}{dJ_i} > 0 \ . \tag{5.11}$$

Hence the minimum theorem holds also for ferromagnetic bodies. It is however doubtful for diamagnetic bodies since α_{rs} is negative because $\chi < 0$.
It is worth remarking that Duhem assumes this minimum property as a principle justified by thermodynamics in order to infer from it the constitutive equations (2.9). In the first part of this section we have in some sense inverted the Duhem theorem; this result is however obvious considering how (2.9) is inferred by (2.8) in section 2. In the second part we have shown, under plausible assumptions, the minimum properties both for paramagnetic bodies and for ferromagnetic ones, provided the hysteresis can be neglected.
As for the diamagnetic bodies are concerned, Duhem concluded that they could not exist because the minimum principle is not always verified. However in further investigation ([11] p. 368) by means of a more complete examination of the thermodynamical equilibrium condition, he was able to overcome the above difficulties.

6) To conclude this talk, I will show a further minimum theorem valid under conditions less restrictive than the former ones. To this end, let us recall that Coleman and Dill introduce the so called free enthalpy whose volume density is given by the formula:

$$\zeta = \psi - \vec{H} \cdot \vec{B} - \vec{E} \cdot \vec{D} \ . \tag{6.1}$$

Under the usual assumptions of invariability and uniformity for the temperature ϑ, and of time independence of $\vec{D}$ and $\vec{E}$ and expressing by (2.16) $\vec{B}$ as a function of $\vec{H}$, we may write:

$$\zeta = \zeta(\vec{H}) \ . \tag{6.2}$$

The total free enthalpy is therefore:

$$G(\vec{H}) = \int_{S_\infty} \zeta(\vec{H})\, dS_\infty \quad . \tag{6.3}$$

And taking into account (1.9) and (1.11) we have:

$$\dot{\zeta} = -\vec{B}(\vec{H}) \cdot \dot{\vec{H}} \implies d\zeta = -\vec{B}(\vec{H}) \cdot d\vec{H} \quad . \tag{6.4}$$

Let us consider, as in the preceding sections, the magnetic body magnetized through induction by some stationary electric currents flowing on a conductor exterior to the body and non magnetic (i.e. with $\chi = 0$ such as, with fairly accurate approximation, in copper); these last hypothesis as well the exclusion of permanent magnet are stated only to semplify the exposition. Let $\vec{H}_c$ be the field generated by the currents due to the magnetization of body, we have to add to $\vec{H}_c$ a field $\vec{H}_i$ of potential W; let us therefore write:

$$\vec{H} = \vec{H}_c - \nabla W \quad . \tag{6.5}$$

By (2.16) there corresponds an induction vector, generally speaking, non solenoidal:

$$\vec{B} = \vec{B}(\vec{H}_c - \nabla W) \quad . \tag{6.6}$$

Now the field expressed by (6.5), called the field compatible with the current (or, as sect. 5, the admissible field), coincides with the field of equilibrium only when W coincides with the value of V for which the corresponding $\vec{B}$ given (6.6) is solenoidal, i.e. fulfills (3.4) and (3.5). Now remark that inserting (6.4) in (6.3) since $\vec{H}_i$ is given, the expression of the total free enthalpy for all admissible fields in a functional of W. Let us show the following theorem [12],[13]. The total free enthalpy is a maximum, and hence the complementary energy (defined in [11] as minus the free enthalpy) is a minimum when W coincides with V, i.e. when the field is of equilibrium.

To prove this result let we first set:

$$\vec{H}_o = \vec{H}_c - \nabla V \tag{6.7}$$

$$\vec{B}_o = \vec{B}(H_o) \tag{6.8}$$

and the vector $\vec{B}_o$ is solenoidal by what mentioned above.
Furthermore we set:

$$\vec{H}(\lambda) = \vec{H}_o + \lambda(\vec{H} - \vec{H}_o) \ , \qquad \lambda \in [0,1] \ . \tag{6.9}$$

Obviously $\vec{H}(0) = \vec{H}_o$, $\vec{H}(1) = \vec{H}$; $\vec{H}$ is supposed fixed as $\vec{H}_o$. Inserting (6.9) in (6.4) we have:

$$d\xi = - \vec{B}(\vec{H}_o + \lambda(\vec{H} - \vec{H}_o)) \cdot (\vec{H} - \vec{H}_o) \ dh =$$

$$= - \frac{\vec{B}(\vec{H}_o + \lambda(\vec{H} - \vec{H}_o)) - \vec{B}(\vec{H}_o)}{\lambda} \cdot \lambda(\vec{H} - \vec{H}_o) d\lambda + \vec{B}_o (\vec{H} - \vec{H}_o) d\lambda \ . \tag{6.10}$$

By inserting in (2.17) $\vec{H}_o = \vec{H}$, $\vec{H}' = \vec{H}_o + \lambda(\vec{H} - \vec{H}_o)$, the first terms in the r.h.s. of (6.10) result negative. Hence integrating (6.9) with respect λ from 0 to 1, one has:

$$\zeta(\vec{H}) - \zeta(\vec{H}_o) < \vec{B}_o \cdot (\vec{H} - \vec{H}_o) \tag{6.11}$$

i.e. inserting in (6.3):

$$G(\vec{H}) - G(\vec{H}_o) \leq - \int_{S_\infty} \vec{B}_o \cdot \nabla(W - V) \ dS_\infty \ . \tag{6.12}$$

Since $\vec{B}_o$ is solenoidal and $\vec{B}_o$ convergent to infinity as $W - V$, the last term of (6.12) vanishes. Hence

$$G(\vec{H}) < G(\vec{H}_o)$$

in agreement with the statement of the theorem.
Remark that this proof is that of [12] semplified by Guido Ascoli (see [13]). Le us finally remark that the maximum is taken over all solutions fulfilling the determinacy condition of problem, including the constitutive equations, but excluding the solenoidality of $\vec{B}$, while in the former section, on the contrary, only the constitutive equation was excluded.

REFERENCES

[1] B. D. Coleman and E. H. Dill, Thermodynamics restrictions on the constitutive equations of electromagnetic theory, *Zeitschrift für angewandte Mathematik und Physik (ZAMP)*, XXII (1971), 691-702.

[2] B. D. Coleman and E. H. Dill, On the thermodynamics of electromagnetic fields in materials with memory, *Archive of Rational Mechanics and Analysis*, XLI (1971), 121-172.

[3] M. Fabrizio, Una teoria fenomenologica per la termodinamica del campo elettromagnetico, *Bollettino U.M.I.*, (4) IX (1971), 708-719.

[4] M. Fabrizio, Teoremi di approssimazione e restrizioni termodinamiche per le equazioni costitutive del campo elettromagnetico, *Atti Accademia delle Scienze di Bologna*, (12) X (1972-73), 97-110.

[5] M. Fabrizio, Convessità dei potenziali termodinamici nell'elettromagnetismo, *Rendiconti Accademia Lincei*, LV (1973), 241-248.

[6] D. Graffi, Questioni di elettromagnetismo. Liguori, Napoli, 1981.

[7] P. Duhem, Leçons sur l'electricité et le magnetisme. Gauthier Villars,Paris, 1892.

[8] M. Jouguet, Traité d'electricité théorique. Gauthier Villars, Paris, 1955.

[9] J. C. Maxwell, Treatise of electricity and magnetism (traduzione italiana di Evandro Agazzi, U.T.E.T., Torino, 1973).

[10] D. Graffi, Non linear partial differential equations in physical problems. Pitman, London, 1980.

[11] L'oeuvre scientifique de Pierre Duhem. Blanchard Paris et Teret Bordeau, 1928.

[12] D. Graffi, Su una legge di minimo della magnetostatica, *Annali dell'Università di Ferrara*, III (1954), 25-29.

[13] D. Graffi, Alcuni problemi non lineari della Fisica Matematica, *Rendiconti del Seminario Matematico dell'Università e del Politecnico di Torino*, XIV (1955), 75-86.

RELAXATION PHENOMENA VIA HIDDEN VARIABLE THERMODYNAMICS

by
Angelo Morro
Istituto di Matematica dell'Università
16132 Genova, Italy

1. Introduction

There are various phenomenological models accounting for dissipative effects in continuous media. They, though, may be framed in one of the following three schemes. First, *materials with instantaneous response*: the response at time t is determined by a suitable set of state variables at the same time t. Such is the case, for instance, of thermoelasticity and of the Navier-Stokes-Fourier (NSF) theory of viscosity and heat conduction. Second, *materials with (fading) memory*: the response at time t is determined by the history of certain state variables up to time t. Viscoelasticity is an outstanding topic where this scheme is applied. Third, *materials with hidden variables*: the response at time t is determined by a suitable set of observable variables and of hidden (or internal) variables (named also additional parameters), all at the same time t; the rate of change of the hidden variables is governed by a function on the observable variables and on the hidden variables as well.

It is evident that the first scheme is so poor as to justify the search for generalizations. Yet, in view of the comprehensiveness of the fading memory theories and in view of the rather special nature of the hidden variable model as a way of accounting for memory, one might well ask if anything motivates the recourse to hidden variables. One motivation is as follows. The explicit structure of memory functionals, which is needed to make a model operative, is obtained by a nonlinear multiple integral representation. In practice, one has to be content with only the first few integrals; even so, however, it is a formidable task to evaluate the corresponding kernels. That is why the relatively simple character of the memory introduced by hidden variables is likely to be more profitable in trying to set up a satisfactory model of a material behaviour. That this is so may be ascertained through the rapidly growing literature on materials with hidden variables; references [1-26] are only a few of the papers which I am aware of.

To my knowledge, the paper of Coleman and Gurtin [1], appeared in 1967, marks the beginning of thermodynamics with hidden variables. Starting from the physical motivation of setting up a model for the transfer of molecular motion from the translational mode to one or several internal modes in polyatomic gases [27], Coleman and Gurtin elaborated quite a general theory of thermoelastic materials with hidden variables. While they did not claim to present a theory of reacting materials, there are formal similarities between the two theories and thus results obtained from one theory have application to the other. Later, hidden variable thermodynamics was applied in other contexts such as visco-plasticity [11], rate type plasticity [13], dielectric relaxation phenomena [7], magnetic relaxation phenomena [14].

Apart from direct comparison with possible experimental results, the validity of a model is examined against some general requirements of an acceptable description of material behaviour. In continuum thermodynamics such requirements embrace compatibility with wave propagation. Now, it is well known that the NSF theory does not allow wave front propagation; among the attempts to overcome this drawback I mention that of Kosiński and Perzyna [5] who described heat conduction through a hidden variable thus accounting for thermal waves at finite speed. An account of heat conduction and viscosity through hidden variables has been set up by myself which generalizes the NSF theory and, meanwhile, is compatible with wave propagation. Lately, more refined models have been performed which show how fruitful the hidden variable scheme really is. It is just the aim of this paper to outline some general ideas pertaining to the hidden variable thermodynamics and then to emphasize some successful models of dissipative phenomena in solids and fluids.

The plan of the paper is summarized as follows. Section 2 gives a résumé of the main features characterizing a system with hidden variables; particular attention is focused on the asymptotic stability of the evolution equation. Next, section 3 delivers a relatively simple model of a heat conducting and viscous fluid with hidden variables; after having derived the main consequences of the entropy inequality, a comparison is made with the theory elaborated by Coleman, Fabrizio, and Owen [28]. More involved models of fluids and solids are examined in section 4; such models are based on evolution equations where gradients of the hidden variables occur and then they are likely to be adequate for many phenomena involving steep gradients or rapid variations. In connection with the analysis of the general features of the hidden variable thermodynamics (HVT), a detailed comparison with the extended irreversible thermodynamics (EIT) is exhibited in section 5; the main differences and analogies are emphasized through the example of ultrasonic waves in metals. Finally, in section 6, it is pointed out how the objectivity and the hyperbolicity may provide a priori restrictions on the evolution equations.

2. Essentials of systems with hidden variables

Henceforth $\mathbf{Y}$, $\mathbf{A}$, and Φ denote finite dimensional real normed vector spaces, subject to the requirement $\dim \mathbf{A} \leq \dim \mathbf{Y}$, while $\mathbf{L}(\cdot,\cdot)$ stands for the normed vector space of all linear maps from a vector space into another. The symbol $|\cdot|$ is adopted to denote the usual norm - $|\mathbf{p}| = (\mathbf{p}\cdot\mathbf{p})^{1/2}$ - both in $\mathbf{Y}$, $\mathbf{A}$ and in $\mathbf{L}$. A superposed dot designates (material) time differentiation.

Following Day [12], we define a *system with hidden variables* $\{y_0, a_0, \mathbf{Y}, \phi, f\}$ on $\mathbf{Y}\times\mathbf{A}$ to consist of a ground value (y_0, a_0) of the variables $(y, a) \in \mathbf{Y}\times\mathbf{A}$, y representing the set of *observable variables* and a the set of *hidden variables*, together with the maps

$$\phi \in C^2(Y\times A, \Phi) , \qquad f \in C^2(Y\times A, \Phi) .$$

It is usual to assume that Y is an open connected neighbourhood of y_0. A pair (y, a) is a *state*, ϕ is the *response* of the system and f governs the growth of the hidden variables through the ordinary differential equation

$$\dot{a}(t) = f(y(t), a(t)), \quad t \geqslant t_0, \quad a(t_0) = \hat{a} , \tag{2.1}$$

t being the time. We assume that, corresponding to each $y \in Y$ - or, at least, $y \in U$, U being a suitable open connected neighbourhood of y_0, $U \subset Y$ - there is just one hidden variable $E(y) \in A$ such that

$$f(y, E(y)) = 0 ;$$

moreover we assume that $a_0 = E(y_0)$. The inverse map $E^{-1}: A \to Y$ is supposed to exist. Although improperly, the pairs $(y, E(y))$ are termed *equilibrium states*. Further hypotheses on the evolution function f allow the solution to (2.1) to be endowed with some significant properties. Such hypotheses may be given the following form.

I. *There is a map* $\Lambda \in L(A, A)$ *and a positive constant* δ *such that*

$$|f(y,a+b) - f(y,a) - \Lambda b| \leqslant \delta |b|, \quad y \in Y, \quad a, a+b \in A,$$

and each eigenvalue of $\Lambda + \delta I_A$ *has a negative real part.*

II. *There is a positive constant* ε *such that*

$$|f(y+z,a) - f(y,a)| \leqslant \varepsilon |z|, \quad y, y+z \in Y, \quad a \in A.$$

It is a direct consequence of I and II that

$$f(y,\cdot) \in \mathrm{Lip}(|\Lambda| + \delta), \quad y \in Y,$$
$$f(\cdot,a) \in \mathrm{Lip}\,\varepsilon , \quad a \in A.$$

To show how the behaviour of the solution to (2.1) is influenced by the conditions I, II, it is convenient to have recourse to the following

Theorem [25]. *Let* g, k, h *be (real) continuous functions on the interval* $[t_0, \hat{t}]$, $t_0, \hat{t} \in \mathbb{R}$, $\hat{t} > t_0$, $kh > 0$. *If a continuous function* v *has the property that*

$$v(t) \leqslant g(t) + k(t) \int_{t_0}^{t} h(s)\, v(s)\, ds, \quad t \in [t_0, \hat{t}],$$

then

$$v(t) \leqslant g(t) + k(t) \int_{t_0}^{t} h(s)\, g(s) \left[\exp \int_{s}^{t} h(u)\, k(u)\, du\right] ds, \quad t \in [t_0, \hat{t}] .$$

Consider now the hidden variables $a, a+b : [t_0, \hat{t}] \to A$ corresponding to the histories y, $y+z : [t_0, \hat{t}] \to Y$, namely

$$\dot{a}(t) = f(y(t), a(t)), \quad t \in [t_0, \hat{t}], \quad a(t_0) = a^*,$$
$$\dot{a}(t) + \dot{b}(t) = f(y(t)+z(t), a(t)+b(t)), \quad t \in [t_0, \hat{t}], \quad a(t_0) + b(t_0) = a^* + b^* .$$

Then, letting $-m < -\delta$ be the real part of the eigenvalue of Λ with the greatest real part, appeal to I, II allows to obtain

$$|b(t)| \leqslant \exp[-m(t-t_0)]\, |b(t_0)| + \varepsilon \int_{t_0}^{t} \exp[-m(t-s)]\, |z(s)|\, ds$$
$$+ \delta \int_{t_0}^{t} \exp[-m(t-s)]\, |b(s)|\, ds .$$

Thus, upon the identifications

$$v(t) = |b(t)|,$$
$$g(t) = \exp[-m(t-t_0)]\, |b(t_0)| + \varepsilon \int_{t_0}^{t} \exp[-m(t-s)]\, |z(s)|\, ds ,$$
$$k(t) = \delta \exp(-mt) ,$$
$$h(t) = \exp(mt) ,$$

the previous theorem and some rearrangement lead to the estimate

$$|b(t)| \leq \exp[-(m-\delta)(t-t_0)]\,|b(t_0)| + \varepsilon\int_{t_0}^{t}\{\exp[-m(t-s)]\,|z(s)| + \delta\exp[-(m-\delta)(t-s)]\int_{t_0}^{s}\exp[-m(s-u)]\,|z(u)|\,du\}\,ds\,.$$

For practical purposes, often it is convenient to employ the less accurate estimate

$$|b(t)| \leq \exp[-(m-\delta)(t-t_0)]\,|b(t_0)| + \frac{\varepsilon}{m-\delta}\{m - \delta\exp[-(m-\delta)(t-t_0)]\}\int_{t_0}^{t}|z(u)|\,du \qquad (2.2)$$

which follows from the previous one by considering 1 as an upper bound for $\exp[-m(t-s)]$ and $\exp[-m(s-u)]$.

We are now in a position to exhibit two significant properties of the solution to (2.1). First, according to the estimate (2.2) we may have $z(t)$ arbitrary and, nevertheless, $|b(t)|$ as small as we please. This behaviour may be phrased by saying that the present value of the hidden variable $a(t)$ is independent of the present value of the observable variable $y(t)$. Second, two hidden variables a, $a+b$ corresponding to the same observable variable differ, at most, by

$$|b(t)| \leq \exp[-(m-\delta)(t-t_0)]\,|b(t_0)|,$$

i. e. the effect of different initial values of the hidden variables fades in time and then asymptotic stability holds.

The asymptotic stability property is important in many respects. On the one hand, it provides an operative procedure to have a take any value $\hat{a}\in\mathbf{A}$ at time t: independently of the initial value of a at $t_0\to-\infty$, we need only keep $y = E^{-1}(\hat{a})$ until time t. On the other, asymptotic stability allows a precise thermodynamic interpretation. Suppose that we make the following experience: the observable variables are changed at time t_0 as

$$y = \begin{cases} y_1\,, & t<t_0\,,\\ y_2\,, & t>t_0\,.\end{cases}$$

We have $a = E(y_1)$ up to time t_0 and $a = E(y_2)$ as $t\to\infty$. More specifically, for times $t>t_0$ close enough to t_0, the hidden variables are close enough to $E(y_1)$ thus playing the role of constant parameters. On the largest time scale $(t\to\infty)$ $a = E(y_2)$; the hidden variables assume the new equilibrium values. So the system, starting from the equilibrium state $(y_1, E(y_1))$, eventually reaches the new equilibrium state $(y_2, E(y_2))$ after having passed through nonequilibrium states of the form (y_2, a), $a \neq E(y_2)$; a similar viewpoint about materials with hidden variables is given by Becker [29].

It is currently accepted that hidden variables may be a useful tool for describing nonequilibrium phenomena. However, since they could seem somewhat unphysical in character, hidden variables would be certainly acceptable if they would not be necessary when dealing with equilibrium states. Asymptotic stability allow these two viewpoints to coexist; essentially this is the physical motivation for the introduction of the requirements I, II.

We end this section with a further remark about stability. Following Coleman and Gurtin [1] we say that the domain of attraction, at constant observable variables, of an equilibrium state (y^*, a^*) is the set $D(y^*, a^*)$ of all $\hat{a}$ such that the solution $a(t)$ of the initial value problem

$$\dot{a} = f(y^*, a), \qquad a(0) = \hat{a}\ ,$$

exists for all $t \geq 0$ and tends to a^* as $t \to \infty$. The estimate (2.2), which is a consequence of I and II, ensures that $D = \mathbf{Y}$, namely any equilibrium state is asymptotically stable.

3. A constitutive theory for heat-conducting viscous fluids

The key role of thermodynamics is to reduce postulated constitutive equations via the restrictions placed by the second law. Unfortunately, as it stands this assertion is in fact meaningless because there is not a unique statement of the second law [30]. It is not the purpose of this work to investigate the various formulations of the second law of thermodynamics and then I content myself with mentioning some outstanding contributions to the subject.

Undoubtedly, a cornerstone is the work of Coleman and Noll [31] supporting the Clausius-Duhem inequality as a statement of the second law and then as a restriction on the constitutive equations. Later, a more refined statement has been given by Müller [32]: the entropy flux need not be the heat flux divided by the temperature but, rather, it must be specified through a constitutive equation. Moreover Müller analysis takes into account the balance equations through Lagrange multipliers. In Day's approach [33] the second law consists in the validity of the Clausius inequality for infinite processes which start and end at equilibrium states. According to Coleman and Owen's formulation [34] of the second law there is one (not necessarily equilibrium) state at which a preassigned action has the Clausius property, i. e. is approximately negative on processes which start at that state and return near to it. Serrin's conceptual analysis of the second law [35] resulted in the statement that, for every cycle of classical thermodynamic systems, the accumulation inequality

$$\int_0^\infty \frac{Q(P,\theta)}{\theta^2}\, d\theta \leq 0$$

holds; it is worth mentioning that the accumulation inequality has been exploited by Iannece, Romano and Starita [36] in conjunction with systems with fading memory. Finally, Coleman, Owen, and Serrin [37] gave a generalization of the accumulation inequality so as to incorporate approximate cycles: for each hotness level L, for each state σ and for each $\varepsilon > 0$ there is a neighbourhood $\mathcal{O}(\sigma,\varepsilon,L)$ of σ for which

$$\int_0^\infty \frac{Q(P,\sigma;\theta)}{\theta^2}\, d\theta < \varepsilon$$

whenever the process (P,σ) operates at or below L and has its final state in $\mathcal{O}$.

The purpose of this section is to exhibit a thermodynamic scheme of heat-conducting viscous fluids via hidden variables [15-17]. The scheme exploits as a statement of the second law a dissipation inequality involving a Müller-like entropy flux.

Let y be the array $(\rho,\theta,\langle \mathbf{D}\rangle,\mathrm{tr}\,\mathbf{D},\mathbf{g})$ of the mass density ρ, the (absolute) temperature θ, the traceless part of the stretching tensor $\mathbf{D}$, the trace of $\mathbf{D}$, and the temperature gradient $\mathbf{g} = \nabla\theta$. Choosing the hidden variables a as the triple $(\Sigma, \Theta, \Lambda)$ of a second rank symmetric traceless tensor Σ, a scalar Θ, and a vector Λ, we allow the evolution function f to make the equations (2.1) into the form

$$\dot{\Sigma} = \sigma_1 \langle \mathbf{D} \rangle - \sigma_2 \Sigma,$$
$$\dot{\Theta} = \omega_1 \operatorname{tr} \mathbf{D} - \omega_2 \Theta, \tag{3.1}$$
$$\dot{\Lambda} = \lambda_1 \mathbf{g} - \lambda_2 \Lambda,$$

where σ_1, σ_2, ω_1, ω_2, λ_1, λ_2 are functions on ρ and θ. Note first that compatibility of such evolution equations with the requirement I holds provided σ_2, ω_2, λ_2 are positive; accordingly σ_2^{-1}, ω_2^{-1}, λ_2^{-1} may be viewed as relaxation times. The requirement II holds as well and then all properties examined in section 2 are true for the system (3.1). Indeed, it is a trivial matter to obtain the solution of the evolution equations (3.1); for example,

$$\Sigma(t) = \Sigma(t_0) \exp[-\int_{t_0}^{t} \sigma_2(\tau)\, d\tau] + \int_{t_0}^{t} \exp[-\int_{\tau}^{t} \sigma_2(\xi)\, d\xi]\, \sigma_1(\tau) \langle \mathbf{D}(\tau) \rangle\, d\tau \tag{3.2}$$

where $\sigma_{1,2}(\tau)$ are shorthands for $\sigma_{1,2}(\rho(\tau),\theta(\tau))$. As a consequence $\Sigma(t)$, as well as $\Theta(t)$ and $\Lambda(t)$, turns out to be independent of the present values $\rho(t)$, $\theta(t)$, $\dot{\theta}(t)$, $\mathbf{g}(t)$, $\mathbf{D}(t)$. Observe now that we lose no generality by assuming that σ_1, ω_1, λ_1 be positive because, otherwise, we could simply let $\Sigma \to -\Sigma$, $\Theta \to -\Theta$, $\Lambda \to -\Lambda$ thereby obtaining the desired form of the evolution equations.

To complete the constitutive assumptions we must spacify the response functions and the statement for the second law. As to the response functions it is convenient to identify ϕ with the array $(\psi,\eta,\mathbf{T},\mathbf{q},\mathbf{F})$ of the free energy ψ, the entropy η, the stress $\mathbf{T}$, the heat flux $\mathbf{q}$, and the entropy flux $\mathbf{F}$; these functions are assumed to depend on the observable variables ρ, θ and on the hidden variables Σ, Θ, Λ. So the dependence on the histories of $\langle \mathbf{D} \rangle$, $\operatorname{tr} \mathbf{D}$, $\mathbf{g}$ is included through the hidden variables Σ, Θ, Λ while the dependence on the present values of $\langle \mathbf{D} \rangle$, $\operatorname{tr} \mathbf{D}$, $\mathbf{g}$ is dropped out. As to the second law we assume that the inequality

$$\rho\dot{\eta} + \nabla\cdot\mathbf{F} - \rho\theta^{-1} r \geq 0 \tag{3.3}$$

hold identically, the term r representing the heat supply. On account of the balance of energy, the inequality (3.3) may be written as

$$-\rho(\dot{\psi} + \eta\dot{\theta}) + \mathbf{T}\cdot\mathbf{D} - \nabla\cdot\mathbf{q} + \theta\nabla\cdot\mathbf{F} \geq 0. \tag{3.4}$$

We are now in a position to derive some restrictions placed by the entropy inequality. Let $\mathbf{N} = \mathbf{F} - \theta^{-1}\mathbf{q}$ be the entropy extra-flux. Since $\psi = \psi(\rho,\theta,\Sigma,\Theta,\Lambda)$ and $\mathbf{N} = \mathbf{N}(\rho,\theta,\Sigma,\Theta,\Lambda)$ the inequality (3.4) becomes

$$-\rho(\psi_\theta + \eta)\dot{\theta} + (\rho^2\psi_\rho \mathbf{1} - \rho\sigma_1\psi_\Sigma - \rho\omega_1\psi_\Theta \mathbf{1} + \mathbf{T})\cdot\mathbf{D} - (\rho\lambda_1\psi_\Lambda + \theta^{-1}\mathbf{q})\cdot\mathbf{g}$$
$$+\rho(\sigma_2\psi_\Sigma\cdot\Sigma + \omega_2\psi_\Theta\Theta + \lambda_2\psi_\Lambda\cdot\Lambda) + \theta[\mathbf{N}_\rho\cdot\nabla\rho + \mathbf{N}_\theta\cdot\mathbf{g} + \mathbf{N}_\Sigma\cdot(\nabla\Sigma) + \mathbf{N}_\Theta\cdot\nabla\Theta + \mathbf{N}_\Lambda\cdot(\nabla\Lambda)] \geq 0$$

the subscripts θ, ρ, Σ, Θ, Λ denoting partial derivatives. Owing to the arbitrariness of $\nabla\rho$, $\nabla\Sigma$, $\nabla\Theta$, $\nabla\Lambda$, this inequality holds provided

$$\mathbf{N}_\rho = \mathbf{0}, \quad \mathbf{N}_\Sigma = \mathbf{0}, \quad \mathbf{N}_\Theta = \mathbf{0}, \quad \mathbf{N}_\Lambda = \mathbf{0}$$

and hence

$$\mathbf{N} = \mathbf{N}(\theta).$$

The isotropy of the fluid implies that $\mathbf{N}$ must vanish. Then standard arguments [15,23] give

$$\eta = -\psi_\theta, \tag{3.5}$$

$$\mathbf{T} = -p\mathbf{1} + \rho\sigma_1\psi_\Sigma + \rho\omega_1\psi_\Theta\mathbf{1}, \qquad p := \rho^2\psi_\rho \ , \tag{3.6}$$

$$\mathbf{q} = -\rho\theta\lambda_1\psi_\Lambda \ , \tag{3.7}$$

$$\sigma_2\psi_\Sigma\cdot\Sigma + \omega_2\psi_\Theta\Theta + \lambda_2\psi_\Lambda\cdot\Lambda \geq 0 \ . \tag{3.8}$$

Accordingly, any free energy function ψ satisfying the reduced dissipation inequality (3.8) yields response functions η, $\mathbf{T}$, $\mathbf{q}$ compatible with the second law of thermodynamics (3.3); based on physical arguments, we will find shortly an expression for ψ satisfying (3.8). It seems natural to require that the present scheme leads to the NSF theory when $\mathbf{D}, \mathbf{g}, \rho, \theta$ are constant in time. Now, if $\mathbf{D}$, $\mathbf{g}$ and ρ, θ are constant, asymptotically we have

$$\Sigma = \sigma_1\sigma_2^{-1}\langle\mathbf{D}\rangle,$$

$$\Theta = \omega_1\omega_2^{-1}\mathrm{tr}\,\mathbf{D}\,,$$

$$\Lambda = \lambda_1\lambda_2^{-1}\mathbf{g}\ .$$

Then (3.6), (3.7) reduce to the NSF scheme provided we identify the hidden variables as

$$\Sigma = \frac{\sigma_1}{2\mu\sigma_2}\langle\mathbf{T}\rangle\ , \qquad \Theta = \frac{\omega_1}{3\zeta\omega_2}(\mathrm{tr}\,\mathbf{T} + 3p)\ , \qquad \Lambda = -\frac{\lambda_1}{\kappa\lambda_2}\mathbf{q}\ ;$$

the shear viscosity μ, the bulk viscosity ζ, and the heat conductivity κ may depend on ρ, θ. As a consequence, (3.6), (3.7) yield

$$\psi_\Sigma = \frac{\mu\sigma_2}{\rho\sigma_1^2}\Sigma\,, \qquad \psi_\Theta = \frac{\zeta\omega_2}{\rho\omega_1^2}\Theta\,, \qquad \psi_\Lambda = \frac{\kappa\lambda_2}{\rho\theta\lambda_1^2}\Lambda$$

whence

$$\psi = \Psi(\rho,\theta) + \frac{1}{\rho}\left(\frac{\mu\sigma_2}{\sigma_1^2}\Sigma\cdot\Sigma + \frac{\zeta\omega_2}{2\omega_1^2}\Theta^2 + \frac{\kappa\lambda_2}{2\theta\lambda_1^2}\Lambda\cdot\Lambda\right)\ .$$

Substitution into (3.8) and the mutual independence of Σ, Θ, Λ allow us to find that the usual inequalities

$$\mu \geq 0\ , \qquad \zeta \geq 0\ , \qquad \kappa \geq 0$$

are necessary and sufficient for the identical validity of (3.8).

As an aside, it is worth looking briefly at the dependence of $\sigma_{1,2}$, $\omega_{1,2}$, $\lambda_{1,2}$ and μ, ζ, κ on ρ, θ. As already noticed, σ_2^{-1}, ω_2^{-1}, λ_2^{-1} play the role of relaxation times; accordingly the scheme allows the relaxation times associated with viscosity and heat conduction to be functions on density and temperature. Moreover, these dependences may be in order so as to account for required thermodynamic properties. For example, the specific heat $c = -\theta\psi_{\theta\theta}$ is expressed by

$$c = -\theta\Psi_{\theta\theta} - \frac{\theta}{\rho}\left[\left(\frac{\mu\sigma_2}{\sigma_1^2}\right)_{\theta\theta}\Sigma\cdot\Sigma + \tfrac{1}{2}\left(\frac{\zeta\omega_2}{\omega_1^2}\right)_{\theta\theta}\Theta^2 + \tfrac{1}{2}\left(\frac{\kappa\lambda_2}{\theta\lambda_1^2}\right)_{\theta\theta}\Lambda\cdot\Lambda\right]\,.$$

If we assume, as usual, that the thermostatic term $\Psi_{\theta\theta}$ is negative, the inequalities

$$\left(\frac{\mu\sigma_2}{\sigma_1^2}\right)_{\theta\theta} \leq 0\,, \qquad \left(\frac{\zeta\omega_2}{\omega_1^2}\right)_{\theta\theta} \leq 0\,, \qquad \left(\frac{\kappa\lambda_2}{\theta\lambda_1^2}\right)_{\theta\theta} \leq 0\,,$$

are sufficient for the specific heat to be positive. Should, instead, κ, λ_1, λ_2 be nonvanishing constants, the last inequality could not hold.

In ending this section a comparison with some results obtained recently by Coleman, Fabrizio, and Owen [28] is in order. They consider an evolution equation of the form

$$\dot{\alpha} = \nu(e,\alpha) - \mathbf{H}(e,\alpha)\,\mathbf{g}$$

e being the internal energy and α the (vector) hidden variable, and choose

$$\theta = \hat{\theta}(e,\alpha) \ ,$$
$$\mathbf{q} = \hat{\mathbf{q}}(e,\alpha)$$

as response functions. Moreover they specify the contents of the second law by asserting that the action

$$s(P_t \, , \sigma_0) = \int_0^t [\frac{r}{\theta} - \nabla\cdot(\frac{\mathbf{q}}{\theta})] \, d\zeta$$

has the Clausius property at least at one state σ_0, namely $s(P_t, \sigma_0)$ is approximately negative when $(P_t \, , \sigma_0)$ is approximately cyclic. As a result they find that there exists an entropy function $\hat{\eta}$ obeying the relations

$$\hat{\theta}(e,\alpha) = \hat{\eta}_e(e,\alpha) \ , \tag{3.9}$$
$$\hat{\mathbf{q}}(e,\alpha) = -\,[\hat{\theta}(e,\alpha)]^2 \, \mathbf{H}(e,\alpha) \, \hat{\eta}_\alpha(e,\alpha) \ , \tag{3.10}$$
$$\hat{\eta}_\alpha(e,\alpha)\cdot\nu(e,\alpha) \geq 0 \ , \tag{3.11}$$

the (invertible) tensor $\mathbf{H}$ being symmetric and the subscripts e, α denoting partial derivatives. Moreover, upon identifying α with $\mathbf{q}$, they arrive at the expression

$$\psi(\theta,\mathbf{q}) = \psi_0(\theta) + \frac{1}{\theta}\int_0^1 s\,\mathbf{q}\cdot\mathbf{H}(\theta,s\mathbf{q})^{-1}\,\mathbf{q}\,ds$$

for the free energy. If $\mathbf{H}$ is independent of $\mathbf{q}$ it follows at once that

$$\psi(\theta,\mathbf{q}) = \psi_0(\theta) + \frac{1}{2\theta}\mathbf{q}\cdot\mathbf{H}^{-1}\mathbf{q} \ . \tag{3.12}$$

Apart from the dependence of $\mathbf{H}$ on the hidden variable α, one realizes immediately the wide agreement between (3.9)-(3.11) and (3.5), (3.7), (3.8). Differences occur because of e, α, rather than θ, α, being the independent variables. Since e cannot be a function on θ alone the transformation between the two sets of independent variables is not trivial; this aspect is considered in [38].

4. Further thermodynamic models via hidden variables

The work of Müller [39], appeared in 1967, has rightly influenced the subsequent research on nonequilibrium thermodynamics especially in connection with the physical aspects. The central idea of Müller's paper is that, for a complete description of the state of a system out of equilibrium, additional quantities are required besides those specifying equilibrium states. Moreover, Müller's analysis of the entropy balance leads to the result that the entropy flux need not coincide with the heat flux divided by the absolute temperature. Should we succeed in setting up a hidden variable scheme embracing Müller's one as a particular case, the recourse to hidden variables in nonequilibrium thermodynamics would gain further credit. In fact such a scheme has been accomplished in [23] and is now outlined.

Look at a viscous and heat-conducting fluid. For the purpose we have in mind it is again convenient to imagine the hidden variables as the triple $(\Sigma, \Theta, \Lambda)$ and to choose f so as to make the evolution equation into the coupled system

$$\dot{\Sigma} = \tau_1^{-1}(\langle\mathbf{D}\rangle - \Sigma) + a\langle\nabla\Lambda\rangle \ , \tag{4.1}$$
$$\dot{\Theta} = \tau_2^{-1}(\mathrm{tr}\,\mathbf{D} - \Theta) + b\nabla\cdot\Lambda \ , \tag{4.2}$$
$$\dot{\Lambda} = \tau_3^{-1}(\mathbf{g} - \Lambda) + c\nabla\Theta + d\nabla\cdot\Sigma \ , \tag{4.3}$$

where a, b, c, d are phenomenological coefficients and τ_1, τ_2, τ_3 (constant) relaxation times. For usual evolution equations the present value of the hidden variables is independent of the present value of the observable variables. To show that this feature occurs here as well consider, for example, (4.3); at any particle of the fluid a trivial integration yields

$$\Lambda(t) = \Lambda(t_0)\exp[-\tau_3^{-1}(t-t_0)] + \int_{t_0}^{t}\exp[-\tau_3^{-1}(t-\xi)]\,[\tau_3^{-1}\mathbf{g} + c\nabla\Theta + d\nabla\cdot\Sigma](\xi)\,d\xi$$

whence it follows the desired result. Letting again $\mathbf{N}$ be the entropy extra-flux, the isotropy of the fluid and the analysis developed in section 3 suggest that we make the ansatz

$$\mathbf{N} = K\Sigma\Lambda + L\Theta\Lambda\ ; \tag{4.4}$$

a connection between the phenomenological coefficients K, L and a, b, c, d will be given shortly. Accordingly, upon substitution of (4.1)-(4.4) into the entropy inequality (3.4), the arbitrariness of the present values $\theta(t)$, $\mathbf{D}(t)$, $\mathbf{g}(t)$ and the obvious independence of $\nabla\cdot\Sigma$, $\nabla\Theta$, $\nabla\Lambda$ of Σ, Θ, Λ allow us to find that the restrictions

$$\eta = -\psi_\theta\ , \tag{4.5}$$

$$\mathbf{T} = -p\mathbf{1} + \frac{\rho}{\tau_1}\psi_\Sigma + \frac{\rho}{\tau_2}\psi_\Theta\mathbf{1}\ , \qquad p := \rho^2\psi_\rho\ , \tag{4.6}$$

$$\mathbf{q} = -\frac{\rho\theta}{\tau_3}\psi_\Lambda\ , \tag{4.7}$$

and

$$\theta K\Sigma = \rho a\psi_\Sigma\ , \tag{4.8}$$

$$\theta L\Theta = \rho b\psi_\Theta\ , \tag{4.9}$$

$$\theta L\Lambda = \rho c\psi_\Lambda\ , \tag{4.10}$$

$$\theta K\Lambda = \rho d\psi_\Lambda\ , \tag{4.11}$$

$$\tau_1^{-1}\psi_\Sigma\cdot\Sigma + \tau_2^{-1}\psi_\Theta\Theta + \tau_3^{-1}\psi_\Lambda\cdot\Lambda \geqslant 0\ , \tag{4.12}$$

must hold. If now, by analogy with section 3, we select the particular free energy function given by

$$\psi = \Psi(\rho,\theta) + \frac{1}{\rho}(\mu\tau_1\Sigma\cdot\Sigma + \tfrac{1}{2}\zeta\tau_2\Theta^2 + \frac{\kappa\tau_3}{2\theta}\Lambda\cdot\Lambda) \tag{4.13}$$

the conditions (4.8)-(4.11) become

$$2\mu\tau_1 a = \theta K, \qquad \zeta\tau_2 b = \theta L, \qquad \kappa\tau_3 c = \theta^2 L, \qquad \kappa\tau_3 d = \theta^2 K, \tag{4.14}$$

while (4.12) gives the customary inequalities

$$\mu \geqslant 0\ , \qquad \zeta \geqslant 0\ , \qquad \kappa \geqslant 0\ .$$

The conditions (4.14) reduce the effective number of the phenomenological coefficients as a, b, c, d turn out to be determined by K, L.

The scheme so outlined may be identified with Müller's one (in the particular case (4.13)); this is easily seen by noting that, in view of (4.13), the conditions (4.6), (4.7) give

$$\mathbf{T} = -p\,\mathbf{1} + 2\mu\Sigma + \zeta\Theta\,\mathbf{1}\ ,$$

$$\mathbf{q} = -\kappa\Lambda\ .$$

Then we need only identify the components Σ_{ij}, Θ, and Λ_i with $-\bar{p}_{ij}/2\mu$, $-p_{jj}/\zeta$, and q_i/κ of Müller [39]; the sought connection is established.

As the above models show, once the thermodynamic theory is developed, the properly chosen hidden variables acquire a clear physical meaning. Furthermore, the hidden variable scheme has the advantage of leading straightaway to the desired thermodynamic model provided only the evolution equations are introduced. There are a number of examples corroborating this assertion (see, e. g., [14,15,18,19,24]); a further example will be given shortly.

Since the early experiments performed by Bömmel and Mackinnon around 1954, numerous measurements of the ultrasonic attenuation in metals have been made [40]. Meanwhile, several theoretical explanations have been formulated; nevertheless the first macroscopic (thermodynamic) model accounting for ultrasonic attenuation in metals has been accomplished very recently by Jou, Bampi, and myself [26] within the context of EIT. Here I will try to set up the corresponding model within HVT.

The starting point consists in the following two observations. First, kinetic theory arguments [41] show that, owing to collisions with ions, electrons in metals may be viewed as a viscous fluid with a negligible bulk viscosity. Second, solid state physics [42] indicates that a macroscopic model for ultrasonic waves in metals should incorporate relaxation effects along with nonlocal ones.

The simplest way of accounting for the electronic motion in metals is the MFD approximation; if, for simplicity, we assume that heat conduction is negligible then the entropy inequality may be written as

$$-\rho(\dot{\psi}+\eta\dot{\theta}) + \mathbf{T}\cdot\mathbf{D} + \mathbf{J}\cdot\mathbf{e} + \theta\nabla\cdot\mathbf{N} \geqslant 0 \tag{4.15}$$

$\mathbf{J}$ being the conduction current and $\mathbf{e}$ the electromagnetic intensity. Now, describe the states of the system through the observable variables $(\rho, \theta, \langle\mathbf{D}\rangle, \mathbf{e})$ and the hidden variables (Σ, Γ), Γ being a vector. Upon disregarding cross effect coupling terms and assuming that nonlocality affects Σ only, we set

$$\dot{\Sigma} = \tau_1^{-1}(\langle\mathbf{D}\rangle - \Sigma) + \gamma\nabla^2\Sigma\,,$$
$$\dot{\Gamma} = \tau_2^{-1}(\mathbf{e} - \Gamma),$$

γ being a phenomenological parameter. Then, letting ψ, η, $\mathbf{T}$, $\mathbf{J}$, $\mathbf{N}$ be functions on the state variables eventually we arrive at

$$\eta = -\psi_\theta\,,$$
$$\mathbf{T} = -p\,\mathbf{1} + \rho\tau_1^{-1}\psi_\Sigma\,, \qquad p := \rho^2\psi_\rho\,,$$
$$\mathbf{J} = \rho\tau_2^{-1}\psi_\Gamma\,,$$
$$\theta\nabla\cdot\mathbf{N} - \rho\psi_\Sigma\cdot\gamma\nabla^2\Sigma + \rho(\tau_1^{-1}\psi_\Sigma\cdot\Sigma + \tau_2^{-1}\psi_\Gamma\cdot\Gamma) \geqslant 0\,. \tag{4.16}$$

The reduced dissipation inequality (4.16) suggests that we set $\mathbf{N} = \frac{1}{2}\lambda\nabla(\Sigma\cdot\Sigma)$, λ being a constant; substitution gives

$$(\lambda\theta\Sigma - \gamma\rho\psi_\Sigma)\cdot\nabla^2\Sigma + \lambda\theta(\nabla\Sigma)\cdot(\nabla\Sigma) + \rho(\tau_1^{-1}\psi_\Sigma\cdot\Sigma + \tau_2^{-1}\psi_\Gamma\cdot\Gamma) \geqslant 0\,. \tag{4.17}$$

The expression

$$\psi = \Psi(\rho,\theta) + \rho^{-1}(\tfrac{1}{2}\sigma\tau_2\Gamma\cdot\Gamma + \mu\tau_1\Sigma\cdot\Sigma)$$

allows us to get the direct generalization of Navier-Stokes' and Ohm's laws; specifically (4.17) implies that

$$\lambda \geqslant 0\,, \qquad \mu \geqslant 0\,, \qquad \sigma \geqslant 0\,,$$
$$\gamma = \frac{2\mu\tau_1}{\lambda\theta} \geqslant 0\,.$$

It is of interest to remark that the analysis of ultrasonic waves based on the previous model and the corresponding analysis based on the kinetic description enable us to obtain a precise estimate for the coefficients λ, γ. In fact, on account of the results in [26] we have

$$\gamma = \frac{9}{35}\frac{\ell^2}{\tau_1}$$

ℓ being the electronic mean-free-path.

In ending this section, devoted to models of continuous media, I mention that dissipative constrained solids with hidden variables are investigated in [43].

5. HVT versus EIT

As the literature shows, at present there are essentially two approaches to nonequilibrium thermodynamics, namely EIT and HVT. EIT has been recently developed to a large extent and applied extensively by Jou, Lebon, Casas-Vázquez et al. [44,45]; the physical ideas and the associated procedures trace back to Müller [39]. HVT, some developments of which have been described above, originates mainly from the paper of Coleman and Gurtin [1]. Since both approaches aim to account for nonequilibrium thermodynamics, it is of interest to compare them and to emphasize similarities and differences. Similarities are apparent: both theories involve nonequilibrium variables, besides equilibrium ones, and show how relaxation phenomena are natural ingredients of nonequilibrium thermodynamics. Differences appear at various stages.

As to the independent variables, EIT deals with observable variables only. HVT, instead, involves hidden variables besides observable variables; however, if we look at hidden variables as solutions of the evolution equation then HVT involves in fact observable variables as EIT does.

Concerning the evolution equations, in HVT they are specified on the basis of hints arising from the observation of the behaviour of the body at hand. The structure of the entropy function and of the entropy flux is then derived as a consequence of the second law. In EIT, instead, the entropy function is assumed to be given explicitly in that the starting point of the thermodynamic analysis is the generalized Gibbs equation. Next the evolution equation and the entropy flux are derived as a consequence of the second law.

A comparison between EIT and HVT has been performed in [24] in connection with a wavelength-dependent description of heat conduction. As a further example emphasizing the analogies between the two approaches, the thermodynamic description of ultrasonic attenuation in metals is now considered; since the HVT version has been delivered in the previous section here its EIT counterpart is outlined [26].

Letting $\mathbf{T} = -p\,\mathbf{1} + \boldsymbol{\Upsilon}$, $\mathrm{tr}\,\boldsymbol{\Upsilon} = 0$, the generalized Gibbs equation for $\eta = \eta(\varepsilon, \rho, \boldsymbol{\Upsilon}, \mathbf{J})$ is assumed to take the form

$$\dot{\eta} = \frac{1}{\theta}\dot{\varepsilon} - \frac{p}{\rho^2\theta}\dot{\rho} - \frac{1}{\rho\theta}\mathbf{A}\cdot\dot{\boldsymbol{\Upsilon}} - \frac{1}{\rho\theta}\mathbf{b}\cdot\dot{\mathbf{J}}$$

where ε is the internal energy while $\mathbf{A}$ and $\mathbf{b}$ are unspecified as yet. Then, in view of the energy equation

$$\rho\dot{\varepsilon} = \mathbf{T}\cdot\mathbf{D} + \mathbf{J}\cdot\mathbf{e} + \rho r\ ,$$

the entropy inequality

$$\rho\dot{\eta} + \nabla\cdot\mathbf{N} - \frac{1}{\theta}\rho r \geqslant 0$$

leads to

$$\Upsilon\cdot\langle \mathbf{D}\rangle + \mathbf{J}\cdot\mathbf{e} - \mathbf{A}\cdot\dot{\Upsilon} - \mathbf{b}\cdot\dot{\mathbf{J}} + \theta\nabla\cdot\mathbf{N} \geqslant 0.$$

Introduce now the further assumption

$$\mathbf{A} = \Lambda\Upsilon , \qquad \mathbf{b} = \beta\mathbf{J} \tag{5.2}$$

where the phenomenological coefficients Λ, β may depend on ρ, θ. It follows that

$$-(\Lambda\dot{\Upsilon} + \langle \mathbf{D}\rangle)\cdot\Upsilon - (\beta\dot{\mathbf{J}} - \mathbf{e})\cdot\mathbf{J} + \theta\nabla\cdot\mathbf{N} \geqslant 0. \tag{5.3}$$

Physical arguments, together with the inequality (5.1), suggest that we put

$$\Lambda\dot{\Upsilon} + \langle \mathbf{D}\rangle = -\frac{1}{2\mu}\Upsilon + \gamma\theta\nabla^2\Upsilon , \tag{5.4}$$

$$\beta\dot{\mathbf{J}} - \mathbf{e} = -\frac{1}{\sigma}\mathbf{J} , \tag{5.5}$$

μ, σ, γ being phenomenological parameters (γ = constant). Substitution of (5.4),(5.5) shows that (5.3) holds provided

$$\mathbf{N} = \tfrac{1}{2}\gamma\nabla(\Upsilon\cdot\Upsilon)$$

and

$$\mu \geqslant 0 , \qquad \sigma \geqslant 0 , \qquad \gamma \geqslant 0 .$$

Accordingly it seems natural to require that

$$\Lambda > 0 , \qquad \beta > 0$$

and to view

$$\tau_1 = 2\mu\Lambda , \qquad \tau_2 = \sigma\beta$$

as (positive) relaxation times.

The scheme so outlined allows us to restate the typical differences between EIT and HVT. First, according to (5.1), (5.2) the entropy function is assumed to be known explicitly at the outset while in HVT the entropy function (or the free energy function) is determined through compatibility with the entropy inequality. Second, the evolution equations (5.4), (5.5) are so chosen as to satisfy the entropy inequality (5.3) while in HVT they have to be known at the outset. In both theories the entropy flux $\mathbf{F}$ is determined in such a way that the entropy inequality hold identically.

In conclusion, a question arises naturally: which approach is preferable? Setting aside any consideration about the formal structure, it seems that HVT is to be preferred when we can rely upon precise physical hints concerning the evolution of certain nonequilibrium variables.

6. Remarks

In setting up a thermodynamic model with hidden variables the starting point is the evolution equation. However, usually the evolution equation is not completely known; in fact, there may be experimental observations and hints of physical character which suggest a suitable structure of the evolution equation. General rules, if any, could provide profitable guidelines for making the available suggestions into a precise form of the evolution equation. Here two possible general rules are outlined.

a) *Objectivity*

In the last decade there has been much controversy about objectivity as a principle (see, e. g., [46,47]); without entering such a controversy, here the principle of objectivity is assumed to hold and then a consequence is derived which should constitute one of the aforementioned guidelines.

According to the principle of objectivity [48] constitutive equations must be form invariant under the Euclidean group (objective). Since they are a particular case of constitutive equations, the evolution equations as well must be objective. Now, if an evolution equation involves objective quantities then the time derivative characterizing the evolution equation must be objective. In ref. [49] the proof is given that the most general objective time derivative $\mathring{\mathbf{A}}$ of an objective vector $\mathbf{A}$ is expressed as

$$\mathring{A}^i = \dot{A}^i + S^i_{\ j} A^j + H^i_{\ j} A^j \tag{6.1}$$

where $\mathbf{S}$ is an arbitrary objective tensor and $\mathbf{H}$ a (nonobjective) skewsymmetric tensor. We realize immediately that customary (objective) time derivatives are particular cases of (6.1). For example, the choice $\mathbf{H} = -\mathbf{W}$, $\mathbf{W}$ being the spin tensor, and $\mathbf{S} = \mathbf{0}$ yields the co-rotational derivative while setting $\mathbf{H} = -\mathbf{W}$ and $\mathbf{S} = -\mathbf{D} + w(\mathrm{tr}\,\mathbf{D})\mathbf{1}$, w being the weight of the vector, yields the convected time derivative.

In passing I mention the wide use of the co-rotational derivative in plasticity. For example, according to Tokuoka [13] the rate type plastic material is described by the Cauchy stress $\mathbf{T}$, the scalar hidden variable α, the tensor hidden variable β, and the translated stress $\check{\mathbf{T}} = \mathbf{T} - \beta$ through the constitutive equations

$$\mathring{\mathbf{T}} = \mathcal{H}(\langle \check{\mathbf{T}} \rangle, \alpha)\, \mathbf{D} ,$$

$$\dot{\alpha} = \phi(\langle \check{\mathbf{T}} \rangle, \alpha) \cdot \mathbf{D} ,$$

$$\mathring{\beta} = \psi(\langle \check{\mathbf{T}} \rangle, \alpha)\, \mathbf{D} ,$$

where "$\circ$" denotes the co-rotational derivative, namely

$$\mathring{\mathbf{T}} = \dot{\mathbf{T}} - \mathbf{W}\mathbf{T} + \mathbf{T}\mathbf{W} .$$

b) *Hyperbolicity*

Constitutive theories are usually expected to be compatible with wave propagation, i. e. they are required to allow the propagation of discontinuity waves at finite speed. Sometimes, it is rather required that the system of equations governing the behaviour of the continuum under consideration is hyperbolic, namely that there exists a complete set of propagation modes with finite speeds.

On adopting this more severe point of view, recently Ruggeri [50] succeeded in describing viscous and heat-conducting fluids. Specifically, in connection with a quasi-linear first order conservative system, associated with a supplementary conservation law, the main field $\mathbf{u}'$ is introduced which makes the original system of equations into a hyperbolic one. Then, on assuming for simplicity that the transition to off-equilibrium affects the Gibbs function only, a generalization of the NSF theory is obtained where $\theta^{-1}\langle \mathbf{T} \rangle$, $\theta^{-1}(\mathrm{tr}\,\mathbf{T} + 3p)$, $\theta^{-2}\mathbf{q}$, instead of $\langle \mathbf{T} \rangle$, $\mathrm{tr}\,\mathbf{T} + 3p$, $\mathbf{q}$, are in fact the most suitable hidden variables (or relaxed observable variables). The reader interested in a detailed account of the procedure is referred to the contribution of Ruggeri in this volume.

In ending this paper I remark that no mention has been given of the schemes describing viscosity and heat conduction without having recourse to hidden variables. The interested reader is advised to consult the papers [50,51] both to compare them with the hidden variable scheme and for their references to alternative approaches.

Acknowledgment

I am indebted to Professors B. D. Coleman, M. Fabrizio, D. R. Owen and to Professor T. Ruggeri for providing me with a copy of their papers [28], [50] before the publication. The research leading to this work was performed under the auspices of GNFM-CNR.

References

1. B. D. Coleman and M. E. Gurtin, J. Chem. Phys. **47**, 597 (1967).
2. B. D. Coleman and M. E. Gurtin, Phys. Fluids **10**, 1454 (1967).
3. J. Lubliner, Acta Mechanica **8**, 75 (1969).
4. P. J. Chen and M. E. Gurtin, Phys. Fluids **14**, 1091 (1971).
5. W. Kosiński and P. Perzyna, Arch. Mech. **24**, 629 (1972).
6. R. M. Bowen and P. J. Chen, Arch. Mech. **25**, 703 (1973).
7. G. A. Kluitenberg, Physica **68**, 75 (1973).
8. J. Lubliner, Acta Mechanica **17**, 109 (1974).
9. W. Kosiński, Int. J. Non-Linear Mechanics **9**, 481 (1974).
10. W. Kosiński, Arch. Mech. **27**, 733 (1975).
11. K. C. Valanis, Arch. Mech. **27**, 857 (1975).
12. W. A. Day, Arch. Rational Mech. Anal. **62**, 367 (1976).
13. T. Tokuoka, Int. J. Non-Linear Mechanics **13**, 199 (1978).
14. G. A. Maugin, J. Mécan. **18**, 541 (1979).
15. A. Morro, Arch. Mech. **32**, 145 (1980).
16. A. Morro, Arch. Mech. **32**, 193 (1980).
17. A. Morro, Rend. Sem. Mat. Univ. Padova **63**, 169 (1980).
18. A. Morro, Phys. Lett. A **79**, 84 (1980).
19. A. Morro, Acta Phys. Hung. **48**, 369 (1980).
20. F. Bampi and A. Morro, J. Math. Phys. **21**, 1201 (1980).
21. F. Bampi and A. Morro, Wave Motion **2**, 153 (1980).
22. E. Matsumoto, Acta Mechanica **39**, 241 (1981).
23. F. Bampi and A. Morro, J. Phys. A **14**, 631 (1981).
24. F. Bampi, A. Morro, and D. Jou, Physica A **107**, 393 (1981).
25. A. Morro, Boll. Un. Mat. Ital. B **1**, 553 (1982).
26. D. Jou, F. Bampi, and A. Morro, J. Non-Equilib. Thermodyn., to appear.
27. B. D. Coleman and M. E. Gurtin, Meccanica **2**, 135 (1967).
28. B. D. Coleman, M. Fabrizio, and D. R. Owen, Arch. Rational Mech. Anal., to appear.

29. E. Becker, in *Trends in Solid Mechanics 1979*, Delft University Press, 1979.
30. K. Hutter, Acta Mechanica **27**, 1 (1977).
31. B. D. Coleman and W. Noll, Arch. Rational Mech. Anal. **13**, 167 (1963).
32. I. Müller, *Thermodynamik - Grundlagen der Materialtheorie*, Bertelsmann Universitatsverlag, Düsseldorf, 1973.
33. W. A. Day, *The Thermodynamics of Simple Materials with Fading Memory*, Springer, Berlin, 1972.
34. D. D. Coleman and D. R. Owen, Arch. Rational Mech. Anal. **54**, 1 (1974).
35. J. Serrin, Arch. Rational Mech. Anal. **70**, 355 (1979).
36. D. Iannece, A. Romano, and G. Starita, Arch. Rational Mech. Anal. **75**, 373 (1981).
37. B. D. Coleman, D. R. Owen, and J. Serrin, Arch. Rational Mech. Anal. **77**, 103 (1981).
38. F. Bampi and A. Morro, Phys. Lett. A **79**, 156 (1980).
39. I. Müller, Zeit. Phys. **198**, 329 (1967).
40. H. Stoltz, Phys. Status Solidi **3**, 1153 (1963).
41. W. P. Mason, Phys. Rev. **97**, 557 (1955).
42. J. M. Ziman, *Electrons and Phonons*, Clarendon Press, Oxford, 1979; p. 218.
43. A. Morro, Arch. Mech., to appear.
44. D. Jou, J. Casas-Vázquez, and G. Lebon, J. Non-Equilib. Thermodyn. **4**, 349 (1979).
45. G. Lebon, D. Jou, and J. Casas-Vázquez, J. Phys. A **13**, 275 (1980).
46. I. Müller, Arch. Rational Mech. Anal. **45**, 241 (1972).
47. C. Truesdell, Meccanica **11**, 196 (1976).
48. C. Truesdell and W. Noll, in *Encyclopedia of Physics*, S. Flügge ed., Springer, Berlin, 1965; vol. III/3.
49. F. Bampi and A. Morro, Found. Phys. **10**, 905 (1980).
50. T. Ruggeri, Acta Mechanica, to appear.
51. G. Grioli, Atti Accad. Naz. Lincei Rend. **67**, 332 (1980); **67**, 426 (1980).

CONSERVATION LAWS AND SYMMETRIC HYPERBOLIC SYSTEM FOR A THERMOVISCOUS FLUID

by

T.RUGGERI

Istituto di Matematica Applicata - Università di Bologna

Via Vallescura, 2 - 40136 BOLOGNA (ITALY)

The field equations for the motion of a thermoviscous fluid, in conservative form, are :

$$\begin{cases} \dfrac{\partial \rho}{\partial t} + \dfrac{\partial}{\partial x_i}(\rho\, u_i) = 0 \\[2ex] \dfrac{\partial(\rho\, u_k)}{\partial t} + \dfrac{\partial}{\partial x_i}(\rho\, u_i\, u_k + t_{ik}) = 0 \\[2ex] \dfrac{\partial \varepsilon}{\partial t} + \dfrac{\partial}{\partial x_i}(\varepsilon\, u_i + q_i + t_{ij}\, u_j) = 0\ ; \end{cases} \tag{1}$$

$t_{ik} = p\ \delta_{ik} - \sigma_{ik}$ being the pressure tensor; p the pressure, $\underset{\sim}{\sigma} \equiv (\sigma_{ik})$ the viscous tensor, $\varepsilon = \rho(e + u^2/2)$ the total energy, e the internal energy and ρ, $\underline{u}$, $\underline{q}$ respectively the mass density, the velocity of the fluid particle and the heat flux.

To the eqs. (1), the constitutive equation must be added; in particular one has to determine the constitutive relations for the heat flux and the viscous tensor. In the classical approach for a Newtonian fluid these are the Fourier and Navier-Stokes ones :

$$\begin{cases} \nabla \theta = - \underline{q}/\chi \\ \frac{1}{2}(\nabla \underline{u} + \nabla^T \underline{u}) - \frac{1}{3} \underset{\sim}{I}\, div\, \underline{u} = \frac{1}{2\eta} \underset{\sim}{\sigma}^D \\ div\, \underline{u} = \frac{1}{3\zeta} tr\, \underset{\sim}{\sigma}\, . \end{cases} \tag{2}$$

They satisfy the entropy inequality:

$$\frac{\partial}{\partial t}(\rho S) + \frac{\partial}{\partial x_i}(\rho S u_i + \frac{q_i}{\theta}) = s = - \frac{\underline{q} \cdot \nabla \theta}{\theta^2} + \frac{\underset{\sim}{\sigma} : \nabla \underline{u}}{\theta} \geqslant 0 \tag{3}$$

provided that the scalar functions χ, η and ζ are positive (χ is the thermal conductivity and η and ζ the usual viscous coefficients; $\underset{\sim}{\sigma}^D = \underset{\sim}{\sigma} - \frac{1}{3} \underset{\sim}{I}\, tr\, \underset{\sim}{\sigma}$ denoting the deviator stress corresponding to $\underset{\sim}{\sigma}$ and θ is the absolute temperature).
The system (1) - (2) (denoted in the following as FNS system), with the associated constitutive relations for all thermodynamic variables is a first order system of 14 partial differential equations for 14 unknowns (for example two thermodynamic variables (ρ, θ), the three components of $\underline{u}$, the three components of $\underline{q}$, the five components of $\underset{\sim}{\sigma}^D$ and $tr\, \underset{\sim}{\sigma}$).

As it is known, the system is not hyperbolic because of the form of constitutive relation (2); the consequence is an infinite wave propagation speed. Several authors proposed alternative equation to replace the FNS system in order to eliminate the paradox and hyperbolize the system. Starting from Maxwell's idea and from the well known paper by C.Cattaneo |1| (in the case of a rigid heat conductor), a large body of literature exists.
Substantially two points of view are predominant: i) the "*extended irreversible thermodynamics*" (Müller |2|) based on the hypothesis that the non-equilibrium thermodynamic variables (in particular the

entropy density) are different from equilibrium and depend also on $\underline{q}$ and $\underset{\sim}{\sigma}$. So the relaxation effects are introduced by a simple modification of Gibbs relation.

In this context see also the works by Carassi-Morro |3|, Lebon and Coworkers |4| - |6| in the non relativistic frame and Müller |7|, Kranyš |8|, Israel |9|, Israel-Stewart |10| for relativistic theories.

ii) The "*rational thermodynamics*" in which the entropy inequality is to be regarded as a constraint to the constitutive equations and not as an identification of a privileged time orientation : Coleman-Noll |11|, Müller |12|; see also Gurtin |13|, Gurtin-Pipkin |14|, Green-Lindsay |15|, the Truesdell book |16| and the review article by Hutter |17|.

Different approaches to similar problems are due to Boillat |18|, Massa-Morro |19|, Dixon |20|, Morro |21|, Grioli |22|, Bressan |23|. In |24| (to which we refer for the details of this lecture), we have exposed some critical considerations on the models alternative to the F.N.S. system, in particular on the question of hyperbolicity (generally not guaranteed for all times |25|) and on the lack of "*conservative form*" for the new heat and viscous stress equations, that prevents us from establishing the usual weak formulation for the Cauchy problem, therefore preventing us from studying shock waves. Starting from this two objections we expose an attempt to set up a model that have the following requests:

R1) *Unification of the two type of approach* ("extended" and "rational thermodynamics") in the sense that we think the heat flux and the viscous stress as field components that appear in the non-equilibrium entropy as in the extended thermodynamics approach, but we determine the evolution of $\underline{q}$ and $\underset{\sim}{\sigma}$ directely from an "entropy principle" in the frame of "rational thermodynamics".

R2) *Conservation laws:* we adopt a restrictive "entropy principle" so that also the evolution equations for $\underline{q}$ and $\underset{\sim}{\sigma}$ are in the form of conservation laws as the field equations (1).

R3) *Hyperbolicity*. We find constraints only on the constitutive relations so that the hyperbolicity holds for all time.

The technique used in order to arrive to the request R1) - R3) is based on some recent results for the structure of a quasi linear partial differential system in conservative form compatible with a supplementary conservation law.

Let us shortly sketch the methodology.

We consider a quasi-linear first order conservative system

$$\partial_\alpha \underline{F}^\alpha(\underline{u}) = \underline{f}(\underline{u}) \tag{4}$$

where $\underline{F}^\alpha$ and $\underline{f}$ are column vectors $\in \mathbb{R}^N$ and $\underline{u}(x^\alpha)$ is the unknown $\mathbb{R}^N$ - vector of the system (α=0,i; i=1,2,3; $\partial_o = \partial/\partial t$).

One looks for the functional dependence of $\underline{F}^\alpha \equiv \underline{F}^\alpha(\underline{u})$ so that the system (4) become compatible with the supplementary conservation law

$$\partial_\alpha h^\alpha(\underline{u}) = g(\underline{u}) \tag{5}$$

Observing the quasi linearity of (4) and (5), the compatibility is ensured by the existence of a vector $\underline{u}' \in R^N$, the components of which are function of $\underline{u}$ such that

$$\underline{u}' \cdot d\,\underline{F}^\alpha \equiv d\,h^\alpha \tag{6}$$

$$\underline{u}' \cdot \underline{f} = g \tag{7}$$

Then, by supposing local invertibility of the mapping relating $\underline{u}$ and $\underline{u}'$, it is easy to see that four scalar functions h'^α exist, defined as

$$h'^\alpha = \underline{u}' \cdot \underline{F}^\alpha - h^\alpha \tag{8}$$

so that

$$\underline{F}^{\alpha} = -\frac{\partial h'^{\alpha}}{\partial \underline{u}'} \tag{9}$$

and therefore the system assumes in the field $\underline{u}'$ (*"main field"*) the form :

$$\partial_{\alpha} \left(\frac{\partial h'^{\alpha}}{\partial \underline{u}'}\right) = \underline{f} \iff \frac{\partial^2 h'^{\alpha}}{\partial \underline{u}' \partial \underline{u}'} \partial_{\alpha} \underline{u}' = \underline{f} \, . \tag{10}$$

If $\partial \underline{F}^{0}/\partial \underline{u}$ is not singular (so that it is possible to choose $\underline{F}^{0}$ as field $\underline{u}$) and h^{0} is a convex function of $\underline{F}^{0}$ then it is possible to show from (8) that h'^{0} is the Legendre transformation of h^{0} and therefore h'^{0} is also a convex function of $\underline{u}'$ so that $\partial^2 h'^{0}/\partial u' \partial u'$ is definite positive and the system (10) is symmetric hyperbolic in the sense of Friedrichs definition.

Systems of the form (10) have been considered first by Godunov |26| who has observed that the system of perfect fluid and the Euler-Lagrange equations are suscettible of this form but without the use of compatibility from (4) and (5). Successively Godunov |27| noted that systems of form (10) admit always an equation (5) as consequence. In 1971 Friedrichs and Lax |28| proved that all system (4) compatible with (5) with h^{0} convex function of $\underline{F}^{0}$ are symmetric if we multiply (4) by an opportun matrix $\underset{\sim}{H}(\underline{u})$. The disavantage of this method is the lack of the conservative form of the new symmetric system.

The sketch here presented is due substantially to Boillat |29| with some generalization due to Ruggeri and Strumia |30| and myself in |24|.

Returning to our problem we observe the following.

i) *Systems of type (4) with the supplementary conservation law are completely determined if we know the 2N + 4 objects:* $\{\underline{u}', h'^{\alpha}, \underline{f}\}$. *Infact from (9) we determine* $\underline{F}^{\alpha}$ *and from (8) and (7)* h^{α} *and* g. *We call these quantities "Generators".*

ii) *The system (4) is the wanted form of the system that we seek, given by eq. (1) and the new equation of evolution for* $\underline{q}$ *and* $\underset{\sim}{\sigma}$; *(5) plays the role of an "entropy principle" once a definite sign for the source term is taken.*

Therefore we would know the system satisfying the request R1), R2) if we could determine the generators $\underline{u}'$, h'^{α} and $\underline{f}$. Furtbermore if we could prove that h^{o} is a convex function of $\underline{F}^{o}$ also R3) would be satisfied. Then the problem now is to identify the generators.
The generators can be found for non-hyperbolic system also (obviously in this case it is not possible that $\underline{F}^{o}$ is a field and h^{o} a convex function of $\underline{F}^{o}$) and so the simplest idea we will follow is based on the following steps :

I°) *evaluation of the generators of FNS systems.*

II°) *modification of the generators by substituting the entropy density and related thermodynamic quantities with new functions to be determined, playing the role of the same quantities but evaluated at non-equilibrium and to be taken as dependent on the heat flux and the viscous stress tensor beside the usual variables.*

III°) *for the modified generators determination of the new system and the new entropy law.*

IV°) *prove that the new h^{o} is a convex function of the new $\underline{F}^{o}$ such a convexity condition is sufficient to guarantee that the new system obtained in this way is a "symmetric" one and then conservative and hyperbolic for any field.*

I° STEP : GENERATORS OF F.N.S. SYSTEM

The system (1), (2) assume the form (4) and the entropy law (3) the form (5) when we choose :

$$\underline{F}^{o} \equiv (\rho,\ \rho u_k,\ \varepsilon,\ 0_3,\ 0_5,\ 0_1)^T \tag{11}$$

$$\underline{F}^{i} \equiv (\rho\, u_i,\ \rho u_i u_k + t_{ik},\ \varepsilon\, u_i + q_i + t_{ij}\, u_j,\ \theta\, \delta_{ik},\ \frac{1}{2}(u_j\, \delta_{ki} + u_k\, \delta_{ji}) - \frac{1}{3}\ \delta_{jk}\, u_i,\ u_i)^T \tag{12}$$

$$\underline{f} \equiv (0_1, 0_3, 0_1 - k\, q_k, \alpha\, \sigma^D_{jk} ;\ \beta\, tr\, \underset{\sim}{\sigma})^T \tag{13}$$

$$h^o = -\rho\, S; \quad h^i = -(\rho\, S\, u_i + q_i/\theta) \tag{14}$$

$$g = -s = -(k\, \frac{q^2}{\theta^2} + \alpha\, \frac{\sigma^D : \sigma^D}{\theta} + \beta\, \frac{(tr\sigma)^2}{\theta}) \tag{15}$$

where $k = 1/\chi$; $\alpha = 1/(2\eta)$; $\beta = 1/(3\zeta)$; $0_1, 0_3, 0_5$ denotes respectively the scalar, vector and traceless symmetric matrix null elements. From (6), (7) and (8) we obtain the generators of FNS system :

$$\underline{u}' \equiv \frac{1}{\theta}\,(G - \frac{u^2}{2}, u_k, -1, \frac{q_k}{\theta} ;\ -\sigma^D_{jk} ;\ \frac{-tr\,\sigma}{3})^T \tag{16}$$

$$h'^o = \frac{p}{\theta} ; \qquad h'^i = \frac{1}{\theta}\,(t_{ik}\, u_k + q_i) ; \tag{17}$$

in which G is the chemical potential (free entalpy):

$$G = e + p\, V - \theta\, S , \qquad (V = 1/\rho) . \tag{18}$$

II° STEP : MODIFICATIONS OF THE GENERATORS -

As we pointed out in the previous step the mathematical quantities that determine the FNS system are the main field $\underline{u}'$ supplied by (16), the generating functions h'^α given by (17) and the source $\underline{f}$ given by (13).

Following Müller's approach, we may introduce relaxation terms by modifying some thermodynamic variables, e.g. $S_N(\rho, e, \underline{q}, \underset{\sim}{\sigma})$ different from the corresponding equilibrium quantities. The simplest idea is then to modify h'^α, $\underline{u}'$ and $\underline{f}$ of the classical theory by substituting into them the new non-equilibrium variables. Since in this formulation only the absolute temperature and the chemical potential appear as thermodynamic variables and since it does not seem to be reasonable to redefine the former, the chemical potential alone (and with it the entropy density) will be changed.

In other words, we are looking for a system of type (4), if it exists, such that the new generating function and the new main field are :

$$h'^{\alpha}_{N} = h'^{\alpha} \tag{19}$$

$$\underline{u}'_N \equiv \frac{1}{\theta}\left(G_N - \frac{u^2}{2},\ u_k,\ -1,\ \frac{q_k}{\theta},\ -\sigma^D_{jk},\ -\frac{1}{3}\,tr\,\underset{\sim}{\sigma}\right)^T ; \tag{20}$$

$G_N(\rho, e, \underline{q}, \underset{\sim}{\sigma})$ is a function yet to be determined, which we shall interpret as the non-equilibrium chemical potential.

As a consequence of the previous discussion the advantages of this simple criterion are the following : i) the new system is conservative; ii) a supplementary conservation equation (for the entropy) is garanteed: ii) the entropy inequality is also satisfied; in fact the "entropy source" $s = -g$ doesn't change from the F.N.S. theory since the last 9 components of $\underline{u}'_N$ and the source $\underline{f}$ are also unchanged [see (7), (3), (13), (16) and (20)] .

(The index N will denote quantities in the new model, to be determined (e.g. $\underline{u}'_N$, h'^{α}_N, etc...), while quantities without on index are the ones of the FNS theory).

III° STEP. THE NEW SYSTEM AND THE NEW ENTROPY INEQUALITY -

By differentiating h'^{α}_N with respct to $\underline{u}'_N$ we obtain $\underline{F}'^{\alpha}_N$ i.e. the new conservative system, then through (8) we reach h^{α}_N and from (7) g_N, i.e. the new supplementary law and the new entropy density.

Summarizing the final results are: $\exists$ a constitutive function $\mathcal{S}$ ·

$$\mathcal{S} \equiv \mathcal{S}(q_*, \sigma^D_*, \sigma_*); \qquad (q_* = q/\theta^2_*; \ \sigma^D_* = \sigma^D/\theta; \ \sigma_* = \frac{tr\,\sigma}{\theta})$$

so that the new system is formed by the same eqs. (1) and the following new equations replacing the FNS constitutive relations :

$$\begin{cases} \rho \dfrac{d}{dt} \left(\dfrac{\partial \mathcal{S}}{\partial q_*}\right) + \nabla\theta = -k\, q \\ \rho \dfrac{d}{dt} \left(\dfrac{\partial \mathcal{S}}{\partial \sigma^D_*}\right) - \dfrac{1}{2}(\nabla u + \nabla^T u) + \dfrac{1}{3} I\, div\, u = -\alpha\, \sigma^D \\ \rho \dfrac{d}{dt} \left(\dfrac{\partial \mathcal{S}}{\partial \sigma_*}\right) - div\, u = -\beta \ \ tr\, \sigma \end{cases} \tag{21}$$

with the supplementary law :

$$\rho \frac{d\, S_N}{dt} + div\, (q/\theta) = k\left(\frac{q^2}{\theta^2}\right) + \frac{\alpha}{\theta}\left(\sigma^D : \sigma^D\right) + \frac{\beta}{3\theta}(tr\, \sigma)^2 \geq 0 \tag{22}$$

(d/dt) denotes the substantial derivative).
We note that in this approach all the other thermodynamical quantities e.g. e, ρ, θ, p are the same as in the classical one.
Furthermore we have the new entropy density and the new chemical potential by :

$$S_N = S + \mathcal{S} - \left\{ q_* \cdot \frac{\partial \mathcal{S}}{\partial q_*} + \sigma^D_* : \frac{\partial \mathcal{S}}{\partial \sigma^D_*} + \sigma_* \frac{\partial \mathcal{S}}{\partial \sigma_*} \right\} \tag{23}$$

$$G_N = G - \theta \mathcal{S}. \tag{24}$$

IV STEP. CONVEXITY AND HYPERBOLICITY OF THE SYSTEM -

Until now we have shown that the new system eqs. (1), (21) is conservative, the entropy principle is satisfied and S_N and G_N may be determined from (23) and (24) starting only from one

constitutive function $\mathcal{S}$.
The crucial point, i.e. the hyperbolicity of the system, must now be checked.
To prove hyperbolicity using the usual definition would be a considerable task; however as a consequence of the construction of the system, it is enough to show that h_N^o is a convex function of the field $\underline{F}_N^o$, since this condition ensure hyperbolicity (through the properties of symmetric systems). Thus, we investigate whether the quadratic form

$$Q = \delta^2 h_N^o = \frac{\partial^2 h_N^o}{\partial \underline{F}_N^o \, \partial \underline{F}_N^o} \; \delta \underline{F}_N^o \cdot \delta \underline{F}_N^o$$

is positive for any non vanishing variation $\delta \underline{F}_N^o$.
Straightforward calculations yield for our system to

$$Q = Q_f + \rho \left\{ \delta \underline{q}_* \cdot \delta\left(\frac{\partial \mathcal{S}}{\partial \underline{q}_*}\right) + \delta \overset{D}{\underset{\sim}{\sigma}}_* : \delta\left(\frac{\partial \mathcal{S}}{\partial \overset{D}{\underset{\sim}{\sigma}}_*}\right) + \delta\sigma_* \delta\left(\frac{\partial \mathcal{S}}{\partial \sigma_*}\right) \right\} \qquad (25)$$

where Q_f denotes the analogous quantity for an inviscid fluid at thermal equilibrium and it is known to be positive if the specific heat at constant pressure c_p and $(\partial p/\partial \rho)_S$ are both positive. Therefore from (25), *we conclude that* Q *is positive, if and only if* $\mathcal{S}$ *is a convex function of its arguments.*
Furthermore by choosing $\mathcal{S}$ as a convex function of $\underline{q}_*$, $\overset{D}{\underset{\sim}{\sigma}}_*$, and σ_* one finds that h_N^o is also a convex function of $\underline{F}_N$ and then obtains a symmetric-hyperbolic system.
A question arises on the physical meaning for the convexity condition for the function $\mathcal{S}$. We recall that the convexity of a generic function ψ with respect to a vector field $\underline{X} \in \mathbb{R}^n$ is expressed also by the condition

$$\psi(\underline{X}_o) - \psi(\underline{X}) + (\underline{X} - \underline{X}_o) \cdot \frac{\partial \psi}{\partial \underline{X}} > 0$$

for any $\underline{X}$, $\underline{X}_o$ $(\underline{X} \neq \underline{X}_o)$ belonging to a convex subset of $\mathbb{R}^n$. Let us identify ψ with $\mathcal{S}$, $\underline{X}$ with $(\underline{q}_*, \overset{D}{\underset{\sim}{\sigma}}_*, \sigma_*) \in R^9$ and $\underline{X}_o$ with the

equilibrium state $\left[\underset{-o}{X} \equiv (0_3, 0_5, 0_1) \right]$ in which $\mathcal{S}$ vanishes.
Then the convexity condition becomes

$$\mathcal{S} - q_* \cdot \frac{\partial \mathcal{S}}{\partial q_*} - \overset{D}{\underset{\sim}{\sigma}}_* : \frac{\partial \mathcal{S}}{\partial \overset{D}{\underset{\sim}{\sigma}}_*} - \sigma_* \frac{\partial \mathcal{S}}{\partial \sigma_*} < 0 \, .$$

From (23)$^{\prime}$ the above condition is equivalent to the physically relevant result :

$$S_N < S \qquad \forall \, (q_*, \overset{D}{\underset{\sim}{\sigma}}_*, \sigma_*) \neq (0_3, 0_5, 0_1) \, . \tag{26}$$

Condition (26), which expresses the physical condition of thermodynamic stability (maximum entropy) at equilibrium (required also by Müller in |2|), ensures hyperbolicity of our system.

CONSEQUENCES

Interesting consequence, both from the mathematical and physical point of view, arise from the conservative and symmetric hyperbolic structure of our system. Let us examine some problems :

i) Cauchy problem - For a symmetric system a general theorem given by Fischer and Marsden |31| on the well-position (locally) of the Cauchy problem, holds, ensuring existence and uniqueness of the solution with the same regularity of the initial data in a neighborhood of the initial manifold, when the initial data are chosen in a Sobolev space $\mathcal{H}^s$, with $s \geq 4$.

ii) Shock waves - The conservative form of our system makes possible to define weak solution in the usual way and, in particular, to study shock waves. Moreover the properties of shoch waves shown in |29|, |30|, |32|, hold :

- The "entropy" increases across the shock wave front :

In fact it is proved that across a shock front Γ while the field equations (4) satisfy the Rankine-Hugoniot relations:

$-s\,[\underline{F}^{o}] + [\underline{F}^{i}]\,n_i = 0$ on $\underline{\Gamma}$, for the supplementary equations (5) the quantity

$$\eta = -s[h^{o}] + [h^{i}]\,n_i$$

is generally non vanishing. Furthermore it is possible to show that η is non negative. (s is the velocity of shock, $\underline{n}$ is the unity normal to the shock front and the brackets denoting the jump across Γ). In our case taking in account also the Rankine-Hugoniot conditon for the mass balance, we have :

$$\eta = \rho_o\,(s - u_{on})\,[S_N] - [\frac{q_n}{\theta}]\ .$$

where the o denotes the imperturbated field and $u_n = \underline{u}\cdot\underline{n}$; $q_n = \underline{q}\cdot\underline{n}$.

- <u>The knowledge of</u> η <u>as a function of the imperturbed field</u> $\underline{u}_o = \underline{F}^{o}_o$ <u>determine the jump of the main field</u> $\underline{u}'$:

$$[\underline{u}'] = \left\{-s\,(\frac{\partial F^{o}}{\partial \underline{u}})_o + n_i\,(\frac{\partial F^{i}}{\partial \underline{u}})_o\right\}^{-1}\frac{\partial\eta}{\partial \underline{u}_o}\ .$$

- <u>The shock propagation speeds are bounded between the smallest and the largest characteristic speeds</u> $\lambda^{(k)}$ $(k = 1, 2, \ldots, N)$:

$$\inf_{\substack{\underline{u}'\in\mathcal{D}' \\ \underline{n}\in\|n\|=1}}\ \min_{k}\ \{\lambda^{(k)}\} < |s| < \sup_{\substack{\underline{u}'\in\mathcal{D}' \\ \underline{n}\in\|n\|=1}}\ \max_{k}\ \{\lambda^{(k)}\}\ ;$$

where $\mathcal{D}'$ is on open convex set $\subseteq \mathbb{R}^N$.

iii) *Special Case*

Our model is specified by the constitutive function $\mathcal{S}$ which depends on the variables $\underline{q}_*$, $\overset{D}{\underset{\vee}{\sigma}}{}_*$, σ_*. Such a function is a measure of the "distance" between our model and the classical FNS theory. In fact when $\mathcal{S}$ vanishes, the model becomes the FNS theory.

It is reasonable to assume that this function is not very large. If it is required to be convex, it seems interesting to examine the special case where $\mathcal{S}$ is a quadratic function of the variables :

$$\mathcal{S} = \frac{1}{2} (\tau_o \, q_*^2 + \tau_1 \overset{D}{\underset{\sim}{\sigma}}_* : \overset{D}{\underset{\sim}{\sigma}}_* + \tau_2 \, \sigma_*^2) \qquad (27)$$

in which τ_i are positive constants. The equations (21) then become

$$\begin{cases} \chi \{\tau_o \, \rho \, \frac{d}{dt} (\underline{q}/\theta^2) + \nabla\theta\} = - \underline{q} \\ 2\eta \, \{\tau_1 \, \rho \, \frac{d}{dt} (\overset{D}{\underset{\sim}{\sigma}}/\theta) - \frac{1}{2} (\nabla\underline{u} + \overset{T}{\nabla}\underline{u}) + \frac{1}{3} \, I \, div \, \underline{u} \} = - \overset{D}{\underset{\sim}{\sigma}} \\ 3 \, \zeta \{\tau_2 \, \rho \, \frac{d}{dt} (tr \underset{\sim}{\sigma}/\theta) - div \, \underline{u} \} = - tr \, \underset{\sim}{\sigma} \end{cases} \qquad (28)$$

and (24), (23), become :

$$G_N = G - \theta\mathcal{S} \, ; \qquad S_N = S - \mathcal{S} \, . \qquad (29)$$

Gibbs relation is modified into :

$$d \, S_N = d \, S - \tau_o \, \frac{\underline{q}}{\theta^2} \cdot d \, (\frac{\underline{q}}{\theta^2}) - \tau_1 \, \frac{\overset{D}{\underset{\sim}{\sigma}}}{\theta} : d \, (\frac{\overset{D}{\underset{\sim}{\sigma}}}{\theta}) - \tau_2 \, \frac{tr \, \underset{\sim}{\sigma}}{\theta} \, d \, (\frac{tr \, \underset{\sim}{\sigma}}{\theta}) . \quad (30)$$

We point out that in (30) the measure of "nearess" is quantitatively defined by the squares of $\underline{q}/\theta^2$, $\overset{D}{\underset{\sim}{\sigma}}/\theta$ and $tr \, \underset{\sim}{\sigma}/\theta$, instead of the squares of $\underline{q}$ $\overset{D}{\underset{\sim}{\sigma}}$, $tr \, \underset{\sim}{\sigma}$ as it happens in Müller's theory. Our measure involves also the "coldness" $1/\theta$.

iv) *Acceleration waves.*

Acceleration waves (weak discontinuities) propagate with real and finite speeds.

Furthermore it is possible to show that anisotropy appears because of

the presence of heat flow (the velocity are different if we change the normal $\underline{n}$ in $-\underline{n}$). This properties are usually not present when Cattaneo-Müller type equations is employed.
In particular in the special case treated before with $\zeta = 0$; $p = R\rho\theta$; $e = c_v\theta$ we obtain the following 13 velocities :

$v_{(i)} = 0$ *for* $i = 1,\ldots,5$ *(contact waves)* ;

$v_{(i)}$ $(i = 6,\ldots,11)$ *roots of the polinomial of* 6° *degree:*

$$v^2(\mu_4^2 - \mu_5^2)\,(R\theta + \frac{1}{4}\mu_3^2) + (v^2 - \frac{3}{4}\mu_3^2)\Big\{(v^2 - 2\mu_1 v - \mu_2^2)\cdot$$

$$\cdot\,(v^2 - \mu_3^2 - R\theta) - v^2\mu_4^2\Big\}\;; \qquad (31)$$

$v_{(i)}$ $(i=12,13)$ *roots* of $v^2 = \frac{3}{4}\mu_3^2$.

Where $v = \lambda - u_n$ (λ is the characteristic velocity and $u_n = \underline{u}\cdot\underline{n}$) and μ_j $(j = 1,\ldots,5)$ are the following quantities having dimensions of velocities :

$$\mu_1 = \frac{q_n}{\rho\theta\, c_v}\;;\quad \mu_2^2 = \frac{\theta^2}{\rho^2\tau_o c_v}\;;\quad \mu_3^2 = \frac{2\theta}{3\,\tau_1\rho^2}$$

$$\mu_4^2 = \frac{t_{nn}^2}{\rho^2\,\theta\, c_v}\;;\qquad \mu_5^2 = \frac{t_n^2}{\rho^2\,\theta c_v}\;;$$

$$(q_n = \underline{q}\cdot\underline{n};\quad \underline{t}_n = \underset{\sim}{t}\,\underline{n}\;;\; t_{nn} = \underline{n}^T\underset{\sim}{t}\,\underline{n}\;;\; \underset{\sim}{t} = p\,\underset{\sim}{I} - \underset{\sim}{\sigma}).$$

The mentioned anisotropy appears as μ_1 and $\underline{n}$ become respectively $-\mu_1$ and $-\underline{n}$.
In the case of principal waves $\mu_4 = \mu_5$ and (31) give the double velocities $v^2 = \frac{3}{4}\mu_3^2$ and a polynomial of 4° degree that in a

constant state ($\underline{q} = \underset{\sim}{\sigma} = 0$) becomes biquadratic.

ν) *Objectivity principle* - The heat equation (21.1), as the analogous law for the viscous stress (21.2-3), do not fulfill the objectivity principle, generally required for constitutive equations; however our equations are invariant under Galilean trasnformations. It is known that such circumstance characterizes the model of the extended thermodynamics.

Several authors introduced "ad hoc" terms to restore objectivity (e.g. |5|).

We think that to be coherent with our approach the principle must not be fulfilled by the heat equation (and viscous stress). In fact, by requiring the heat equation to be conservative, and the heat flux to be a field variable, in order to have a hyperbolic conservative system, we require that the heat law is a field equation, playing the same role of the remaining equations of the system. This claim is equivalent to the assumption for the conservative heat equation to be of the same type, e.g., of momentum equation (but with a dissipative source), i.e. of the form:

$$\rho_o \frac{d\underline{w}}{dt} + div\,\underset{\sim}{\Theta} = -\underline{q}/\chi .$$

in which $\rho_o\, \underline{w}$ plays the role of "thermal momentum" and $\underset{\sim}{\Theta}$ the one "thermal stress". This starting hypotesis is equivalent to consider the heat equation as a balance equation (which is a resonable assumption, coming it from a principle); the quantities $\underline{w}$ and $\underset{\sim}{\Theta}$ are related to the fields variables through constitutive relations which must fulfill the principle of objectivity. The entropy principle here adopted shows the special structure :

$$\underline{w} = \partial\mathcal{S}/\partial\,\underline{q}_* , \qquad \underset{\sim}{\Theta} = \theta\,\underset{\sim}{I} .$$

Analogous considerations are for the viscous stress tensor $\underset{\sim}{\sigma}$;

The only claim is that $\mathcal{S}$ is a scalar function of their arguments. Even if in a differente frame and with different conclusions the consideration on the objectivity principle, seem to be analogous to the ones by Bressan in |23|.
An analogous treatment is obtained in |33| for a thermoelastic body with finite deformations. An attempt to extend this methodology to the relativistic case is in preparation by A. Strumia.

R E F E R E N C E S

|1| C.CATTANEO :Atti Sem.Mat. Fis. Modena, 3, 83-101 (1948).
C.R. Acad. Sc. Paris 247, 431-433 (1958).

|2| I., MÜLLER : Zeitschrift für Physik 198, 329-344 (1967).

|3| M. CARASSI and A., MORRO : Nuovo Cimento 9B 321-343 (1972).

|4| G.,LEBON : Acad. Roy. Soc. Belgique 64, 456-472 (1978).

|5| G.LEBON, D.JOU, J.CASAS VASQUEZ : J. of Physics 13A, 275-290 (1980). J. Non-Eq. Thermodyn. 4, 349-362 (1979).

|6| G.LEBON, J.M.RUBI : J. Non-Eq. Thermodyn. 5, 285-300 (1980).

|7| I.MÜLLER : Doctor thesis, Darmstadt (1966).

|8| M.KRANYŠ : Nuovo Cimento 48B, 51-70 (1966); 8B, 417-441 (1972).

|9| W.ISRAEL : Ann. Physics 100, 310-331 (1976).

|10| W.ISRAEL, J.M.STEWART : Ann. Physics 118, 341-372 (1979).

|11| B.D. COLEMAN, W.NOLL : A.R.M.A. 13, 167-178 (1963).

|12| I. MÜLLER : A.R.M.A. 40, 319-332 (1971). Entropy in non equilibrium. A challenge to mathematiciancs. In Trend in application of pure Mathematics to Mechanics, VoL.II Ed. H.Zorski. 281-295 Pitman London (1979).

|13| M. GURTIN : A.R.M.A. 28, 40-50 (1968).

|14| M. GURTIN, A.C. PIPKIN : A.R.M.A. 31, 113-126 (1968).

|15| A.E. GREEN, K.A. LINDSAY : J. Elasticity 2, 1-7 (1972).

|16| C. TRUESDELL : Rational Thermodynamics. Mac Graw-Hill, New York (1969).

|17| K. HUTTER : Acta Mech. 27, 1-54 (1977).

|18| G. BOILLAT : Nuovo Cimento 10, 295-300 (1974); 10, 352-354 (1974). Lett. math. Phys. 2, 257-263 (1978).

|19| E. MASSA, A. MORRO : Ann. Inst. H.Poincaré 24, 423-454 (1978).

|20| G. DIXON : Special relativity, the foundation of macroscopic physics. Cambridge (1978).

|21| A. MORRO : Arch. Mech. 32, 145-161 (1980).

|22| G. GRIOLI : Atti Acc. Naz. Lincei I, 67 (5), 332-339; II,67 (6), 426-432 (1980).

|23| A. BRESSAN : Lett. Nuovo Cimento, 33 (4), 108)-112; 34 (2), 63. (1982).

|24| T. RUGGERI : Symmetric hyperbolic system of conservative equations for a viscous heat conducting fluid. In press in Acta Mech. (1982).

|25| F.BAMPI, A.MORRO : Wave Motion, 2, 153-157 (1980).

|26| S.K.GODUNOV : Sov. Math. 2, 947-948 (1961).

|27| S.K.GODUNOV : Russ. Math. Surv. 17, 145-156 (1962)

|28| K.O. FRIEDRICHS, P.D. LAX : Proc. Nat. Acad. Sc. USA 68, 1686-1688 (1971).

|29| G. BOILLAT : C.R. Acad. Sc. Paris 278A, 909-912 (1974); 283A, 409-412 (1976). "Urti" in Wave Propagation, Corso C.I.M.E., 1980, coord. G. Ferrarese, Liguori Ed., Napoli, 1982.

|30| T.RUGGERI, A.STRUMIA : Ann. Inst. H.Poincaré 34, 65-84 (1981). See also: Corso C.I.M.E., 1980 (op.cit.).

|31| A.FISCHER, D.P.MARSDEN : Comm. Math. Phys. 28, 1-38 (1972).

|32| G.BOILLAT, T.RUGGERI : C.R. Acad. Sc. Paris 289A, 257-258 (1979).

|33| T.RUGGERI : "Generators of hyperbolic heat equation in non-linear thermoelasticity".
In press in Rend. Sem. Padova (vol. dedicato a G.Grioli nel suo 70° compleanno - 1982).

C.I.M.E. Session on "Thermodynamics and constitutive equations"

List of Participants

G. ALBANO, Via Arturo Graf 4, 90145 Palermo

F. BAMPI, Istituto Matematico Università, Via L.B.Alberti 4, 16132 Genova

M. BARTOLOZZI, Via Sciuti 71, 90144 Palermo

G. BOILLAT, Univ. de Clermont, Dept. Math., 63170 Aubière, France

A. BRESSAN, Via Pasubio 20, 35141 Padova

S. BUSENBERG, Dept. of math., Harvey Mudd College, Claremont, Cal. 91711, USA

P. CARBONARO, Via Portella 3, 98011 Bordonaro (Messina)

F. CARDIN, Via Palesa 33, 35100 Padova

G. CARICATO, Via Piero Foscari 40, 00139 Roma

G. CAVIGLIA, Istituto Matematico Università, Via L.B. Alberti 4, 16132 Genova

B.D. COLEMAN, Dept. of Math., Schenley Park, Pittsburgh, PA 15213, USA

C.W. DAFERMOS, Division of Appl. Math., Brown Univ., Providence, RI 02912, USA

W.G. DIXON, Churchill College, Cambridge, England

A. DONATO, Istituto Matematico Università, Via C. Battisti 90, 98100 Messina

P. FERGOLA, Istituto Matematico Università, Via Mezzocannone, 8, 80134 Napoli

F. FERRAIOLI, Istituto Matematico Università, Città Universitaria, 00185 Roma

G. FERRARESE, Istituto Matematico Università, Città Universitaria, 00185 Roma

F. FRANCHI, Via Enrico Toti 36, 40051 Altedo (Bologna)

D. FUSCO, Viale Europa 83/E, 98100 Messina

S. GIAMBO', Istituto Matematico Università, Via C. Battisti 90, 98100 Messina

D. GRAFFI, Via A. Murri 9, 40137 Bologna

A. GRECO, Istituto Matematico Università, Via Archirafi 34, 90123 Palermo

G. GRIOLI, Via Luzzatti 16, 35100 Padova

P. MARCATI, Dipartimento di Matematica, Università di Trento, 38050 Povo (Trento)

V. MOAURO, Dipartimento di Matematica, Università di Trento, 38050 Povo (Trento)

M.S. MONGIOVI' TELARETTI, Via Nairobi 31, 90129 Palermo

A. MORRO, Istituto Matematico Università, Via L.B. Alberti 4, 16132 Genova

I. MULLER, T.U. Berlin, Sekr HF.1, Str. des 17 Juni 135, D 1000 Berlin

I. MURDOCH, Math. Dept., Univ. of Strathclyde, Livingstone Tower,
26 Richmond St., Glasgow G1 1XH

M.C. NUCCI, Istituto Matematico, Via Vanvitelli 1, 06100 Perugia

A. PALUMBO, Istituto Matematico Università, Via C. Battisti 80, 98100 Messina

P. PANTANO, Dipartimento di Matematica, Università della Calabria, 87036 Arcavacata (Cosenza)

M. PITTERI, Via Bonazza 60, 35109 Padova

S. PLUCHINO, Via G. Leopardi 60, 95127 Catania

P. PODIO GUIDUGLI, Facoltà di Ingegneria, Università, 56100 Pisa

M.A. POZIO, Dipartimento di Matematica, Università di Trento, 38050 Povo (Trento)

L. RESTUCCIA, Via Damabianca is.263, 98100 Messina

T. RUGGERI, Ist. Mat. Appl. Univ., Via Vallescura 2, 40136 Bologna

A. SCALIA, Via Lombardia 7, 95014 Giarre (Catania)

M. STUDNIARSKI, ul. 11-go Listopada 47 n.29, 91-371 Lodz (Poland)

C. TENNERIELLO, Istituto Matematico Università, Via Mezzocannone 8, 80134 Napoli

M. TORRISI, Via Fusco 6, 95128 Catania

A. VALENTI, Via Ala 61, 95123 Catania

N. VIRGOPIA, Istituto Matematico Università, Città Universitaria, 00185 Roma

FONDAZIONE C.I.M.E.
CENTRO INTERNAZIONALE MATEMATICO ESTIVO
INTERNATIONAL MATHEMATICAL SUMMER CENTER

"Some Problems in Nonlinear Diffusion"

is the subject of the Second 1985 C.I.M.E. Session.

The Session, sponsored by the Consiglio Nazionale delle Ricerche and the Ministero della Pubblica Istruzione, will take place under the scientific direction of Prof. ANTONIO FASANO and Prof. MARIO PRIMICERIO (Università di Firenze, Italy) at Villa «La Querceta», Montecatini Terme (Pistoia), Italy, ***from June 10 to June 18, 1985.***

Courses

a) ***The Porous Medium Equation.*** (8 lectures in English).
Prof. D. G. ARONSON (University of Minnesota, USA).

— Physical background, elementary properties, basic existence and uniqueness theory.
— Regularity in one space dimension.
— Properties of the interface in one space dimension: smoothness, nonsmoothness, waiting time.
— The relationship between the porous medium pressure equation and a Hamilton-Jacobi equation: limiting behavior as $m \to 1$.
— Regularity in R^d for $d > 1$: Hölder continuity and quasi convexity.
— Initial trace, existence and uniqueness in R^d: Harnack-type estimates and growth at infinity.
— Asymptotic behavior of solutions of the initial-boundary value problem.
— Stabilization results.

References

- D.G. ARONSON and P. BENILAN, Regularité des solutions de l'équation des milieux poreux dans R^n. C.R. Acad. Sci. Paris, Série A-B, 288 (1979), 103-105.
- D.G. ARONSON and L.A. CAFFARELLI, The initial trace of a solution of the porous medium equation. Trans. Amer, Math. Soc. 280 (1983), 351-366.
- D.G. ARONSON, L.A. CAFFARELLI and J.L. VAZQUEZ, Interfaces with a corner point in one-dimensional porous medium flow. Comm. Pure Appl. Math., to appear.
- D.G. ARONSON, M.G. CRANDALL and L.A. PELETIER, Stabilization of solutions of a degenerate nonlinear diffusion problem. Nonlinear Analysis 6 (1982), 1002-1022.
- D.G. ARONSON, S. KAMIN and L.A. CAFFARELLI, How an initially stationary interface begins to move in porous medium flow. SIAM J. Math. Anal. 14 (1983), 639-658.
- D.G. ARONSON and L.A. PELETIER, Large time behaviour of solutions of the porous medium equation. J. Diff. Eq. 39 (1981), 378-412.
- P. BENILAN, M.G. CRANDALL and M. PIERRE, Solutions of the porous medium equation in R^n under optimal conditions on initial values. Indiana Univ. Math. J. 33 (1984), 51-87.
- L.A. CAFFARELLI and A. FRIEDMAN, Continuity of the density of a gas flow in a porous medium. Trans. Amer. Math. Soc. 252 (1979), 99-113.
- L.A. CAFFARELLI and A. FRIEDMAN, Regularity of the free boundary of a gas flow in an n-dimensional porous medium. Indiana U. Math. J. 29 (1980), 361-391.
- B.E.J. DAHLBERG and C.E. KENIG, Nonnegative solutions of the porous medium equation. Comm. P.D.E., to appear.
- B.F. KNERR, The porous medium equation in one dimension. Trans. Amer. Math. Soc. 234 (1977), 381-415.
- O.A. OLEINIK, S.A. KALASHNIKOV and CHZOU YUI-LIN, The Cauchy problem and boundary problems for equations of the type of nonstationary filtration. Izv. Akad. Nauk SSSR Ser. Mat., 22 (1958), 667-704.

b) ***Qualitative Methods in Reaction-Diffusion Equations.*** (8 lectures in English).
Prof. Joel SMOLLER (University of Michigan).

Outline.

Topological Methods, the Conley Index, Applications to Reaction-Diffusion Equations, the Fitz-Hugh Nagumo equations, Stability and Bifurcation of Stationary Solutions to Predator-Prey Equations, Symmetry-Breaking Solutions of Semilinear Elliptic equations.

References

- J. SMOLLER, Shock Waves and Reaction-Diffusion Equations. Springer Verlag, 1983.

c) ***Reaction-Diffusion Problems in Chemistry.*** (8 lectures in English).
Prof. I. STAKGOLD (University of Delaware).

— The quasilinear parabolic system of equations for the concentration and temperature of a simple, irreversible reaction.
— The Frank-Kamenetskii, Semenov and Gelfand approximations, The notion of criticality.
— The method of upper and lower solutions.
— Existence, uniqueness and multiplicity of solutions.
— The case of nonlinearities that are not piecewise smooth. Extinction, dead cores and quenching.
— Free boundary problems and related estimates.
— Two-phase problems, particularly gas-solid interactions.
— The pseudo-steady-state approximation and estimates for the conversion.

Seminars

A number of seminars and special lectures will be offered during the Session.

LIST OF C.I.M.E. SEMINARS

		Publisher
1954 -	1. Analisi funzionale	C.I.M.E.
	2. Quadratura delle superficie e questioni connesse	"
	3. Equazioni differenziali non lineari	"
1955 -	4. Teorema di Riemann-Roch e questioni connesse	"
	5. Teoria dei numeri	"
	6. Topologia	"
	7. Teorie non linearizzate in elasticità, idrodinamica, aerodinamica	"
	8. Geometria proiettivo-differenziale	"
1956 -	9. Equazioni alle derivate parziali a caratteristiche reali	"
	10. Propagazione delle onde elettromagnetiche	"
	11. Teoria della funzioni di più variabili complesse e delle funzioni automorfe	"
1957 -	12. Geometria aritmetica e algebrica (2 vol.)	"
	13. Integrali singolari e questioni connesse	"
	14. Teoria della turbolenza (2 vol.)	"
1958 -	15. Vedute e problemi attuali in relatività generale	"
	16. Problemi di geometria differenziale in grande	"
	17. Il principio di minimo e le sue applicazioni alle equazioni funzionali	"
1959 -	18. Induzione e statistica	"
	19. Teoria algebrica dei meccanismi automatici (2 vol.)	"
	20. Gruppi, anelli di Lie e teoria della coomologia	"
1960 -	21. Sistemi dinamici e teoremi ergodici	"
	22. Forme differenziali e loro integrali	"
1961 -	23. Geometria del calcolo delle variazioni (2 vol.)	"
	24. Teoria delle distribuzioni	"
	25. Onde superficiali	"
1962 -	26. Topologia differenziale	"
	27. Autovalori e autosoluzioni	"
	28. Magnetofluidodinamica	"
1963 -	29. Equazioni differenziali astratte	"
	30. Funzioni e varietà complesse	"
	31. Proprietà di media e teoremi di confronto in Fisica Matematica	"

1964 -	32.	Relatività generale	C.I.M.E.
	33.	Dinamica dei gas rarefatti	"
	34.	Alcune questioni di analisi numerica	"
	35.	Equazioni differenziali non lineari	"
1965 -	36.	Non-linear continuum theories	"
	37.	Some aspects of ring theory	"
	38.	Mathematical optimization in economics	"
1966 -	39.	Calculus of variations	Ed. Cremonese, Firenze
	40.	Economia matematica	"
	41.	Classi caratteristiche e questioni connesse	"
	42.	Some aspects of diffusion theory	"
1967 -	43.	Modern questions of celestial mechanics	"
	44.	Numerical analysis of partial differential equations	"
	45.	Geometry of homogeneous bounded domains	"
1968 -	46.	Controllability and observability	"
	47.	Pseudo-differential operators	"
	48.	Aspects of mathematical logic	"
1969 -	49.	Potential theory	"
	50.	Non-linear continuum theories in mechanics and physics and their applications	"
	51.	Questions of algebraic varieties	"
1970 -	52.	Relativistic fluid dynamics	"
	53.	Theory of group representations and Fourier analysis	"
	54.	Functional equations and inequalities	"
	55.	Problems in non-linear analysis	"
1971 -	56.	Stereodynamics	"
	57.	Constructive aspects of functional analysis (2 vol.)	"
	58.	Categories and commutative algebra	"
1972 -	59.	Non-linear mechanics	"
	60.	Finite geometric structures and their applications	"
	61.	Geometric measure theory and minimal surfaces	"
1973 -	62.	Complex analysis	"
	63.	New variational techniques in mathematical physics	"
	64.	Spectral analysis	"

1974 -	65. Stability problems	Ed. Cremonese, Firenze
	66. Singularities of analytic spaces	"
	67. Eigenvalues of non linear problems	"
1975 -	68. Theoretical computer sciences	"
	69. Model theory and applications	"
	70. Differential operators and manifolds	"
1976 -	71. Statistical Mechanics	Ed. Liguori, Napoli
	72. Hyperbolicity	"
	73. Differential topology	"
1977 -	74. Materials with memory	"
	75. Pseudodifferential operators with applications	"
	76. Algebraic surfaces	"
1978 -	77. Stochastic differential equations	"
	78. Dynamical systems	Ed. Liguori, Napoli and Birkhäuser Verlag
1979 -	79. Recursion theory and computational complexity	Ed. Liguori, Napoli
	80. Mathematics of biology	"
1980 -	81. Wave propagation	"
	82. Harmonic analysis and group representations	"
	83. Matroid theory and its applications	"
1981 -	84. Kinetic Theories and the Boltzmann Equation	(LNM 1048) Springer-Verlag
	85. Algebraic Threefolds	(LNM 947) "
	86. Nonlinear Filtering and Stochastic Control	(LNM 972) "
1982 -	87. Invariant Theory	(LNM 996) "
	88. Thermodynamics and Constitutive Equations	(LN Physics 228) "
	89. Fluid Dynamics	(LNM 1047) "
1983 -	90. Complete Intersections	(LNM 1092) "
	91. Bifurcation Theory and Applications	(LNM 1057) "
	92. Numerical Methods in Fluid Dynamics	(LNM 1127) "
1984	93. Harmonic Mappings and Minimal Immersions	to appear "
	94. Schrödinger Operators	to appear "
	95. Buildings and the Geometry of Diagrams	to appear "
1985 -	96. Probability and Analysis	to appear "
	97. Some Problems in Nonlinear Diffusion	to appear "
	98. Theory of Moduli	to appear "

Note: Volumes 1 to 38 are out of print. A few copies of volumes 23,28,31,32,33,34,36,38 are available on request from C.I.M.E.

Lecture Notes in Physics

Vol. 195: Trends and Applications of Pure Mathematics to Mechanics. Proceedings, 1983. Edited by P. G. Ciarlet and M. Roseau. V, 422 pages. 1984.

Vol. 196: WOPPLOT 83. Parallel Processing: Logic, Organization and Technology. Proceedings, 1983. Edited by J. Becker and I. Eisele. V, 189 pages. 1984.

Vol. 197: Quarks and Nuclear Structure. Proceedings, 1983. Edited by K. Bleuler. VIII, 414 pages. 1984.

Vol. 198: Recent Progress in Many-Body Theories. Proceedings, 1983. Edited by H. Kümmel and M. L. Ristig. IX, 422 pages. 1984.

Vol. 199: Recent Developments in Nonequilibrium Thermodynamics. Proceedings, 1983. Edited by J. Casas-Vázquez, D. Jou and G. Lebon. XIII, 485 pages. 1984.

Vol. 200: H.D. Zeh, Die Physik der Zeitrichtung. V, 86 Seiten. 1984.

Vol. 201: Group Theoretical Methods in Physics. Proceedings, 1983. Edited by G. Denardo, G. Ghirardi and T. Weber. XXXVII, 518 pages. 1984.

Vol. 202: Asymptotic Behavior of Mass and Spacetime Geometry. Proceedings, 1983. Edited by F. J. Flaherty. VI, 213 pages. 1984.

Vol. 203: C. Marchioro, M. Pulvirenti, Vortex Methods in Two-Dimensional Fluid Dynamics. III, 137 pages. 1984.

Vol. 204: Y. Waseda, Novel Application of Anomalous (Resonance) X-Ray Scattering for Structural Characterization of Disordered Materials. VI, 183 pages. 1984.

Vol. 205: Solutions of Einstein's Equations: Techniques and Results. Proceedings, 1983. Edited by C. Hoenselaers and W. Dietz. VI, 439 pages. 1984.

Vol. 206: Static Critical Phenomena in Inhomogeneous Systems. Edited by A. Pękalski and J. Sznajd. Proceedings, 1984. VIII, 358 pages. 1984.

Vol. 207: S. W. Koch, Dynamics of First-Order Phase Transitions in Equilibrium and Nonequilibrium Systems. III, 148 pages. 1984.

Vol. 208: Supersymmetry and Supergravity/Nonperturbative QCD. Proceedings, 1984. Edited by P. Roy and V. Singh. V, 389 pages. 1984.

Vol. 209: Mathematical and Computational Methods in Nuclear Physics. Proceedings, 1983. Edited by J. S. Dehesa, J. M. G. Gomez and A. Polls. V, 276 pages. 1984.

Vol. 210: Cellular Structures in Instabilities. Proceedings, 1983. Edited by J. E. Wesfreid and S. Zaleski. VI, 389 pages. 1984.

Vol. 211: Resonances – Models and Phenomena. Proceedings, 1984. Edited by S. Albeverio, L. S. Ferreira and L. Streit. VI, 369 pages. 1984.

Vol. 212: Gravitation, Geometry and Relativistic Physics. Proceedings, 1984. Edited by Laboratoire "Gravitation et Cosmologie Relativistes", Université Pierre et Marie Curie et C.N.R.S., Institut Henri Poincaré, Paris. VI, 336 pages. 1984.

Vol. 213: Forward Electron Ejection in Ion Collisions. Proceedings, 1984. Edited by K. O. Groeneveld, W. Meckbach and I. A. Sellin. VII, 165 pages. 1984.

Vol. 214: H. Moraal, Classical, Discrete Spin Models. VII, 251 pages. 1984.

Vol. 215: Computing in Accelerator Design and Operation. Proceedings, 1983. Edited by W. Busse and R. Zelazny. XII, 574 pages. 1984.

Vol. 216: Applications of Field Theory to Statistical Mechanics. Proceedings, 1984. Edited by L. Garrido. VIII, 352 pages. 1985.

Vol. 217: Charge Density Waves in Solids. Proceedings, 1984. Edited by Gy. Hutiray and J. Sólyom. XIV, 541 pages. 1985.

Vol. 218: Ninth International Conference on Numerical Methods in Fluid Dynamics. Edited by Soubbaramayer and J. P. Boujot. X, 612 pages. 1985.

Vol. 219: Fusion Reactions Below the Coulomb Barrier. Proceedings, 1984. Edited by S. G. Steadman. VII, 351 pages. 1985.

Vol. 220: W. Dittrich, M. Reuter, Effective Lagrangians in Quantum Electrodynamics. V, 244 pages. 1985.

Vol. 221: Quark Matter '84. Proceedings, 1984. Edited by K. Kajantie. VI, 305 pages. 1985.

Vol. 222: A. García, P. Kielanowski, The Beta Decay of Hyperons. Edited by A. Bohm. VIII, 173 pages. 1985.

Vol. 223: H. Saller, Vereinheitlichte Feldtheorien der Elementarteilchen. IX, 157 Seiten. 1985.

Vol. 224: Supernovae as Distance Indicators. Proceedings, 1984. Edited by N. Bartel. VI, 226 pages. 1985.

Vol. 225: B. Müller, The Physics of the Quark-Gluon Plasma. VII, 142 pages. 1985.

Vol. 226: Non-Linear Equations in Classical and Quantum Field Theory. Proceedings, 1983/84. Edited by N. Sanchez. VII, 400 pages. 1985.

Vol. 227: J.-P. Eckmann, P. Wittwer, Computer Methods and Borel Summability Applied to Feigenbaum's Equation. XIV, 297 pages. 1985.

Vol. 228: Thermodynamics and Constitutive Equations. Proceedings, 1982. Edited by G. Grioli. V, 257 pages. 1985.

Selected Issues from

Lecture Notes in Mathematics

Vol. 932: Analytic Theory of Continued Fractions. Proceedings, 1981. Edited by W.B. Jones, W.J. Thron, and H. Waadeland. VI, 240 pages. 1982.

Vol. 934: M. Sakai, Quadrature Domains. IV, 133 pages. 1982.

Vol. 935: R. Sot, Simple Morphisms in Algebraic Geometry. IV, 146 pages. 1982.

Vol. 936: S.M. Khaleelulla, Counterexamples in Topological Vector Spaces. XXI, 179 pages. 1982.

Vol. 937: E. Combet, Intégrales Exponentielles. VIII, 114 pages. 1982.

Vol. 938: Number Theory. Proceedings, 1981. Edited by K. Alladi. IX, 177 pages. 1982.

Vol. 942: Theory and Applications of Singular Perturbations. Proceedings, 1981. Edited by W. Eckhaus and E.M. de Jager. V, 363 pages. 1982.

Vol. 953: Iterative Solution of Nonlinear Systems of Equations. Proceedings, 1982. Edited by R. Ansorge, Th. Meis, and W. Törnig. VII, 202 pages. 1982.

Vol. 956: Group Actions and Vector Fields. Proceedings, 1981. Edited by J.B. Carrell. V, 144 pages. 1982.

Vol. 957: Differential Equations. Proceedings, 1981. Edited by D.G. de Figueiredo. VIII, 301 pages. 1982.

Vol. 963: R. Nottrot, Optimal Processes on Manifolds. VI, 124 pages. 1982.

Vol. 964: Ordinary and Partial Differential Equations. Proceedings, 1982. Edited by W.N. Everitt and B.D. Sleeman. XVIII, 726 pages. 1982.

Vol. 968: Numerical Integration of Differential Equations and Large Linear Systems. Proceedings, 1980. Edited by J. Hinze. VI, 412 pages. 1982.

Vol. 970: Twistor Geometry and Non-Linear Systems. Proceedings, 1980. Edited by H.-D. Doebner and T.D. Palev. V, 216 pages. 1982.

Vol. 972: Nonlinear Filtering and Stochastic Control. Proceedings, 1981. Edited by S.K. Mitter and A. Moro. VIII, 297 pages. 1983.

Vol. 978: J. Ławrynowicz, J. Krzyż, Quasiconformal Mappings in the Plane. VI, 177 pages. 1983.

Vol. 979: Mathematical Theories of Optimization. Proceedings, 1981. Edited by J.P. Cecconi and T. Zolezzi. V, 268 pages. 1983.

Vol. 982: Stability Problems for Stochastic Models. Proceedings, 1982. Edited by V.V. Kalashnikov and V.M. Zolotarev. XVII, 295 pages. 1983.

Vol. 989: A.B. Mingarelli, Volterra-Stieltjes Integral Equations and Generalized Ordinary Differential Expressions. XIV, 318 pages. 1983.

Vol. 994: J.-L. Journé, Calderón-Zygmund Operators, Pseudo-Differential Operators and the Cauchy Integral of Calderón. VI, 129 pages. 1983.

Vol. 999: C. Preston, Iterates of Maps on an Interval. VII, 205 pages. 1983.

Vol. 1000: H. Hopf, Differential Geometry in the Large, VII, 184 pages. 1983.

Vol. 1003: J. Schmets, Spaces of Vector-Valued Continuous Functions. VI, 117 pages. 1983.

Vol. 1005: Numerical Methods. Proceedings, 1982. Edited by V. Pereyra and A. Reinoza. V, 296 pages. 1983.

Vol. 1007: Geometric Dynamics. Proceedings, 1981. Edited by J. Palis Jr. IX, 827 pages. 1983.

Vol. 1015: Equations différentielles et systèmes de Pfaff dans le champ complexe – II. Seminar. Edited by R. Gérard et J.P. Ramis. V, 411 pages. 1983.

Vol. 1021: Probability Theory and Mathematical Statistics. Proceedings, 1982. Edited by K. Itô and J.V. Prokhorov. VIII, 747 pages. 1983.

Vol. 1031: Dynamics and Processes. Proceedings, 1981. Edited by Ph. Blanchard and L. Streit. IX, 213 pages. 1983.

Vol. 1032: Ordinary Differential Equations and Operators. Proceedings, 1982. Edited by W.N. Everitt and R.T. Lewis. XV, 521 pages. 1983.

Vol. 1035: The Mathematics and Physics of Disordered Media. Proceedings, 1983. Edited by B.D. Hughes and B.W. Ninham. VII, 432 pages. 1983.

Vol. 1037: Non-linear Partial Differential Operators and Quantization Procedures. Proceedings, 1981. Edited by S.I. Andersson and H.-D. Doebner. VII, 334 pages. 1983.

Vol. 1041: Lie Group Representations II. Proceedings 1982–1983. Edited by R. Herb, S. Kudla, R. Lipsman and J. Rosenberg. IX, 340 pages. 1984.

Vol. 1045: Differential Geometry. Proceedings, 1982. Edited by A.M. Naveira. VIII, 194 pages. 1984.

Vol. 1047: Fluid Dynamics. Seminar, 1982. Edited by H. Beirão da Veiga. VII, 193 pages. 1984.

Vol. 1048: Kinetic Theories and the Boltzmann Equation. Seminar, 1981. Edited by C. Cercignani. VII, 248 pages. 1984.

Vol. 1049: B. Iochum, Cônes autopolaires et algèbres de Jordan. VI, 247 pages. 1984.

Vol. 1054: V. Thomée, Galerkin Finite Element Methods for Parabolic Problems. VII, 237 pages. 1984.

Vol. 1055: Quantum Probability and Applications to the Quantum Theory of Irreversible Processes. Proceedings, 1982. Edited by L. Accardi, A. Frigerio and V. Gorini. VI, 411 pages. 1984.

Vol. 1057: Bifurcation Theory and Applications. Seminar, 1983. Edited by L. Salvadori. VII, 233 pages. 1984.

Vol. 1058: B. Aulbach, Continuous and Discrete Dynamics near Manifolds of Equilibria. IX, 142 pages. 1984.

Vol. 1059: Séminaire de Probabilités XVIII, 1982/83. Proceedings. Edité par J. Azéma et M. Yor. IV, 518 pages. 1984.

Vol. 1063: Orienting Polymers. Proceedings, 1983. Edited by J.L. Ericksen. VII, 166 pages. 1984.

Vol. 1065: A. Cuyt, Padé Approximants for Operators: Theory and Applications. IX, 138 pages. 1984.

Vol. 1066: Numerical Analysis. Proceedings, 1983. Edited by D.F. Griffiths. XI, 275 pages. 1984.

Vol. 1071: Padé Approximation and its Applications, Bad Honnef 1983. Prodeedings Edited by H. Werner and H.J. Bünger. VI, 264 pages. 1984.

Vol. 1072: F. Rothe, Global Solutions of Reaction-Diffusion Systems. V, 216 pages. 1984.

Vol. 1085: G.K. Immink, Asymptotics of Analytic Difference Equations. V, 134 pages. 1984.

Vol. 1086: Sensitivity of Functionals with Applications to Engineering Sciences. Proceedings, 1983. Edited by V. Komkov. V, 130 pages. 1984

Vol. 1100: V. Ivrii, The Precise Spectral Asymptotics for Elliptic Operators Acting in Fiberings over Manifolds with Boundary. V, 237 pages. 1984.